Biological Modeling and Simulation

Computational Molecular Biology

Sorin Istrail, Pavel Pevzner, and Michael Waterman, editors

Computational molecular biology is a new discipline, bringing together computational, statistical, experimental, and technological methods, which is energizing and dramatically accelerating the discovery of new technologies and tools for molecular biology. The MIT Press Series on Computational Molecular Biology is intended to provide a unique and effective venue for the rapid publication of monographs, textbooks, edited collections, reference works, and lecture notes of the highest quality.

Computational Molecular Biology: An Algorithmic Approach
Pavel A. Pevzner, 2000

Computational Methods for Modeling Biochemical Networks
James M. Bower and Hamid Bolouri, editors, 2001

Current Topics in Computational Molecular Biology
Tao Jiang, Ying Xu, and Michael Q. Zhang, editors, 2002

Gene Regulation and Metabolism: Postgenomic Computation Approaches
Julio Collado-Vides, editor, 2002

Microarrays for an Integrative Genomics
Isaac S. Kohane, Alvin Kho, and Atul J. Butte, 2002

Kernel Methods in Computational Biology
Bernhard Schölkopf, Koji Tsuda and Jean-Philippe Vert, editors, 2004

Immunological Bioinformatics
Ole Lund, Morten Nielsen, Claus Lundegaard, Can Keşmir and Søren Brunak, 2005

Ontologies for Bioinformatics
Kenneth Baclawski and Tianhua Niu, 2005

Biological Modeling and Simulation
Russell Schwartz, 2008

BIOLOGICAL MODELING AND SIMULATION

A Survey of Practical Models, Algorithms, and Numerical Methods

Russell Schwartz

The MIT Press
Cambridge, Massachusetts
London, England

MIT Press books may be purchased at special quantity discounts for business or sales promotional use. For information, please email special_sales@mitpress.mit.edu or write to Special Sales Department, The MIT Press, 55 Hayward Street, Cambridge, MA 02142.

This book was set in Times New Roman and Syntax on 3B2 by Asco Typesetters, Hong Kong. Printed and bound in the United States of America.

Library of Congress Cataloging-in-Publication Data

Schwartz, Russell.
Biological modeling and simulation : a survey of practical models, algorithms, and numerical methods / Russell Schwartz.
    p.   cm. — (Computational molecular biology)
Includes bibliographical references and index.
ISBN 978-0-262-19584-3 (hardcover : alk. paper) 1. Biology—Simulation methods. 2. Biology—Mathematical models. I. Title.
QH323.5.S364   2008
570.1′1—dc22   2008005539

10 9 8 7 6 5 4 3 2 1

Contents

Preface   xi

1   Introduction   1
1.1   Overview of Topics   1
1.2   Examples of Problems in Biological Modeling   2
    1.2.1   Optimization   2
    1.2.2   Simulation and Sampling   4
    1.2.3   Parameter-Tuning   8

I   MODELS FOR OPTIMIZATION   13

2   Classic Discrete Optimization Problems   15
2.1   Graph Problems   16
    2.1.1   Minimum Spanning Trees   16
    2.1.2   Shortest Path Problems   19
    2.1.3   Max Flow/Min Cut   21
    2.1.4   Matching   23
2.2   String and Sequence Problems   24
    2.2.1   Longest Common Subsequence   25
    2.2.2   Longest Common Substring   26
    2.2.3   Exact Set Matching   27
2.3   Mini Case Study: Intraspecies Phylogenetics   28

3   Hard Discrete Optimization Problems   35
3.1   Graph Problems   36
    3.1.1   Traveling Salesman Problems   36
    3.1.2   Hard Cut Problems   37
    3.1.3   Vertex Cover, Independent Set, and k-Clique   38
    3.1.4   Graph Coloring   39
    3.1.5   Steiner Trees   40
    3.1.6   Maximum Subgraph or Induced Subgraph with Property Π 42
3.2   String and Sequence Problems   42
    3.2.1   Longest Common Subsequence   42
    3.2.2   Shortest Common Supersequence/Superstring   43

3.3   Set Problems   44
      3.3.1   Minimum Test Set   44
      3.3.2   Minimum Set Cover   45
3.4   Hardness Reductions   45
3.5   What to Do with Hard Problems   46

4   Case Study: Sequence Assembly   57
4.1   Sequencing Technologies   57
      4.1.1   Maxam–Gilbert   57
      4.1.2   Sanger Dideoxy   59
      4.1.3   Automated Sequencing   61
      4.1.4   What About Bigger Sequences?   63
4.2   Computational Approaches   64
      4.2.1   Sequencing by Hybridization   64
      4.2.2   Eulerian Path Method   66
      4.2.3   Shotgun Sequencing   67
      4.2.4   Double-Barreled Shotgun   69
4.3   The Future?   71
      4.3.1   SBH Revisited   71
      4.3.2   New Sequencing Technologies   72

5   General Continuous Optimization   75
5.1   Bisection Method   76
5.2   Secant Method   78
5.3   Newton–Raphson   80
5.4   Newton–Raphson with Black-Box Functions   84
5.5   Multivariate Functions   85
5.6   Direct Methods for Optimization   89
      5.6.1   Steepest Descent   89
      5.6.2   The Levenberg–Marquardt Method   90
      5.6.3   Conjugate Gradient   91

6   Constrained Optimization   95
6.1   Linear Programming   96
      6.1.1   The Simplex Method   97
      6.1.2   Interior Point Methods   104
6.2   Primals and Duals   107
6.3   Solving Linear Programs in Practice   107
6.4   Nonlinear Programming   108

II   SIMULATION AND SAMPLING   113

7   Sampling from Probability Distributions   115
7.1   Uniform Random Variables   115
7.2   The Transformation Method   116
      7.2.1   Transformation Method for Joint Distributions   119
7.3   The Rejection Method   121
7.4   Sampling from Discrete Distributions   124

8   Markov Models   129
  8.1   Time Evolution of Markov Models   131
  8.2   Stationary Distributions and Eigenvectors   134
  8.3   Mixing Times   138

9   Markov Chain Monte Carlo Sampling   141
  9.1   Metropolis Method   141
       9.1.1   Generalizing the Metropolis Method   146
       9.1.2   Metropolis as an Optimization Method   147
  9.2   Gibbs Sampling   149
       9.2.1   Gibbs Sampling as an Optimization Method   152
  9.3   Importance Sampling   154
       9.3.1   Umbrella Sampling   155
       9.3.2   Generalizing to Other Samplers   156

10   Mixing Times of Markov Models   159
  10.1   Formalizing Mixing Time   160
  10.2   The Canonical Path Method   161
  10.3   The Conductance Method   166
  10.4   Final Comments   170

11   Continuous-Time Markov Models   173
  11.1   Definitions   173
  11.2   Properties of CTMMs   175
  11.3   The Kolmogorov Equations   178

12   Case Study: Molecular Evolution   185
  12.1   DNA Base Evolution   185
       12.1.1   The Jukes–Cantor (One-Parameter) Model   185
       12.1.2   Kimura (Two-Parameter) Model   188
  12.2   Simulating a Strand of DNA   191
  12.3   Sampling from Whole Populations   192
  12.4   Extensions of the Coalescent   195
       12.4.1   Variable Population Sizes   196
       12.4.2   Population Substructure   197
       12.4.3   Diploid Organisms   198
       12.4.4   Recombination   198

13   Discrete Event Simulation   201
  13.1   Generalized Discrete Event Modeling   203
  13.2   Improving Efficiency   204
  13.3   Real-World Example: Hard-Sphere Model of Molecular Collision Dynamics   206
  13.4   Supplementary Material: Calendar Queues   209

14   Numerical Integration 1: Ordinary Differential Equations   211
  14.1   Finite Difference Schemes   213
  14.2   Forward Euler   214
  14.3   Backward Euler   217

14.4   Higher-Order Single-Step Methods   219
14.5   Multistep Methods   221
14.6   Step Size Selection   223

15   Numerical Integration 2: Partial Differential Equations   227
15.1   Problems of One Spatial Dimension   228
15.2   Initial Conditions and Boundary Conditions   230
15.3   An Aside on Step Sizes   233
15.4   Multiple Spatial Dimensions   233
15.5   Reaction–Diffusion Equations   234
15.6   Convection   237

16   Numerical Integration 3: Stochastic Differential Equations   241
16.1   Modeling Brownian Motion   241
16.2   Stochastic Integrals and Differential Equations   242
16.3   Integrating SDEs   245
16.4   Accuracy of Stochastic Integration Methods   248
16.5   Stability of Stochastic Integration Methods   249

17   Case Study: Simulating Cellular Biochemistry   253
17.1   Differential Equation Models   253
17.2   Markov Models Methods   256
17.3   Hybrid Models   259
17.4   Handling Very Large Reaction Networks   260
17.5   The Future of Whole-Cell Models   262
17.6   An Aside on Standards and Interfaces   263

III   PARAMETER-TUNING   265

18   Parameter-Tuning as Optimization   267
18.1   General Optimization   268
18.2   Constrained Optimization   269
18.3   Evaluating an Implicitly Specified Function   271

19   Expectation Maximization   275
19.1   The "Expectation Maximization Algorithm"   277
19.2   EM Theory   278
19.3   Examples   280

20   Hidden Markov Models   291
20.1   Applications of HMMs   292
20.2   Algorithms for HMMs   295
        20.2.1   Problem 1: Optimizing State Assignments   295
        20.2.2   Problem 2: Evaluating Output Probability   297
        20.2.3   Problem 3: Training the Model   299
20.3   Parameter-Tuning Example: Motif-Finding by HMM   303

21   Linear System-Solving   309
21.1   Gaussian Elimination   310
        21.1.1   Pivoting   312

21.2   Iterative Methods   316
21.3   Krylov Subspace Methods   317
       21.3.1   Preconditioners   319
21.4   Overdetermined and Underdetermined Systems   320

22   Interpolation and Extrapolation   323
22.1   Polynomial Interpolation   326
       22.1.1   Neville's Algorithm   326
22.2   Fitting to Lower-Order Polynomials   329
22.3   Rational Function Interpolation   330
22.4   Splines   331
22.5   Multidimensional Interpolation   334
22.6   Interpolation with Arbitrary Families of Curves   334
22.7   Extrapolation   337
       22.7.1   Richardson Extrapolation   337
       22.7.2   Aitken's δ^2 Process   338

23   Case Study: Inferring Gene Regulatory Networks   341
23.1   Coexpression Models   342
       23.1.1   Measures of Similarity   342
       23.1.2   Finding a Union-of-Cliques Graph   344
23.2   Bayesian Graphical Models   347
       23.2.1   Defining a Probability Function   347
       23.2.2   Finding the Network   349
23.3   Kinetic Models   351

24   Model Validation   355
24.1   Measures of Goodness   355
24.2   Accuracy, Sensitivity, and Specificity   358
24.3   Cross-Validation   361
24.4   Sensitivity Analysis   362
24.5   Modeling and the Scientific Method   363

References   367
Index   377

Preface

This text arose from a class on biological modeling I have been teaching annually at Carnegie Mellon University since 2004. I created the class to fill what I saw as a gap in the available computational biology teaching materials. There are many excellent sources from which one can learn about successful approaches that have been developed for various core problems in computational biology (e.g., building phylogenies, implementing molecular simulations, or inferring DNA binding motifs). What seems to me to have been missing, though, is material to prepare aspiring computational biologists to solve the next problem, the one that no one has studied yet. Too often, computational biology courses assume that if a student is well prepared in biology and in computer science, then he or she can figure out how to apply the one to the other. In my experience, however, a computational biologist who wants to be prepared for a broad range of unexpected problems needs a great deal of specialized knowledge that is not part of the standard curriculum of either discipline. The material included here reflects my attempt to prepare my students for the sorts of unanticipated problems a computational biology researcher is like to encounter by collecting in one place a set of broadly useful models and methods one would ordinarily find scattered across many classes in several disciplines.

Meeting this challenge—preparing students for solving a wide array of problems without knowing what those problems will be—requires some compromises. Many potentially useful tools had to be omitted, and none could be covered in as much depth as I might have liked so that I could put together a "bag of tricks" that is likely to serve the aspiring researcher well on a broad class of biological problems. I have for the most part chosen techniques that have proved useful in diverse biological modeling contexts in the past. In a few cases, I have selected methods that are not yet widely used in biological modeling but that I believe have great potential. For every topic, I have tried to focus on what the practitioner needs to know in order to use these techniques effectively, sacrificing theoretical depth to accommodate greater breadth. This approach will surely grate on some readers, and indeed I feel that this material is best treated not as a way to master any particular techniques, but rather as a set of possible starting points for use in the modeling problems one encounters. My goal is that a reader who learns the material in this text will be able to make at least a first attempt at solving nearly any computational problem he or she will encounter in biology, and will have a good idea where to go to learn more if that first attempt proves inadequate.

This text is designed for readers who already have some familiarity with computational and biological topics. It assumes an introductory knowledge of algorithms and their analysis. Portions of the text also assume knowledge of calculus, linear algebra, and probability at the introductory undergraduate level. Furthermore, though the text teaches computational methods, its goal is to help readers solve biological problems. The reader should therefore be prepared to encounter many toy examples and a few extended case studies showing how the methods covered here have been applied to various real problems in biology. Readers are therefore likely to need a general knowledge of biology at the level of at least an undergraduate introductory survey course. When I teach this material, a key part of the learning experience consists of exercises in which students are presented with biological problems and are expected to formulate, and often implement, models using the techniques covered here. While one need not necessarily use the text in that way, it is written for readers capable of writing their own computer code.

I would like to thank the many people who have made this work possible. Sorin Istrail, one of my mentors in this field, provided very helpful encouragement for this project, as did my editors at the MIT Press, Bob Prior and Katherine Almeida. Mor Harchol-Balter provided valuable advice on clarifying my presentation of continuous-time Markov models. And I am grateful to my many teachers throughout the years in whose classes I picked up bits and pieces of the material of this text. I had the mixed blessing of having realized I wanted to be a computational biologist as a student in the days before computational biology classes were widespread. Many of the topics here are pieced together from subjects I found useful in inventing my own computational biology curriculum with the advice of my graduate mentor, Bonnie Berger. Most important in preparing this work have been the students in my class, who have provided much helpful criticism as this material evolved from handwritten lecture notes to typeset handouts, and finally to its present form. Though all of my students deserve some thanks, the following have been particularly helpful in offering corrections and criticism on various editions of this work and suggesting new topics that made their way into the final version: Byoungkoo Lee, Srinath Sridhar, Tiequan Zhang, Arvind Ramanathan, and Warren Ruder.

This material is based upon work supported by the National Science Foundation under glant no. 0346981. Any opinions, findings, and conclusions or recommendations expressed in this material are those of the author and do not necessarily reflect the views of the National Science Foundation.

1 Introduction

1.1 Overview of Topics

This book is divided into three major sections: models for optimization, simulation and sampling, and parameter-tuning. Though there is some overlap among these topics, they provide a general framework for learning how one can formulate models of biological systems, what techniques one has to work with those models, and how to fit those models to particular systems.

The first section covers perhaps the most basic use of mathematical models in biological research: formulating optimization problems for biological systems. Examples of models used for optimization problems include the molecular evolution models generally used to formulate sequence alignment or evolutionary tree inference problems, energy functions used to predict docking between molecules, and models of the relationships between gene expression levels used to infer genetic regulatory networks. We will start with this topic because it is a good way for those who already have some computational background to get experience in reasoning about how to formulate new models.

The second section covers simulation and sampling (i.e., how to select among possible system states or trajectories implied by a given model). Examples of simulation and sampling questions we could ask are how a biochemical reaction system might change over time from a given set of initial conditions, how a population might evolve from a set of founder individuals, and how a genetic regulatory network might respond to some outside stimulus. Answering such questions is one of the main functions of models of biological systems, and this topic therefore takes up the greatest part of the text.

The third section covers techniques for fitting model parameters to experimental data. Given a data set and a class of models, the goal will be to find the best model from the class to fit the data. A typical parameter-tuning problem would be to estimate the interaction energy between any two amino acids in a protein structure model by examining known protein structures. Parameter-tuning overlaps with optimization, as finding the best-fit parameters for a model is often accomplished by optimizing for some quality metric. There are, however, many specialized optimization methods that frequently recur in parameter-tuning contexts. We will conclude our discussion of parameter-tuning by considering how to evaluate the quality of whatever fit we achieve.

1.2   Examples of Problems in Biological Modeling

To illustrate the nature of each of these topics, we can work through a few simple examples of questions in biology that we might address through computational models. In this process, we can see some of the issues that come up in reasoning about a model.

1.2.1   Optimization

Often, when we examine a biological system, we have a single question we want to answer. A mathematical model provides a way to precisely judge the quality of possible solutions and formulate a method for solving it. For example, suppose I have a hypothetical group of organisms: a bacterium, a protozoan, a yeast, a plant, an invertebrate, and a vertebrate. Our question is "What are the evolutionary relationships among these organisms?" That may seem like a pretty straightforward question, but it hides a lot of ambiguity. By modeling the problem, we can be precise about what we are asking.

The first thing we need is a model of what "evolutionary relationships" look like. We can use a standard model, the evolutionary tree. Figure 1.1 shows a hypothetical (and rather implausible) example of an evolutionary tree for our organisms. Note that by choosing a tree model, we are already restricting the possible answers to our question. The tree leaves out many details that may be of interest to us, for example, which genes are conserved among subsets of these organisms. It also makes assumptions, such as a lack of horizontal transfer of genes between species, that may be inaccurate when understanding the evolution of these organisms. Nonetheless, we have to make some assumptions to specify precisely what our output looks like, and these

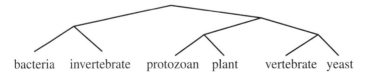

Figure 1.1
Hypothetical evolutionary tree linking our example set of organisms.

are probably reasonable ones. We have now completed one step of formalizing our problem: specifying our *output format*.

We then must deal with another problem: even if our model specifies that our output is a tree, we do not know which one. We cannot answer our question with certainty, so what we really want to find is the best answer, given the evidence available to us. So, what is the evidence available to us? We might suppose that our evidence consists of genetic sequences of some highly conserved gene or genetic region in each organism. That means we assume we are given m strings on an alphabet $\{A, C, T, G\}$. Figure 1.2 is an example of such strings that have been aligned to each other by inserting a gap ("-") in one. We have now completed another step in formalizing our problem: specifying our *input format*.

Now we face another problem. There are many possible outputs consistent with any input. So which is the best one? To answer that, our model needs to include some measure of how well any given tree matches the data. A common way to this is to assume some model of the process by which the input data may have been generated by the process of evolution. This model will then have implications for the probability of observing any given tree. Let us propose some assumptions that will let us define a formal model:

• Our gene is modified only by point mutations, changing one base at a time.
• Mutations are rare.
• Any one mutation (or insertion or deletion) is as likely to occur as any other.
• Mutations are selectively neutral, that is, they do not affect the probability of the organism's surviving and reproducing.

  Those are not exactly correct assumptions, but they may be reasonable approximations, depending on the characteristics of our problem. Given these assumptions, we might propose that the best tree is the one that involves the fewest mutations between organisms. A model that seeks to minimize some measure of complexity of the solution is called a *parsimony* model. Parsimony formulations often lead to recognizable optimization problems. In this case, we can define an *edit distance d* between

```
ACGGTAC
ACCGAAC
AC-GGAC
    .
    .
    .
```

Figure 1.2
A set of strings on the alphabet $\{A, C, T, G\}$ that have been aligned to each other.

two strings s_1 and s_2 to be the minimum number of insertions, deletions, and base changes necessary to convert one string into the other. Then our solution to the problem will consist of a tree with leaves labeled with our input strings and with internal nodes labeled with other strings such that the sum of the edit distances across all edges in the tree is minimized. We have now accomplished the third task in formalizing our problem: specifying a *metric* for it.

Now that we have the three components of our formal specification—an *input format*, an *output format*, and a *metric*—we have specified our model well enough to formulate a well-defined computational optimization problem. We can take the same problem we specified informally above and write it more formally as follows:

Input   A set S of strings on the alphabet $\Sigma = \{A, C, T, G\}$ representing our DNA sequences to be examined

Output   A tree $T = (V, E)$ with $|S|$ leaves $L \subseteq V$ and an assignment of string tags to nodes $t : V \to \Sigma^*$ satisfying the constraint $\forall s \in S \exists l \in L$ s.t. $t(l) = s$ (read as "for all strings s in set S, there exists a leaf node l from set L such that the tag of l, $t(l)$, is the string s")

Metric $\sum_{(u,v) \in E} d(t(u), t(v))$ (read as "the sum over all edges u to v in the edge set E of the edit distance between the tag of u, $t(u)$ and the tag of v, $t(v)$") is minimized over trees T and tag assignments t.

In other words, we want to find the tree whose leaves are labeled with the sequences of our organisms and whose internal nodes are labeled with the sequences of presumed common ancestors such that we minimize the total number of base changes over all pairs of sequences sharing an edge in the tree. This does not yet tell us how to solve the problem, but it does at least tell us what problem to solve. Later in the book, we will see how we might go about solving that problem.

1.2.2   Simulation and Sampling

Another major use of models is for simulation. Usually, we use simulations when we are interested in a process rather than a single outcome. Simulating the process can be useful as a validation of a model or a comparison of two different models. If we have reason to trust our model, then simulation can further be used to explore how interventions in the model might affect its behavior. Simulations are also useful if the long-term behavior of the model is hard to analyze by first principles. In such cases, we can look at how a model evolves and watch for particularly interesting but unexpected properties.

As an example of what one might do with simulation, let us consider an issue motivated by protein structure analysis. Suppose we are given the structure of a protein and we wish to understand whether we can mutate the protein in some way that increases its stability. Simulations can provide a way to answer this sort of question.

Our input can be assumed to be a protein sequence (i.e., a string of amino acids). More formally, our input is a string $s \in \Sigma^*$ ("Σ^*" is a formal notation for a string of zero or more characters from the alphabet Σ), where $\Sigma = \{A, C, D, E, F, G, H, I, K, L, M, N, P, Q, R, S, T, V, W, Y\}$.

If we want to answer this question, we first need a model for the structure of our protein. For the purposes of this illustration, we will use a common form of simplified model called a lattice model. In a lattice model, we treat a protein as a chain of beads sitting at points on a regular grid. To simplify the illustration, we will represent this as a two-dimensional structure sitting on a square grid. In practice, much more flexible lattices are available that better capture the true range of motion of a protein backbone. Lattice models tend to be a good choice for simulations involving protein folding because they are simple enough to allow nontrivial rearrangements to occur on a reasonable time scale. They are also often used in optimizations related to protein folding because of the possibility of enumerating discrete sets of conformations in them. Our model of the protein structure is, then, a self-avoiding chain on a 2-D square lattice (see figure 1.3).

If we want to study protein energetics, we need a model of the energy of any particular structure. Lattice models are commonly used with *contact potentials* that assign a particular energy to any two amino acids that are adjacent on the lattice but not in the protein chain. For example, in the model protein above, we have two contacts, S to L at the top and D to K at the bottom. These are shown as thick dashed lines in figure 1.3. On more sophisticated lattices, these potentials might vary with distance between the amino acids or their orientations relative to one another, but we will ignore that here.

As a first pass at solving our problem, we might simply stop here and say that we can estimate the stability effect of an amino acid change by looking at the change in contact energies it produces. For example, suppose our model specifies a contact energy of $+1$ kcal/mol for contact between S and L and -1 kcal/mol for contact

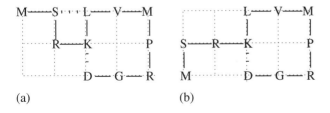

(a) (b)

Figure 1.3
A hypothetical protein folded on a lattice. Solid lines represent the path of the peptide backbone. Thick dashed lines show contacts between amino acids adjacent on the lattice but not on the backbone. Thin dashed lines show the lattice grid. (a) Initial conformation of the protein. (b) Alternative conformation produced by pivoting around the arginine (R) amino acid.

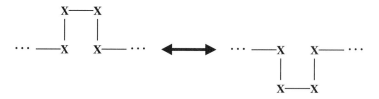

Figure 1.4
An example of a lattice move. X stands for any possible amino acid, and the ellipses stand for any possible conformation of the chain outside of a local region of interest. This move indicates that a 180° bend of four residues can be flipped about the surrounding backbone.

between S and T. Then we might propose that if the conformation in figure 1.3(a) is our protein's native (normal) state, then mutating L to T will increase stability (reduce energy) by 2 kcal/mol. We might then propose to solve the problem by attempting substitutions at all positions until we find the set of amino acids with the lowest possible energy summed over all contacts. This first-pass solution is problematic, though, in that it neglects the fact that an amino acid change which stabilizes the native conformation might also stabilize nonnative conformations. The change might thereby reduce the time spent in the native state even while reducing the native state's intrinsic energy.

We therefore need some way to study how the protein might move under the control of our energy model. There are many *move sets* for various lattices that attempt to capture how a protein chain might bend. A move set is a way of specifying how any given conformation can be transformed into other conformations. Figure 1.4 shows an example of a possible move for a move set. Anywhere we observe a subset of a conformation matching the left pattern, it would be legal to transform it to match the right pattern, and vice versa. This move alone would be insufficient to create a realistic folding model, but it might be part of a larger set allowing more freedom of movement. For this small example, though, we will assume a simpler move set. We will say that a single move of a protein consists of choosing any one bond in the protein and bending it to any arbitrary position that does not produce collisions in the chain. We can get from any chain configuration to any other by some sequence of these single-bond bends. For example, we could legally change our chain configuration in figure 1.3(a) into that in figure 1.3(b) by pivoting 90° at the S-R-K bend. We would not be able to pivot an additional 90°, though, because that would create a collision between the M and D amino acids.

The move set only tells us which moves are allowed, though, not which are likely. We further need a model of *dynamics* that specifies how we select among different legal moves at each point in time. One common method is the *Metropolis criterion*:

1. Pick uniformly at random among all possible moves from the current conformation C_1 to some neighboring conformation C_2.

2. If the energy of C_2 is less than the energy of C_1, *accept* the move and change to conformation C_2.

3. Otherwise, *accept* the move with probability $e^{-(E(C_2)-E(C_1))/k_B T}$, where T is the absolute temperature and k_B is Boltzmann's constant.

4. If the move is not yet accepted, *reject* the move and remain in conformation C_1.

This method produces a sequence of moves with some nice statistical properties that we will cover in more depth in chapter 9. The choice of this model of dynamics once again involves a substantial oversimplification of how a chain would really fold, but it is a serviceable model for this example. This completes a model, if not a very good model, of how a protein chain will move over time.

We are now ready to formulate our initial question more rigorously. We can propose to estimate the stability of the chain as follows:

1. Place the chain into its native configuration.

2. Select the next state according to the Metropolis criterion.

3. If it is in the native configuration, record a *hit*; otherwise, record a *miss*.

4. Return to step 2.

We can run this procedure for some predetermined number of steps and use the fraction of hits as a measure of the stability of the protein. We can repeat this experiment for each mutation we wish to consider. A mutation that yields a higher percentage of hits than the original sequence over a sufficiently long simulation run is inferred to be more stable. A mutation that yields a lower percentage of hits is inferred to be less stable. This example thus demonstrates how we might use simulation to solve a biological problem.

An issue closely related to simulation is sampling: choosing a state according to some probability distribution. For example, instead of simulating a trajectory from the native state, we might repeatedly sample from the partition function defined by the energies of the states of our protein sequence. That is, we might have some probability distribution over possible configurations of the protein defined by the relative energies of the folds, then repeatedly pick random configurations from this distribution. We could then ask what fraction of states that we sample are the native state. This is actually closer to what we really want to do to solve our problem, although if we look at a lot of steps of simulation, the two approaches should converge on the same answers. In fact, simulation is often a valid way to perform sampling, although there may be much more efficient ways for some problems. For a short amino acid chain like this, for example, it might be feasible to analytically determine the probability distribution of states, given our model.

1.2.3   Parameter-Tuning

The final area of modeling and simulation we will consider is how to fit a general class of model to a specific set of data. Whether we are using a model for simulation or optimization, we will commonly have a general format for input and output, but some unknown parameters are needed to translate one to the other. We may also have a set of examples from which to learn the missing parameters. We then wish to establish the function relating inputs to outputs. A model lets us constrain the space of possible functions and judge which among the allowed ones are better explanations than others. That in turn lets us formulate a precise computational problem.

For example, suppose we want to learn about the function of a novel protease we have identified. A protease is a protein that cuts other proteins or peptides. It usually has some specificity in selecting the sites at which it cuts other proteins. That is, if it is presented with many copies of the same protein, there are some sites it will cut frequently and some it will cut rarely or not at all. Suppose we have the following examples of how the protease cleaves some known peptides:

$$SIVVAKSASK \rightarrow SASIVVAK + SASK$$

$$HEPCPDGCHSGCPCAKTC \rightarrow H + EPCPDGCH + SGCPCAKTC.$$

We can treat these examples as the input to a parameter-fitting problem. More formally, we can say our input is a set of strings on the alphabet of amino acids

$$\Sigma = \{A, C, D, E, F, G, H, I, K, L, M, N, P, Q, R, S, T, V, W, Y\}$$

and a set of integer *cut sites* in each string. Our goal is to predict how this protease will act on novel sequences. Typically, we would answer this by assuming a class of models based on prior knowledge about our system, with some unspecified parameters distinguishing particular members of the class. We would then try to determine the parameters of the specific model from our class that best explain our observed data. We can then use the model with that parameter assignment to make predictions about how the protease will act on novel sequences.

We first need to define our class of models. A good way to get started is to ask what we know about proteases in general. Proteases usually recognize a small motif close to the cut site. The closer a residue is to the cut site, the more likely it is to be important to deciding where the cut occurs. A good model then may assume that the protease examines some window of residues around a potential cut site and decides whether or not to cut based on the residues in that window. The parameter-tuning problem for such a model consists of identifying the probability of cutting for any specific window. If we have a lot of training data, we may assume that the protease can consider very complicated patterns. Since our data are very sparse, though, we probably need to assume the motif it recognizes is short and simple. That assumption is not necessarily true, and if it is not, then we will not be able to learn our model without more data. Many known proteases cut exclusively on the basis of the residue immediately N-terminal of the cut site, so for this example we will assume that the window examined consists only of that one residue.

Using these basic assumptions, we can create a formal model for cut-site prediction. As a first pass, we can assume that the probability of cutting at a given site is a function of the amino acid immediately N-terminal from that site. More formally, then, our class of models is the set of mappings from amino acids to cut probabilities,

$$f : \{A, C, D, E, F, G, H, I, K, L, M, N, P, Q, R, S, T, V, W, Y\} \rightarrow [0, 1].$$

The parameters of the model are then the 20 values $f(A), f(C), \ldots, f(Y)$ defining the function over the amino acid alphabet. This may be an acceptable model if we have sufficient data available to estimate all of these values. In this case, though, our training data are so sparse that we do not have any examples of some amino acids with which to estimate cut probabilities. So how do we predict their behavior? Once again, to answer this sort of question we have to ask what we know about our system. Specifically, what do we know about amino acids that might help us reduce the parameter space? One useful piece of information is that some amino acids are more chemically similar than others, and they can be roughly grouped into categories by chemical properties. Typical categories are hydrophobic (H), polar (P), basic (B), acidic (A), and glycine (G). If we then classify our amino acids into these groups, we end up with the following inputs:

$$PHHHHBPHPB \rightarrow PHHHHB + PHPB$$

$$BAPHPAGHBPGHHHHBPH \rightarrow B + APHPAGHB + PGHHHHBPH$$

We now have five parameters to fit in this model: $f(H)$, $f(P)$, $f(B)$, $f(A)$, and $f(G)$, that is, the probabilities of cutting after each amino acid class. In this simple model, the procedure for fitting our model to the data is straightforward: count the fraction of times a particular residue class is followed by a cut site. This procedure gives us the following parameters:

$$f(H) = 0$$

$$f(P) = 0$$

$$f(B) = 0.75$$

$$f(A) = 0$$

$$f(G) = 0$$

That answers our general question about the rules determining the behavior of this protease. In particular, we have derived what are known as *maximum likelihood estimates* of the parameters, which means these are the parameter values that maximize the probability of generating the observed outputs from our model. If we want to get more sophisticated, we can also consider how much confidence to place in our parameters based on the amount of data used to determine each one. We will also need to consider issues of validating the model, preferably on a different data set than the one we used to train it. We will neglect such issues for now, but return to them in chapter 24.

References and Further Reading

Though I am not aware of any references on the general subject matter of this chapter, the specific examples are drawn from a variety of sources in the literature. Evolutionary tree-building is a broad field, and there are many fine references to the general topic. Three excellent texts for the computationally savvy reader are Felsenstein [1], Gusfield [2], and Semple and Steel [3]. The notion of a parsimony-based tree, as we have examined it here, first appeared in the literature in a brief abstract by Edwards and Cavalli-Sforza [4]. There are many computational methods now available for inferring trees by parsimony metrics, and the three texts cited above ([1], [2], [3]) are all good references for these methods. We will see a bit more about them in chapters 2 and 3.

The use of lattice models for protein-folding applications was developed in a paper by Taketomi et al. [5], the first of a series introducing a general class of these lattice models that became known as Gō models. The specific example of a lattice move presented in figure 1.4 was introduced in a paper by Chan and Dill [6] as part of a move set called MS2. The Metropolis method, which we will cover in more detail in chapter 9, is one of the most important and widely used of all methods for sampling from complicated probability distributions. It was first proposed in an influential paper by Metropolis et al. [7].

The problem of predicting proteolytic cleavage sites is not nearly as well studied as evolutionary tree-building or protein-folding, but nonetheless has its own literature. The earliest reference to the computational problem of which I am aware is a paper by Folz and Gordon [8] introducing algorithms for predicting the cleavage of signal peptides. Much of the current interest in the problem arises from its importance in some specific medical contexts. One of these is understanding the activity of the human immunodeficiency virus (HIV) protease, a protein that is critical to the HIV life cycle and an important target of anti-HIV therapeutics. A review by Chou [9] offers a good discussion of the problem and methods in that context. Another important application is prediction of cleavage by the proteasome, a molecular machine found in all living cells. The proteasome is used for general protein degradation, but has evolved in vertebrates to play a special role in the identification of antigens by the immune system. Its specificity has therefore become important to vaccine design, among other areas. Saxová et al. [10] conducted a survey and comparative analysis of the major prediction methods for proteasome cleavage sites, which is a good place to start learning more about that application.

MODELS FOR OPTIMIZATION

2 Classic Discrete Optimization Problems

Often, an excellent way to begin work on a project involving modeling is to look at an informal representation of the problem we need to solve and ask whether it reminds us of any problem we have seen before. We may have to simplify the problem a bit or distort aspects of it to match it to something we already know how to solve. Nonetheless, this can be a good way to get a first pass at a solution. Later on, we can look at whether the problem can be modified to restore important aspects of the real system that were removed in that first pass. The same basic algorithm can often accommodate fairly large changes in the model.

The purpose of this chapter is to help with this process by examining some of the classic discrete optimization problems that we are likely to see reflected in applied work. I have tried to focus here on problems that may "look hard" if we are not familiar with them, since those are the problems for which it is most valuable to know what is and is not doable. For readers who have taken an introductory algorithms class, this chapter will be largely a review, hopefully a useful one. Some of these problems already have important applications in practical computational biology. Others come up often enough in other applied contexts that we might consider them good guesses when looking for solutions to new problems. When looking at a modeling problem for which we do not yet have a solution, we can try to run through some of the problems in this chapter and see if any of them remind us of the problem we need to solve.

We will begin by looking at graph problems. Graphs have proved to be a very useful abstraction for a broad range of optimization problems in computational biology. We will then consider some important string and sequence problems. These two abstractions are broadly important to computer science in general, and particularly to computational biology. We will then examine some of the issues in adapting classic algorithms to real-world problems through a mini case study on the topic of building evolutionary trees.

2.1　Graph Problems

We will start with a quick review of graphs. A graph G consists of a set of nodes, V, and a set of edges, E. Each edge e_i is defined by a pair of nodes (v_i, v_j). A graph can be directed, meaning $(v_i, v_j) \neq (v_j, v_i)$, or undirected, meaning $(v_i, v_j) = (v_j, v_i)$. It can also be weighted, meaning each edge e_i has an associated numerical weight $w(e_i)$. w is then a function from edges to real numbers, denoted by $w : E \rightarrow \mathscr{R}$. In some cases, it is convenient to treat the weight function as a function from $V \times V$ to real numbers (denoted $w : V \times V \rightarrow \mathscr{R}$). We may sometimes see these two representations used interchangeably even though it is an abuse of notation. One may sometimes be interested in *multigraphs*, in which multiple edges can connect the same pair of nodes, but we will not be using multigraphs here. When the type of graph is not specified, we generally assume that we are speaking about a weighted, directed graph. Figure 2.1 shows a weighted, undirected graph that we will use to illustrate the various graph problems we cover.

We will now briefly survey many solvable graph problems that frequently show up in real-world applications. As a practical matter, it is not that important that one memorize the algorithms which solve these problems. In the unlikely event that we need to use a standard algorithm that is not found in some generally available code library, it is easy enough to look it up. We do, however, need to be able to recognize these problems when they come up in practice, so that when we try to model a novel system, we will have a good idea what algorithmic tools are available to us. Once we identify the existing tools suitable for a given problem, we will also be better able to reason about how we might adapt our initial model to make it more realistic or more tractable for our specific application.

2.1.1　Minimum Spanning Trees

Many graph problems consist of finding a subset of a graph that optimizes for some metric. One common variant of optimal subgraph selection is finding a minimum spanning tree. A tree is a graph that is connected and contains no cycles, and a span-

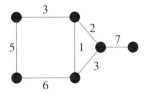

Figure 2.1
A weighted, undirected graph.

ning tree is a tree that includes every node of a given input graph. Informally, the minimum spanning tree problem is the problem of taking a weighted graph and finding the spanning tree of smallest weight (i.e., the tree of minimum total edge weight that connects all nodes of the graph). The minimum spanning tree problem is formalized as follows:

Input   A weighted, undirected, connected graph $G = (V, E)$
Output   A subset of the graph $G' = (V, E')$ such that G' is a tree (a connected, cycle-free graph), and for each vertex $v \in V$, there exists some $u \in V$ such that $(u, v) \in E'$
Metric $\sum_{(u,v) \in E'} w(u, v)$ is minimized.

Figure 2.2 shows the minimum spanning tree for the graph of figure 2.1.

There are two principal algorithms for finding minimum spanning trees in general graphs. Kruskal's algorithm splits a graph into sets, initially one for each node, then repeatedly greedily joins the two sets with the smallest edge weight between them. Figure 2.3 provides pseudocode for the algorithm. The runtime of Kruskal's

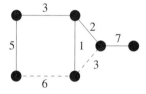

Figure 2.2
A minimum spanning tree for the example graph. Solid edges are those present in the spanning tree, and dashed edges are those absent from it.

---

1. define an initially empty set of edges T
2. for each node v
   A. define a set S_v containing only v
   B. define a label $s(v) = S_v$
3. sort the edges into a list of increasing weight, $e_1, \ldots, e_m$
4. for each edge $e = (u, v)$ in sorted order $e_1 \ldots e_m$
   A. if $s(u) \neq s(v)$
      i. $s(v) \leftarrow s(u) \cup s(v)$
      ii. $s(u) \leftarrow s(v)$
      iii. $T \leftarrow T \cup \{e\}$

Figure 2.3
Pseudocode for Kruskal's algorithm for the minimum spanning tree problem.

---

1. define an initially empty set of edges T
2. pick an arbitrary node s
3. define a set of unassigned nodes $U \leftarrow V - s$
4. for each $u \in U$
   A. $key[u] \leftarrow \infty$
   B. $parent[u] \leftarrow \emptyset$
5. for each edge $(s, u) \in E$
   A. $key[u] \leftarrow w(s, u)$
   B. $parent[u] \leftarrow s$
6. while $(U \neq \emptyset)$
   A. pick the node $u \in U$ for which $key[u]$ is minimum
   B. $U \leftarrow U - \{u\}$
   C. $T \leftarrow T \cup \{(parent[u], u)\}$
   D. for each edge $(u, v) \in E$
      i. if $v \in U$ and $w(u, v) < key[v]$
         a. $key[v] \leftarrow w(u, v)$
         b. $parent[v] \leftarrow u$

Figure 2.4
Pseudocode for Prim's algorithm for the minimum spanning tree problem.

algorithm depends on the sort algorithm and the data structure used for maintaining and merging subsets of the nodes, but in the best case it is $O(|E| \lg|E|)$.

The second general method is Prim's algorithm, which builds the spanning tree outward from a single node, greedily adding in whichever node not in the current tree has the lowest-weight edge connecting it to the tree. Figure 2.4 provides pseudo-code for Prim's algorithm. The runtime of Prim's algorithm depends on the priority queue used to select the node with minimum key. Most standard priority queue algorithms will give $O(|E| \lg|V|)$ runtime, as with Kruskal's algorithm, although Prim's algorithm can be implemented with runtime $O(|E| + |V| \lg|V|)$ by using a sophisticated kind of priority queue called a Fibonacci heap. As is often the case, there are better algorithms for special cases of input. For example, there are algorithms for sparse graphs (graphs with few edges) that can get runtime down to $O(|E| \lg^*|V|)$, which is an improvement over Prim's algorithm if $|E| < |V| \lg|V|/\lg^*|V|$. In applications where runtime is a concern, it may be worth looking into the literature to see whether any special-purpose algorithms apply to the particular problem variant being solved.

There is also a directed version of the minimum spanning tree problem, called a *minimum spanning arborescence*, the algorithms for which are a bit more complicated.

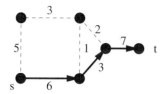

Figure 2.5
A single-pair shortest-path assignment from a source s to a sink t for the graph in figure 2.1. Directed edges on the path are shown as solid arrows. Edges omitted from the path are shown as dashed lines.

2.1.2 Shortest Path Problems

Another common graph problem is to find a shortest path in a graph (i.e., a path of minimum weight or minimum number of edges between a pair of nodes). The simplest version of this problem can be stated as follows:

Input   A graph $G = (V, E)$ (possibly directed), a weight function $w : E \to \mathscr{R}$, a source node $s \in V$, and a sink node $t \in V$
Output   A path $s, v_1, \ldots, v_k, t$ such that $e_0 = (s, v_1) \in E$, $e_i = (v_i, v_{i+1}) \in E$ for all i, and $e_k = (v_k, t) \in E$
Metric $\sum_{i=0}^{k} w(e_i)$ is minimized.

That is, we want a path of minimum weight from s to t in G.

This problem is more specifically known as the *single-pair shortest-path* problem because we want a shortest path between one given pair of nodes. Figure 2.5 shows a single-pair shortest path for the graph of figure 2.1 with a chosen source s and sink t. A special case of this problem is where $w(e) \equiv 1$, meaning we want the path with the minimum number of edges. In an unweighted graph, we can find the single-pair shortest path by an algorithm called breadth-first search in $O(|V| + |E|)$ time. In a breadth-first search, we search through a graph by maintaining a queue of nodes and repeatedly pulling the node off the front of a queue and adding all of its neighbors to the end of the queue. As we process each neighbor of the current node, we update the distance to that neighbor if there is a shorter path through the current node than any of which we were aware before. In a weighted graph, we generally need to use algorithms for the more general *single-source shortest-path* problem, one of several other important shortest-path problem variants:

• Single-source shortest-path: Find the shortest path from a given node to all other nodes.
• Single-destination shortest-path: Find the shortest path from each node to a given node.
• All-pairs shortest-path: Find the shortest path between each pair of nodes.

```
1. define a set of unassigned nodes U = V − s
2. for each u ∈ U
   A. key[u] ← ∞
   B. parent[u] ← ∅
3. for each edge (s, u) ∈ E
   A. key[u] ← w(s, u)
   B. parent[u] ← s
4. while (U ≠ ∅)
   A. pick the node u ∈ U for which key[u] is minimum
   B. U ← U − {u}
   C. for each edge (u, v) ∈ E
      i. if v ∈ U and key[u] + w(u, v) < key[v]
         a. key[v] ← key[u] + w(u, v)
         b. parent[v] ← u
```

Figure 2.6
Pseudocode for Dijkstra's algorithm for the single-source shortest-path problem.

These variants can of course be solved by solving for multiple individual instances of single-pair shortest paths. However, there are in general more efficient methods.

For a general single-source shortest-path problem, we have several options. When all edge weights are nonnegative, we can use Dijkstra's algorithm, which works by successively adding nodes to a growing set of those of known shortest path. It is very similar to Prim's algorithm for finding minimum spanning trees. Figure 2.6 presents pseudocode for Dijkstra's algorithm. Dijkstra's algorithm requires time $O(|V|^2)$.

When edge weights may be negative, we instead need to use the Bellman–Ford algorithm, which performs a repeated "relaxation" operation on all edges to keep updating path costs based on local information until all costs can be guaranteed optimal. Figure 2.7 presents pseudocode for the Bellman–Ford algorithm. Note that the shortest path may not be well defined in a graph with negative-weight edges. In particular, if there is a cycle in the graph for which the total weight is negative, then it is possible to construct pathways of arbitrarily low weight by repeatedly circling around the given cycle. The Bellman–Ford algorithm detects this case and *rejects* its input if a negative-weight cycle is detected. Otherwise, it finds a shortest path and *accepts* the input. The algorithm requires time $O(|V||E|)$. There are also more specialized algorithms for such cases as sparse graphs and restricted domains of edge weights, but we will not cover those here.

The solution to the single-destination shortest path follows trivially from the single-source shortest path. We simply reverse all of the edge directions, and then

```
1. define a set of unassigned nodes U = V − s
2. define key[s] = 0
3. for each u ∈ U
   A. key[u] ← ∞
4. for i = 1 to |V| − 1
   A. for each e = (u, v) ∈ E
      i. if key[v] > key[u] + w(u, v)
         a. key[v] ← key[u] + w(u, v)
5. for each e = (u, v) ∈ E
   A. if key[v] > key[u] + w(u, v)
      i. reject
6. accept
```

Figure 2.7
Pseudocode for the Bellman-Ford algorithm for the single-source shortest-path problem.

the single-destination shortest-path problem becomes a single-source shortest-path problem that we can solve by the Dijkstra or Bellman–Ford algorithm.

The all-pairs shortest path can be solved by multiple runs of a single-source shortest-path algorithm, but there are more efficient ways in practice. One standard algorithm for this problem is the Floyd–Warshall, a form of dynamic programming algorithm that solves for the subproblem of finding the shortest path between each pair of nodes, using only intermediate nodes with index less than k for increasing values of k. It runs in time $O(|V|^3)$. For sparse graphs, the preferred method is Johnson's algorithm, a more complicated method which uses a technique called "reweighting" to eliminate negative-weight cycles and then uses Dijkstra's algorithm to find the shortest path from each node to all of the others. It has runtime $O(|V|^2 \lg |V| + |V| |E|)$. We will omit detailed coverage of those algorithms here.

2.1.3 Max Flow/Min Cut

Two related graph problems of which we should be aware are the maximum flow and minimum cut problems. Maximum flow can be formally stated as follows:

Input A weighted, directed graph $G = (V, E)$ with weight function $w : E \to \mathscr{R}$ (also known as the capacity), a source node $s \in V$, and a sink node $t \in V$

Output An assignment of *flow* to the edges of E, which is a function $f : E \to \mathscr{R}$ satisfying the following properties:

• For all $v \in V$ where $v \neq s$ and $v \neq t$, $\sum_{(u,v) \in E} f((u,v)) = \sum_{(v,u) \in E} f((v,u))$ (i.e., the flow into any node other than the source and sink is equal to the flow out of that node)

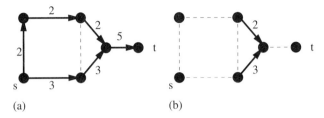

(a)   (b)

Figure 2.8
Maximum flow (a) and minimum cut (b) for the graph in figure 2.1, assuming that we treat edges as bi-directional with the same capacity in each direction. Solid arrows represent edges used by the flow or present in the cut, and dashed lines, edges absent from the flow or cut.

- $f(u, s) = 0$ for all u (i.e., there is no flow into the source)
- $f(t, u) = 0$ for all u (i.e., there is no flow out of the sink)
- $f((u, v)) \leq w((u, v))$ (i.e., the flow through an edge never exceeds its capacity).

Metric $\sum_{(s,u) \in E} f((s, u))$ (or, equivalently, $\sum_{(u,t) \in E} f((u, t))$) is maximized.

We can visualize a flow problem by imagining that the edges of our graph represent pipes joined at the nodes and we are trying to run water through these pipes from the source to the sink. Each pipe has a maximum amount of water it can handle, represented by the edge's capacity. We want to get as much water as we can from the source to the sink, subject to these constraints. We can also make an analogy to electrical current flowing through wires with a maximum current allowable in each. Figure 2.8(a) illustrates a maximum flow assignment for the graph in figure 2.1.

The minimum cut problem can be formally stated as follows:

Input A directed, weighted graph $G = (V, E)$ with weight function $w : E \to \mathcal{R}$, a source $s \in V$, and a sink $t \in V$
Output A *cut*, or set of edges $E' \subseteq E$ such that any path from s to t passes through some $e \in E'$
Metric $\sum_{e \in E'} w(e)$ is minimized.

Informally, a cut is a set of edges we need to remove in order to separate s from t in the graph. The maximum flow and minimum cut problems are closely related in that the maximum flow and the minimum cut in a graph have the same weight. Though we will not cover the proof here, we can intuitively understand why this is so by noting that the minimum cut is the tightest bottleneck through which a flow must pass, and is therefore the limiting factor on the maximum flow. Figure 2.8(b) shows a minimum cut for the graph of figure 2.1.

Because of the relationship between max flow and min cut, they are generally solved simultaneously by the same algorithm. The standard method is the Edmonds–Karp algorithm, an instance of the more general Ford–Fulkerson method. Ford–

Fulkerson constructs a maximum flow iteratively by repeatedly finding a single path from s to t in which all edges have excess capacity, known as an *augmenting path*. The algorithm increases the flow along that one path before searching for a new augmenting path. When no such path is available, the graph will have reached its maximum flow. The Edmonds–Karp algorithm requires $O(|V||E|^2)$ time.

There are many variants on maximum flow problems, some of which are tractable and some of which are not. Some variants can trivially be converted into the form stated above. For example, suppose we want to place capacities on nodes in addition to edges. We can convert such a problem to the above form by splitting each node v with capacity $c(v)$ into v_s and v_t, replacing each edge (u, v) with (u, v_s) and each edge (v, u) with (v_t, u), and adding an edge (v_s, v_t) with capacity $w((v_s, v_t)) = c(v)$. Another variant of this problem is the *multicommodity* flow, in which we have multiple pairs of sources and sinks, and wish to maximize the sum of flows between all pairs, subject to the constraint that the sum of flows along any edge cannot exceed a single capacity for that edge. Some versions of the multicommodity problem can be solved by linear programming, a technique we will cover in chapter 6.

2.1.4 Matching

Another class of graph problem that is somewhat less well known but nonetheless important in practice is a matching problem. Matching frequently comes up in various disguises in problems related to statistical physics, so it is often a good guess for a model when looking at optimization problems in molecular modeling. Given a graph $G = (V, E)$, a *matching* in the graph is a subset of the edges $E' \subseteq E$ such that no vertex is incident to more than one edge. There are several important variants of matching to consider.

The simplest is unweighted bipartite maximum matching. A bipartite graph is one in which the vertex set V can be split into two subsets $V = V_1 \cup V_2$ such that all edges in the graph are between a node in V_1 and a node in V_2 (i.e., $\forall e = (v_1, v_2) \in E$, $v_1 \in V_1$, and $v_2 \in V_2$). Unweighted bipartite maximum matching is simply maximum matching when the input is an unweighted bipartite graph:

Input An unweighted bipartite graph $G = (V, E)$
Output A matching $E' \subseteq E$
Metric $|E'|$ is maximized.

Figure 2.9 is an example of unweighted bipartite maximum matching. Though this variant may seem specialized, it is very useful in practice. It is also easy to solve, since it can be cast as a maximum flow problem. To accomplish this, we divide the the graph into its two parts, V_1 and V_2, and convert all edges into directed edges from V_1 to V_2. We then add an additional source node s and sink node t. Next, we add edges (s, v_1) for each $v_1 \in V_1$ and (v_2, t) for each $v_2 \in V_2$. Finally, we give each

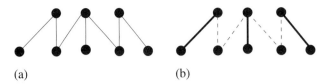

Figure 2.9
Unweighted bipartite maximum matching. (a) A sample unweighted bipartite graph. (b) A maximum matching in the graph. Solid thick edges are those present in the matching, and dashed edges are those absent from the matching.

edge a capacity of 1 and compute the maximum flow from s to t. I will assert without proof that the cost of the maximum flow in this graph is equal to the weight of the maximum matching. Given the weight, it is trivial to find a maximum matching. The best-known algorithm for this matching variant runs in $O(|V||E|)$ time.

A more general but often more useful variant of this problem is weighted bipartite matching. In this case, we have a weighted graph and wish to choose the matching of maximum weight. This problem can be solved in time $O(|V||E|\log_{2+|E|/|V|}|V|)$ by a technique known as the *Hungarian method*. The Hungarian method uses a substantially more complicated conversion into maximum flow problems, which we will not cover here. There are also polynomial algorithms available for matching in non-bipartite graphs, but they are extremely complicated. We will therefore just state some runtimes so we are aware of what can be done with these problems, but we will not cover how to do it. Readers who need to know can refer to the primary sources or more advanced texts covered in the References and Further Reading section. For the fully general case (weighted, nonbipartite) there is an $O(|V|^3)$ algorithm due to Gabow. There are also faster algorithms for various special classes of graphs. Matching algorithms is still an active area of research, and if we actually have to solve a large matching problem, it is often worth our while to do some research on the current state of the art to find the algorithm most specific to the version we are solving.

2.2 String and Sequence Problems

One of the major reasons that computational biology emerged as a distinct field was the need to solve problems that arose as sequence data started to accumulate in large amounts. Important examples include finding frequent patterns in DNA or protein sequences (the motif-finding problem) and finding regions of homology between sequences (the sequence alignment problem). In practice, we often have to modify our normal understanding of the terms "tractable" and "intractable" when dealing with such data. It is common to refer to a problem as "tractable" when it can be solved in runtime polynomial in its input size, but a quadratic algorithm is not generally "tractable" when applied to a 3×10^9 base-pair genome. Though computational biologists often work with very specialized versions of these problems, such specialized problems are frequently solved using methods based on classic string and sequence algorithms.

2.2.1 Longest Common Subsequence

One example of such a classic problem is the longest common subsequence problem. A sequence is an ordered set of characters from some alphabet. A subsequence is an ordered subset of the characters of a sequence in which the characters have the same order as in the original sequence but need not be consecutive in the original sequence. The longest common subsequence problem can be formally stated as follows:

Input   Two sequences A and B
Output   A sequence C that is a subsequence of both A and B
Metric $|C|$ is maximized.

For example, if we take the sequences $A = \text{ABCABCABC}$ and $B = \text{CABBCCBB}$, then the sequence $C = \text{ABBCB}$ is a subsequence of both A and B. We can see this by lining up A and B by the positions of overlap to produce C:

```
  ABCABCA BC
CAB   BC CB B
  AB  BC  B
```

Readers who have already taken an introductory computational biology class may find that this problem looks familiar. It is actually a special case of the global sequence alignment problem, and can be solved by a dynamic programming algorithm similar to the Needleman–Wunsch and Smith–Waterman methods used in sequence alignment. If we define $A[i]$ to be the ith character of A and likewise for $B[i]$, then we can solve for the length of the longest common subsequence by solving for the subproblem $MAX(i, j)$, defined as the size of the longest common subsequence on the first i characters of A and the first j characters of B. This is accomplished by the following recurrences:

$$MAX[i, j] = \max\{MAX(i - 1, j - 1) + 1, MAX(i, j - 1), MAX(i - 1, j)\}$$

$$\text{if } (A[i] = B[j])$$

$$MAX[i, j] = \max\{MAX(i, j - 1), MAX(i - 1, j)\} \quad \text{if } (A[i] \neq B[j]).$$

We can derive the Needleman–Wunsch algorithm, in which we allow gap and mismatch penalties, by some slight modifications of the recurrence equations:

$$MAX[i,j] = \max\{MAX(i-1, j-1) + 1, MAX(i, j-1) - g, MAX(i-1, j) - g\}$$

    if $(A[i] = B[j])$

$$MAX[i,j] = \max\{MAX(i-1, j-1) - m, MAX(i, j-1) - g, MAX(i-1, j) - g\}$$

    if $(A[i] \neq B[j]),$

where m is a penalty for aligning mismatched characters and g is a penalty for inserting a gap in the alignment. Other common sequence alignment variants—including affine gaps, semiglobal alignments, and local alignments—can also be derived with small modifications on the longest common subsequence algorithm. This is thus a fine example of a case where a classic computer science problem can be adapted in various ways to solve important real-world problems in biology.

2.2.2 Longest Common Substring

The longest common substring problem is nearly the same as the longest common subsequence, except that we do not allow gaps in the alignment of the sequences. For the two strings above, $A = $ ABCABCABC and $B = $ CABBCCBB, the longest common substring is $C = $ CAB. We can find this by aligning A and B to one another as follows:

ABCABCABC
   CABBCCBB
   ―――――――――
   CAB

The problem is trivially solvable in quadratic time simply by trying all possible ways of lining one sequence against the other and testing for the longest exact match for each. Though we would traditionally think of a quadratic problem as tractable, for biological problems involving large sequences (e.g., eukaryotic genomes), even quadratic time may be intractable. Perhaps surprisingly, there is a method for solving this problem in linear time in the sum of the sequence lengths. The method uses a data structure called a *suffix tree*. A suffix tree implicitly encodes each suffix of a given sequence or set of sequences. Conceptually, we can think of a suffix tree as if it were similar to the example in figure 2.10, directly encoding every possible suffix of the string. What is surprising about suffix trees, though, is that we can create a data structure that encodes this same information but requires only linear space and can be constructed in linear time. The construction is quite involved and would require a whole chapter to explain, so we omit it here. For our purposes, it is important to know that we can construct suffix trees in linear time in the length of their contents, and that we can then search them as efficiently as we could the abstraction in

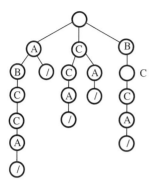

Figure 2.10
Conceptual illustration of a suffix tree encoding the string ABCCA. Each suffix of the string is represented
by a path from the root to a terminal node (/). A true suffix tree would not explicitly encode all of the
nodes in this tree, but can be searched as if it did.

figure 2.10, allowing us to efficiently solve many seemingly difficult string matching
problems.

We can solve the longest common substring problem in time $O(|A| + |B|)$ with suf-
fix trees by creating a suffix tree for A, building another on top of it for B, then finding the deepest node in both trees. The path from the root to that deepest node is the
longest common substring of A and B.

2.2.3   Exact Set Matching

Exact set matching is another classic computer science problem that shows up in
many variations in computational biology applications. It is not an optimization
problem, but is worth covering here for completeness. It is formally stated as follows:

Input   a text T (a large string or possible set of strings) and a set of strings
$S_1, S_2, \ldots, S_m$
Output   the locations of all exact occurrences of any string S_i in T.

We can trivially solve the problem in time $O(|T|(|S_1| + |S_2| + \cdots + |S_m|))$ by
searching for the first pattern, then the second, then the third, and so forth. In practice, though, that is often not sufficient. For example, if our text is a large eukaryotic
genome and we are looking for thousands of patterns representing possible transcription factor binding sites, then this trivial algorithm may be too expensive to be practical. It is less obvious that this problem can be solved much more efficiently by again
using suffix trees. By reading the text into a suffix tree and searching sequentially for
each pattern, we can solve this problem in time $O(|T| + |S_1| + |S_2| + \cdots + |S_m| + k)$,
where k is the number of times the patterns occur.

2.3   Mini Case Study: Intraspecies Phylogenetics

We will now look at an example of how a knowledge of classic computational problems can help one solve a real-world problem in biology: the intraspecies phylogeny problem. A brief history of this problem offers a fine example of how a classic computational problem can provide a beginning from which we can build practical methods for a real-world biological problem. A general phylogeny depicts the evolutionary relationships among a set of sequences of organisms. We saw a brief example of a phylogeny in chapter 1. An intraspecies phylogeny is in principle the same thing, but we take as an assumption of the model that all of the sequences are derived from members of a single species. That assumption has two important consequences. First, it means that we are assuming short time scales compared with what we look at when developing interspecies trees. Second, it means that the ancestral sequences are generally still present in the modern population.

We can see what an intraspecies phylogeny is by looking at an example. Suppose we examine a population which has four different genetic regions, each of which has two variants (called *alleles* in genetics). The first gene comes in alleles A and a, the second in alleles B and b, the third in C and c, and the fourth in D and d. Let us say that we have sequenced chromosomes from five organisms and observed the following sequences of alleles:

- ABCD
- aBCD
- abCD
- abCd
- abcD.

We would like to establish some evolutionary tree describing how the sequences may be related to each other. Figure 2.11 provides a possible answer.

Our problem here is to find some method to produce a "good" phylogeny, given our data. In accomplishing this task, we need to find a way to model the problem consistent with our prior knowledge that draws on the algorithmic tools available to us. As mentioned above, we have two key assumptions that distinguish this intra-

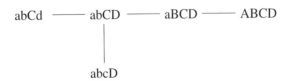

Figure 2.11
Hypothetical intraspecies phylogeny for five input sequences.

species phylogeny problem from the more familiar species tree phylogeny problems. First, common ancestors are usually present in the observed data in an intraspecies phylogeny problem. By contrast, in a species tree, we generally assume the observed data are at the leaves of the tree and that the internal nodes represent extinct, unobserved data points. Suppose we take it as an assumption of the model that all of the common ancestors of our sequences are in fact present in the data set. Then we will say that our input is a set of sequences and our output is a tree each of whose nodes is labeled with one of the input sequences.

Even given that definition, it is unclear which among the many possible trees fitting it is best. We can break the problem down a bit by inferring that we have some metric for evolutionary distance between individual nodes and then embedding that metric into our input. That is, our input becomes a graph of all observed nodes with weighted edges between all pairs of nodes representing the evolutionary distances between them. Our output is then a tree taken from that graph. Which tree to choose is based on how we assign edge weights and how we choose the optimal tree, given those edge weights. We can look at our second assumption, short time spans, to get an idea for a metric in this model. We would like to find a metric that fits that assumption and turns our problem into something we know how to solve. The assumption of short time spans suggests a specific definition of edge weight—the number of allele changes between two nodes—and a specific metric for optimization: parsimony. That is, we want the tree such that the number of mutations between all pairs of adjacent nodes is minimized. By now, we should recognize what we are solving here: we have defined the problem such that its solution is a minimum spanning tree of the input graph. That seems to be a success for us; we have found a reasonable model for our problem and we know how to solve it.

I have been arguing in this chapter that we need to know what tools are available because we want to find models with which we know how to work. In the real world, though, that is just the beginning of crafting a good model. Once we have a theoretically clean model, we then need to ask ourselves what compromises we have made in specifying that model and which of those compromises must be corrected for it to have practical value. This model makes many assumptions that may not be valid. The following are some examples:

• The parsimony criterion assumes all mutations are equally likely and that they occur independently of one another.
• The format of output, a single tree, hides ambiguity in the assignment of the phylogeny to the data.
• The format of input, as we have described it, neglects some important data likely to be available to us, such as population frequencies of individual nodes.

• The method does not include a model of errors in the data, so results may not be robust to the kinds of errors one would expect in sequencing genetic data.
• The method also lacks a model of missing data (e.g., if we have failed to observe some common ancestor of other observed sequences), and so may not be robust to small sample sizes.

We can easily come up with many other hidden assumptions. Model-building almost always involves trade-offs between realism and tractability, and we may decide that some of these are acceptable flaws for a particular application. Sometimes we may have to decide that a weakness of the model is tolerable because resolving it would impose unreasonable difficulties in solving for the model. Some weaknesses may be easy to solve; for example, if we have reason to believe that some alleles mutate at a different rate than others, we can then weight their contributions to edge weight accordingly and still end up with a minimum spanning tree problem. In other cases, though, we may need to augment our basic model in more complicated ways to make it practical.

Let us consider how to adapt our model to address the second flaw above: the loss of ambiguity in the data. For instance, suppose we observe the sequences AB, aB, Ab, and ab. Any of the trees shown in figure 2.12 would be an equally good solution to the problem by our metric because each has cost 3. Whether or not this is a problem depends on the application. In this case, we are generally trying to make a pronouncement of scientific truth—these sequences are related in this way—and it is not satisfactory for us to make such a pronouncement when we know aspects of the tree are completely arbitrary.

To resolve the problem we have observed here, we might propose that we revise our model of the output. Instead of having an edge for each inferred evolutionary relationship, we will have an edge for each relationship we think has a good chance of being present. Suppose we retain our parsimony formulation, but instead of having a minimum spanning tree, we consider the union of all minimum spanning trees. This is now a representation of evolutionary relationships that includes ambiguity. It may contain cycles, which yield multiple possible paths between different nodes. Where

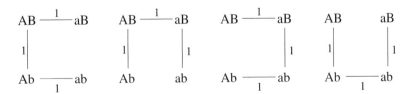

Figure 2.12
Four minimum spanning trees for the input sequences AB, aB, Ab, and ab, each with cost 3.

there are two paths between a given pair of nodes, we can consider them equally good possible explanations of how one evolved from the other.

This seems like a reasonable revision of our model, but it is useful only if we can still solve acceptably for it. It turns out that we can readily adapt our existing algorithms to this problem. Recall that in Kruskal's algorithm, we sort the edges, then successively join together the pair of subtrees not already joined that have the minimum weight edge between them. The algorithm itself can make an arbitrary decision at that point, in that it may have multiple edges of equal weight to choose from. It turns out that if we simply add all such minimum-weight edges connecting disconnected subtrees at each step, rather than just a single arbitrarily chosen edge, we end up with the union of all minimum spanning trees. We have thus found a way to revise our theoretically clean model to make it more realistic without compromising its computational tractability. What we have derived here is in fact a method from the computational genetics literature, originally proposed by Excoffier and Smouse [11], for finding what are called *minimum spanning networks*.

A more insidious problem is that our assumption that all of the common ancestors of observed sequences are also observed may not be valid. For example, suppose we observe the following sequences:

- ABC
- abC
- aBc.

If we see this set of data, we may reasonably infer that we are missing a sequence aBC which lies in between the three observed sequences. Maybe we did not look at a large enough number of samples and missed some sequences. Maybe the missing sequence simply died out through random genetic drift. In any event, we want to be able to infer this "missing" node. This is a case of the model being inadequate to the problem. Our prior model was not wrong to leave out these sequences, but a more useful model would include them.

We want to augment our model again to account for the possibility of these missing intermediates. One hypothetical solution would be to enumerate all of the possible missing nodes (i.e., all possible combinations of our alleles) and then solve the problem for a tree derived from a graph of all of these nodes. If the number of variable sites is not too large, we can construct this graph. The optimization is a little different from the minimum spanning tree, since we really want only a minimum tree that contains all of the observed nodes, rather than a tree containing every possible node. At first glance, it seems that this is a problem not too different from the minimum spanning tree problem, and perhaps may be solved by similar methods. In fact, though, this new variant is an NP-hard problem called the minimum Steiner tree problem, and will not be solvable for realistic data set sizes.

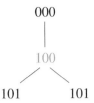

Figure 2.13
Inference of a median node (100) in order to produce an optimal phylogeny on input nodes 000, 101, and 110. Adding the inferred median (in gray) reduces the cost of the optimal phylogeny from 4 to 3.

Since we cannot reasonably neglect the intermediates, and we also cannot rigorously solve for a metric including them, we have to compromise. That is, we have to find some method that is tractable on reasonable data sets and that mostly solves the problem. One possibility is to use the Steiner tree model but use an inexact or heuristic method to solve it. One such heuristic solution to the Steiner tree problem has become widely adopted as a way of solving intraspecies phylogenies in practice. This heuristic relies on resolving one basic subproblem: given a triplet of nodes, find the fourth node that ought to occur between them. For example, given nodes 000, 101, and 110, we can infer that there is likely to be an intermediate node 100 between all of them, as in figure 2.13.

Since we cannot solve the general problem of finding these missing nodes (called *Steiner nodes*), we will try to solve this triplet problem for all triplets in the graph. We can accomplish this for binary (0-1) data by finding, for each triplet of sequences, the *consensus* of those three sequences. The consensus means the sequence with the most common bit value in each position. So, for example, the consensus of 000, 101, and 110 is 100 because two of the three inputs have 1 in the first bit, two of three have 0 in the second bit, and two of three have 0 in the third bit. Using this basic operation, we can create a heuristic method for finding a good solution for the entire graph:

1. Iterate through all triplets of nodes in the graph.
2. For each triplet, find its *consensus* sequence.
3. If the consensus is not already in the graph, add it.
4. If any consensus is missing, return to step 1.

This algorithm will augment the graph with all of the triplet intermediates. We can then find the minimum spanning tree, or minimum spanning network, on the augmented graph. Note that this tree is not in general going to be the most parsimonious tree having all of our sequences and all intermediates, since there may be better sets of intermediates we could derive by considering more than three nodes at a time. But it is an acceptable solution to our problem that is solvable on realistic data sets, and that is what we generally require in a real-world modeling situation.

What we have derived here is a practical method called the reduced median network method, due to Bandelt et al. [12], which has been widely used for this problem. This model is still too simple for some cases—for example, when the variable sites are not binary—and there is an even more complicated algorithm, the median-joining algorithm [13], for handling that case. Nonetheless, this simple study provides a good example of how recognizing common algorithmic problems can help us in practical model-building, and also of how we may have to move beyond our classic problems if we want an approach we can use in the real world.

References and Further Reading

The discussion of graph problems above, and the survey of primary literature below, are heavily indebted to the coverage by Cormen, Leiserson, Rivest, and Stein [14], one of the classic introductory texts on algorithms and an excellent reference for the topic in general. Other recommended texts for introductory discrete algorithms are Aho et al. [15], Knuth [16], and Kozen [17]. The discussion of string problems in this text draws on problems covered in much greater depth in Gusfield [2], a great reference for string algorithms in general.

Most of the problems and algorithms covered here have appeared at one time or another in the primary scientific literature. The introductory reader will almost certainly be better served by learning these methods from the textbooks mentioned above. Curious readers may, however, wish to refer to the following primary sources: Kruskal's algorithm [18], Prim's algorithm [19], Dijkstra's algorithm [20], the Bellman–Ford algorithm [21], [22], the Floyd–Warshall algorithm [23], Johnson's algorithm [24], the Edmonds–Karp algorithm [25], the Ford–Fulkerson method [22], the Hungarian method [26], the Gabow algorithm [27], and the Needleman–Wunsch [28] and Smith–Waterman [29] algorithms for sequence alignment. Suffix tree construction was first shown to be possible in linear time by Weiner [30], although a more memory efficient algorithm by Ukkonen [31] is now generally the preferred method. See Gusfield [2] for more information on applying suffix trees to exact set matching, the longest common subsequence problem, and various other uses in biological data analysis.

The case study on intraspecies phylogenies was based on three papers on the problem from the scientific literature [11], [12], [13]. The interested reader can learn more about this problem in general and other methods for it from Gusfield [2], Felsenstein [1], or Semple and Steel [3].

3   Hard Discrete Optimization Problems

Just as it is important to recognize when a problem we are looking at has been solved, it is also important to recognize a problem we are unlikely to solve. In computer science, problems are conventionally divided into the *tractable*, which admit polynomial-time exact algorithms in input size, and the *intractable*, which do not. Many of the most important problems that come up in practice are not known to have polynomial-time algorithms, but fall into the class of *NP-complete* problems. NP-complete problems can be defined in many ways, one of them being that these are the hardest problems for which it is possible to check the validity of a solution in polynomial time. There is not space in this text for a detailed examination of intractable problems. Readers unfamiliar with NP-completeness or related concepts in complexity theory may benefit by referring to one of the references discussed at the end of this chapter.

There are several important reasons to be able to recognize NP-complete problems when they come up. One is to avoid wasting time trying to solve problems we are not going to be able to solve. It is generally believed that NP-complete problems do not have polynomial-time solutions, so it is probably a good idea to stop looking for one once we know a problem is NP-complete. Tractability is an important consideration in model design, so recognizing that a model leads to hard problems is a good hint that we may need a different model. Subtle changes in model definitions can sometimes make the difference between tractable and intractable models, and one may notice that some of the hard problems we will discuss below seem like minor variations on tractable problems we discussed in chapter 2.

One other important argument for recognizing NP-complete problems is that the split into tractable and intractable is not so clear-cut with real-world problems, and there is likely to be a lot known about some problems that can help us. In practice, it is often less helpful to think of "intractable" problems as unsolvable than to think of them as a class that requires a different set of tools than the "tractable" problems. Depending on the needs of the problem (Do we need optimal solutions? Do we need them quickly?) and the nature of the data (Do we generally have small problem instances? Do they tend to be "easy" cases?), we will often be able to craft an acceptable solution to an "intractable" real-world problem. To make such decisions, though, it helps to know as much as we can about common problems we are likely to encounter.

We will now review some of the frequently occurring intractable problems and see what is known about alternatives for them. As with our quick examination of tractable problems, this chapter is far from exhaustive. It covers only a small selection of problems that are particularly likely to come up in biological modeling contexts. We will cover a set of graph problems and string problems, as well as some problems in set theory. By convention, these hard problems are posed as *decision problems*, for which we must answer a true/false question, rather than as optimization problems. Most have straightforward optimization problems associated with them that can be solved by solving a few instances of the decision problem.

3.1 Graph Problems

3.1.1 Traveling Salesman Problems

One of the best-known NP-hard problems is the traveling salesman problem (TSP). In this problem, we wish to determine whether it is possible to take a *tour* of the graph (a path that traverses each node exactly once) for which the sum of the edge weights is below some bound. The problem derives its name from the fact that it is a model of how a salesman might plan a route to visit a set of cities in which he wishes to sell his goods. The nodes of the graph represent the cities, and the edges represent routes between the cities, weighted by the distance or cost of travel. The salesman wishes to find a path that takes him to each city once while incurring the least time or cost. More formally, we have the following problem:

Traveling Salesman Problem

Input A directed graph $G = (V, E)$ where $V = v_1, \ldots, v_n$ with weight function $f : V \times V \to \mathscr{R}$ and a bound $B \in \mathscr{R}$

Question Does there exist a permutation of the nodes, $\pi_1, \ldots, \pi_n$ such that $f(\pi_n, \pi_1) + \sum_{i=1}^{n-1} f(\pi_i, \pi_{i+1}) \leq B$?

Figure 3.1 provides an example of a graph and a TSP solution for it.

TSP is also related to the Hamiltonian path problem, an NP-complete decision problem that does not have a clear associated optimization problem. The Hamiltonian path problem asks whether an unweighted graph has some path that visits each node in the graph exactly once. The problem can be formally posed as follows:

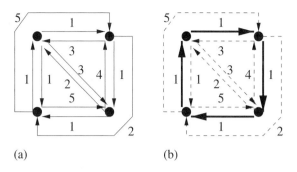

Figure 3.1
A traveling salesman problem. (a) A weighted, directed graph. (b) A minimum weight tour of the graph solving the TSP problem. The edges on the tour are shown with thick solid lines and the edges not on the tour with dashed lines.

Hamiltonian Path Problem

Input   A directed, unweighted graph $G = (V, E)$
Question   Does there exist a permutation of the nodes, $\pi_1, \ldots, \pi_n$ such that $(\pi_i, \pi_{i+1}) \in E$ for all $i = 1, \ldots, n - 1$?

3.1.2   Hard Cut Problems
It is quite common for a tractable problem to have a very similar but intractable variant. Some examples of this principle can be seen with cut problems. Recall that the minimum cut problem is solvable in polynomial time. There are several other problems involving cut inference that are NP-hard. One simple one is maximum cut. Maximum cut is not quite the same as simply flipping the optimization criterion in minimum cut, since that would yield a trivial problem. Rather, in maximum cut, we seek to partition the graph into two sets so as to maximize the cost of the cut separating those sets. The decision variant of this problem can be formally stated as follows:

Maximum Cut

Input   An undirected graph $G = (V, E)$ with weight function $f : V \times V \to \mathcal{R}$ and a bound $B \in \mathcal{R}$
Question   Does there exist a partition of V, $V = V_1 \cup V_2$, such that $\sum_{v_1 \in V_1, v_2 \in V_2} f(v_1, v_2) \geq B$?

Figure 3.2(a) and 3.2(b) provide examples of a graph and a maximum cut in that graph.

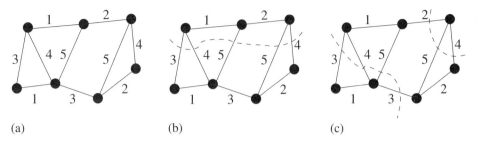

Figure 3.2
Hard cut problems. (a) A weighted, undirected graph. (b) A maximum cut solution for the graph. The
dashed line passes through the edges included in the cut, separating the nodes into two sets. (c) A maximum 3-cut solution for the graph. Two dashed lines pass through edges involved in the 3-cut, separating
the nodes into three sets.

We can also formalize *k-cut* problems, in which we wish to partition nodes into
k sets for some parameter $k > 2$. Figure 3.2(c) provides an example of a max 3-cut
for the graph of figure 3.2(a). Both maximum k-cut and minimum k-cut are NP-
complete.

3.1.3 Vertex Cover, Independent Set, and k-Clique
Two related problems that often come up in practice are vertex cover and independent set. In the vertex cover problem, we want a set of edges of minimum size (or
weight) such that every vertex in the graph has some edge in the cover incident on
it. The formal decision problem for the unweighted version of vertex cover is the
following:

Vertex Cover

Input   An undirected graph $G = (V, E)$ and a bound $B \in \mathcal{R}$
Question   Does there exist a set of vertices $V' \subseteq V$ such that for all $(u, v) \in E$, either
$u \in V'$ or $v \in V'$ and $|V'| \leq B$?

Figure 3.3(a) and 3.3(b) illustrate the vertex cover problem.
   In the independent set problem, we seek a set of nodes in a graph such that no two
nodes share an edge in the graph, where the total number of nodes is maximized. The
formal decision problem is the following:

Independent Set

Input   An undirected graph $G = (V, E)$ and a bound $B \in \mathcal{R}$
Question   Does there exist a set of nodes $V' \subseteq V$ such that there does not exist any
edge $(u, v) \in E$ where $u \in V'$ and $v \in V'$ and such that $|V'| \geq B$?

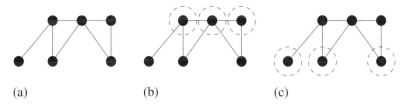

Figure 3.3
Vertex cover and independent set problems. (a) An unweighted, undirected graph. (b) A vertex cover for
the graph, marked by dashed circles around the nodes in the vertex cover. (c) An independent set for the
graph, marked by dashed circles around the nodes in the independent set.

Figure 3.3(c) shows an independent set solution for the graph of figure 3.3(a). It is
not a coincidence that the nodes in the vertex cover of figure 3.3(b) are precisely
those not in the independent set of figure 3.3(c). The independent set and vertex
cover problems are *duals* of one another, meaning that if we are given a vertex cover,
the nodes missing from it are an independent set, and vice versa. That directly
implies that a minimum-size vertex cover is the complement of a maximum-size independent set.

Independent set is also related to a third common problem, maximum clique. Informally, a clique is a subgraph in which all pairs of nodes have edges between them.
The maximum clique problem is the problem of identifying the largest clique in an
input graph. Maximum clique is formally defined as follows:

Maximum Clique

Input An undirected graph $G = (V, E)$ and a bound $B \in \mathcal{R}$
Question Does there exist a set of nodes $V' \subseteq V$ such that $\forall v_1, v_2 \in V'.(v_1, v_2) \in E$,
where $|V'| \geq B$?

Figure 3.4(a) and 3.4(b) show a sample graph and a maximum clique in that graph.
Clique is related to independent set (and thus to vertex cover) because a clique in a
graph corresponds to an independent set in the the complementary graph (that in
which we flip which node pairs have edges). Figure 3.4(c) and 3.4(d) show this relationship by repositioning the nodes of the graph of figure 3.4 to show that it is in fact
the complement graph of that in figure 3.3.

3.1.4 Graph Coloring
Another frequently occurring class of graph problems is called graph coloring.
Graph coloring is applied to an undirected, unweighted graph. The goal of a graph
coloring problem is to assign a distinct "color," which can be thought of as an integer label, to each node in a graph. This needs to be done in such a way as to ensure

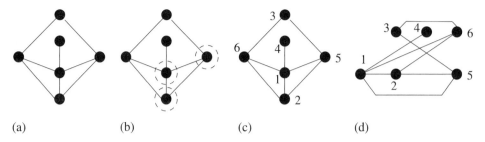

Figure 3.4
A clique problem. (a) An unweighted, undirected graph. (b) A maximum clique in the graph. (c) An arbitrary labeling of nodes in the graph. (d) Repositioning of the labeled nodes to illustrate that the graph in this figure is the complement of that in figure 3.3.

Figure 3.5
A 3-coloring of a graph. The coloring is minimal for this graph.

that no edge connects two nodes of the same color. In the optimization variant, we typically want to solve the problem using as few distinct colors as possible. This minimum number of colors is called the *chromatic number* of the graph. More formally, the decision variant of the problem is the following:

Graph k-Coloring

Input   An undirected graph $G = (V, E)$ and a bound k
Question Does there exist a mapping $c : V \to \{1, \ldots, k\}$ such that $\forall (u, v) \in E.c(u) \neq c(v)$?

Figure 3.5 illustrates a graph coloring for the graph of figure 3.3(a). Graph coloring comes up in practice in many problems in which we are trying to apportion finite resources to several users and need to figure out a way to make sure no two users try to use the same resource at the same time. The problem is easily solvable in polynomial time for $k = 2$ but is hard for any fixed $k \geq 3$. There are many kinds of specialized graphs for which the problem is not hard, though.

3.1.5   Steiner Trees

Another problem that has particular relevance to the biological world is the Steiner tree problem. We saw this problem briefly in chapter 2 in the context of intra-

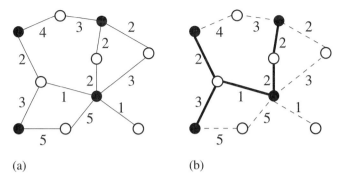

(a)   (b)

Figure 3.6
A Steiner tree problem. (a) An input graph in which nodes required to be in the output tree (terminal nodes) are shown as solid circles, and those that need not be in the output tree (potential Steiner nodes) are shown as open circles. (b) A Steiner tree for the graph.

species phylogenies. Given a graph and a subset of its nodes, called the *terminal nodes*, a *Steiner tree* is a tree containing all of the terminal nodes, as well as possibly others from the graph. The minimum weight Steiner tree problem is defined as follows:

Minimum Weight Steiner Tree

Input   A graph $G = (V, E)$, a weight function $w : E \rightarrow \mathcal{R}$, a subset of the vertices $S \subseteq V$, and a bound $B \in \mathcal{R}$
Question   Does there exist a tree $T = (V', E')$ for which $S \subseteq V' \subseteq V$ and $\sum_{e \in E'} w(e) \leq B$?

In this formalization, the nodes in V' that are not terminal (i.e., $V' - S$) are called *Steiner nodes*. Figure 3.6 gives an example of the Steiner tree problem.

Intraspecies phylogeny and similar problems in molecular evolution are some of the major motivations for studying Steiner tree problems. Often, as in intraspecies phylogeny, we are not explicitly given the Steiner nodes but must infer their existence. When our nodes are labeled with DNA bases, we may assume that any string of bases not in our input is a Steiner node in the graph. In such cases, we need the weight function to be specified in such a way that we can compute it between pairs of nodes not explicitly in the input. This leads to a commonly seen special case of the Steiner tree problem, in which the terminal nodes are bit strings of fixed length n, all other bit strings are presumed to be present as potential Steiner nodes, and the cost of any edge is the number of bit flips between the endpoints (called the *Hamming distance*). This bit-string Steiner tree problem is also NP-hard.

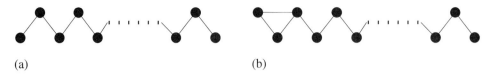

(a)   (b)

Figure 3.7
Demonstration that bipartiteness is nontrivial. (a) A bipartite graph that can be infinitely extended
while remaining bipartite. (b) A nonbipartite graph that can be infinitely extended while remaining nonbipartite. The existence of infinite families of bipartite and nonbipartite graphs proves the property to be
nontrivial.

3.1.6 Maximum Subgraph or Induced Subgraph with Property Π

Finally, we will look at one very broad class of graph problems, all of which are
hard, involving what are called *inheritable* graph properties. A graph property is
called inheritable if, when given a graph possessing the property, removing any
node or edge leaves us with a graph possessing the property. A property is considered
nontrivial if there are an infinite number of graphs that have the property and an
infinite number that do not. An example of a nontrivial, inheritable property is
bipartiteness. Recall that a bipartite graph is defined by the fact that nodes can be
partitioned into two groups (parts) with no edges between any two nodes in the
same group. Figure 3.7 shows that bipartiteness is nontrivial, since there are both bipartite and nonbipartite graphs that can be infinitely extended. Furthermore, removing a node or edge cannot cause an edge to appear between two nodes in the same
part, so bipartiteness is inheritable.

We further need to define a *subgraph* of a graph $G = (V, E)$ to be a graph
$G' = (V, E')$ where $E' \subseteq E$. An *induced subgraph* of a graph $G = (V, E)$ is a graph
$G' = (V', E')$ where $V' \subseteq V$ and $E' = \{(u, v) \in E \mid u, v \in V'\}$. Perhaps surprisingly, it
is NP-hard, given a graph, to find a subgraph or induced subgraph of maximum size
possessing any nontrivial, inheritable property. Some of the problems we have already seen are special cases of one or the other of these two general classes of hard
problem. For example, independent set can be cast as a maximum induced subgraph
problem, where the property Π is independence.

3.2 String and Sequence Problems

3.2.1 Longest Common Subsequence

We discussed a problem by the same name in the context of tractable problems in
chapter 2. This is exactly the same problem, except that now we consider a set of arbitrarily many sequences, not just a pair of sequences. More formally, our problem is
the following:

Longest Common Subsequence

Input A set of sequences $w_1 = (w_{11}, w_{12}, \ldots, w_{1n})$, $w_2 = (w_{21}, w_{22}, \ldots, w_{2n})$, $\ldots$, $w_k = (w_{k1}, w_{k2}, \ldots, w_{kn})$ and a bound B
Question Is there a sequence $s_1, s_2, \ldots, s_m$ such that $s_1, \ldots, s_m$ is a subsequence of each w_i, where $m \geq B$?

For example, the sequences ACGAT, CGAAT, and GCATA can be seen to have the subsequence CAT as follows:

```
ACGA T
 CGAAT
GC A TA
─────────
 C A T
```

The tractable version we have seen is the special case of this problem for $k = 2$. The general case can be solved with dynamic programming, but the runtime will be the product of the lengths of all of the sequences, making it exponential in k. The problem for general k is relevant as a model for multiple sequence alignment problems in computational biology. Because it is intractable, though, it is not so useful as a basis for designing methods for more biologically reasonable variants. The longest common sub*string* problem remains tractable for arbitrary k, and in fact can be solved in linear time in the sum of the sequence lengths by nearly the same suffix tree algorithm as we described for the two-sequence case.

3.2.2 Shortest Common Supersequence/Superstring

Instead of finding the longest sequence contained in a set of input sequences, we can look for the shortest sequence that contains all of a set of input sequences. This is another NP-hard problem called the shortest common supersequence problem. More formally, the problem is the following:

Shortest Common Supersequence

Input A set of sequences $w_1, w_2, \ldots, w_k$ and a bound B
Question Is there a sequence $s_1, s_2, \ldots, s_m$ such that each w_i is a subsequence of $s_1, \ldots, s_m$ and $m \leq B$?

For example, the shortest common supersequence of the inputs ACGAT, GCAAT, and GCATA is AGCGAATA, as shown by the following alignment:

```
A CGA T
  CGAAT
 GC A TA
─────────
AGCGAATA
```

The string analogue to this problem, the shortest common superstring, is also hard. It is formally posed as follows:

Shortest Common Superstring

Input   A set of strings $w_1, w_2, \ldots, w_k$ and a bound B
Question   Is there a string $s_1, s_2, \ldots, s_m$ such that each w_i is a substring of $s_1, \ldots, s_m$ and $m \leq B$?

For example, we can find the shortest common superstring of ACGTTA, TAGCCT, ACAATA, and CCTACA through the following alignment:

```
ACGTTA
   TAGCCT
      CCTACA
         ACAATA
ACGTTAGCCTACAATA
```

It is trivial to find a superstring simply by concatenating all of the strings in the input, but it is hard to find the smallest one. The shortest common superstring problem has particular biological importance because it is a model of sequence assembly, a topic we will cover as an extended case study in chapter 4.

3.3   Set Problems

In addition to graph and sequence problems, we often encounter problems related to sets. Sets are a more general way of representing many of the same kinds of problems as graphs, and thus will sometimes come up as generalizations of graph problems.

3.3.1   Minimum Test Set
Our first problem is called the minimum test set problem, and it is defined as follows:

Minimum Test Set

Input   A set $S = \{s_1, s_2, \ldots, s_n\}$; a collection of subsets of S, $C = \{c_1, c_2, \ldots, c_m\}$ where for all i, $c_i \subseteq S$; and a bound B
Question   Does there exist a set $C' \subseteq C$ such that for each $s_i, s_j \in S$ there exists some $c \in C'$ containing exactly one of s_i and s_j, for which $|C'| \leq B$?

One way to think of what this problem means, and to see a potential connection to biology, is to imagine we are constructing a DNA test. We have a set of n organisms and we look at a piece of DNA that occurs in a different variant in each organism. If we ignore the sites that are the same between them and look only at the sites that differ between at least some organisms, we may see the following:

organism 1 AATAA
organism 2 TCCGG
organism 3 ACCAA
organism 4 TATAG
organism 5 TCTGA

We want to be able to examine a DNA sample and identify which organism provided it, using as few bases as possible. Suppose we define our sets $c_1, \ldots, c_m$ so that c_i is the set of sequences having the more common base at variant site i. So, for example, $c_1 = \{2, 4, 5\}$ because T is the more common base at site 1 and organisms 2, 4, and 5 have that base. Then a minimum test set is the solution to our problem, since it will give us a set of bases that can distinguish between any two organisms. Here, $C' = \{c_1, c_2, c_3\}$ will work, since looking at the first three bases allows us to distinguish any two sequences in the input. This particular example is a real-world problem called "tagging SNP selection" [32].

3.3.2 Minimum Set Cover
The following problem has a statement very similar to the minimum test set, but asks a subtly different question:

Minimum Set Cover

Input   A set S, collection C of subsets of S, and a bound B
Question   Does there exist a $C' \subseteq C$ such that for all $s \in S$ there exists a $c \in C'$ for which $s \in c$, where $|C'| \leq B$?

To think of what this problem is asking, we may think of a different, somewhat contrived medical problem. Imagine we have a sick patient and we have narrowed down the cause of his disease to a set of possible bacteria. Each bacterium is sensitive to certain kinds of antibiotics and not to others. We want to give the patient as few drugs as possible. If we define S to be our set of bacteria and each $c_i \in C$ to be the subset of bacteria affected by antibiotic i, then a minimum cover would give us the smallest possible set of antibiotics that affects every bacterium. The vertex cover problem we saw earlier in the chapter can be thought of as a special case of set cover in which we have a set for each edge containing the two endpoints of the edge.

3.4 Hardness Reductions

Our main goal in studying the preceding problems is to recognize when we have developed a model that leads to a hard optimization problem. When trying to prove a problem is hard, we do not need to show that it is identical to some problem we know to be hard. It is enough to show that it can be *polynomially reduced* from a problem we know to be hard (i.e., that there is a polynomial-time transformation that converts a known hard problem into the new problem). For example, suppose we do not know that the independent set problem is hard. We will be able to show it is hard by constructing a *reduction* from the vertex cover problem to the independent set problem. Suppose we are given a graph $G = (V, E)$ and are asked to find a minimum-size vertex cover in it. Then we can propose that we will accomplish this by first finding a maximum independent set, V', in the graph and then returning $V - V'$ as our vertex cover. If the maximum independent set is efficiently solvable, then this strategy will efficiently solve the vertex cover. Conversely, if the vertex cover is hard, then this strategy cannot work, and thus independent set must be hard as well. We can use similar reductions to show that clique is hard because we can reduce vertex cover or independent set to it, or that set cover is hard because we can reduce vertex cover to it. It is often very difficult to prove that a problem is NP-hard in this way. Those needing to do so would be well advised to study a more thorough text specifically on NP-completeness where they can find less trivial examples of these reductions and various strategies for finding them. Some suggested texts are listed in the section References and Further Reading.

3.5 What to Do with Hard Problems

One reason for studying hard problems is so we can try to develop models that avoid them, or at least recognize when we have a model for which we will not be able to develop provably efficient optimization methods. But that does not mean that if we come up with a model and it turns out to yield a hard problem, we should immediately give up on that model. It is certainly worth considering whether small changes to the model might make it tractable. But even if we cannot make our model tractable, we still have options. This section is an attempt to provide a nonexhaustive but useful guidebook to some of the options we can consider when confronted with a hard problem. There are other, more sophisticated options than those outlined below available, but this is a set that should be reasonably accessible to anyone who can follow the rest of this chapter.

Make Sure the Problem Is Really Hard

One of the most common mistakes in working with NP-hard problems is to assume that because a problem of interest can be cast as an NP-hard problem, it actually is NP-hard. Remember that proofs of NP-hardness actually work in the opposite way: to show our problem is NP-hard, we must show that all instances of a known NP-hard problem can be converted into instances of our problem. The reverse does not follow because many NP-hard problems are not hard if restricted to a subset of possible inputs. If we are more specific about the assumptions of a seemingly hard problem, then the problem may become tractable. Some things that often make a big difference in the tractability of a problem are the following:

• For graph problems, special classes of graphs (e.g., planar, bounded degree, chordal, overlap graph) may yield tractable problems.
• Hard string problems may be tractable with bounded numbers or sizes of strings or with bounded (or nonbounded) alphabets.
• Set problems may be tractable when set sizes are bounded.
• Many error correction problems are tractable when the number of errors is bounded.
• Any intractable problem becomes tractable when the total input size is bounded.

Brute Force

Even if our problem is NP-complete, it is always solvable by trying every possible solution. That approach may be theoretically unsatisfying, but in the real world it is often a perfectly valid solution. For example, with the "tagging SNP" variant of the test set problem mentioned earlier in this chapter, sets with up to about 20 variant bases are likely to be easily solvable in practice by trying all possibilities. In fact, brute force is often the best way to solve a problem in the real world because it is generally easier to code quickly, and may actually run faster than more sophisticated methods on small problem instances. So before going to more advanced methods, we should always ask if our real-world problem involves problem instances that justify those more advanced methods.

Approximation Algorithms

If we have ruled out the two "easy" options above, then we should start considering the more advanced techniques. Approximation algorithms are one such class of methods. An approximation algorithm is a tractable algorithm for an optimization problem that does not exactly solve the problem, but gets a solution that is guaranteed to be close to an optimal solution. For example, we cannot efficiently find the smallest possible vertex cover of a graph, but there is an efficient algorithm to find a vertex cover that is at most twice the size of the smallest one: find a maximal matching in the graph and return all the endpoints of the edges in the matching. (Note that a maximal matching only means that we cannot add any additional edges without the set's no longer being a matching. A maximal matching need not be a maximum matching, but a maximum matching is necessarily maximal.) This set of edges will be a vertex cover because any edge uncovered could have been added to the matching, meaning that the set leaves an edge uncovered only if the matching is not maximal.

The set will not generally be of minimum size, but it cannot be more than twice the minimum size because we add two points per edge of the matching, and we must add at least one point per edge of the matching to cover all edges. This is known as a 2-approximation algorithm.

Different problems are approximable to different degrees. Some problems have α-*approximation algorithms*, in which we have a method which guarantees that we find a solution within some factor of α of the quality of the best possible solution. Others have *approximation schemes*, in which we can solve a problem within some factor ε of the optimum for any possible $\varepsilon > 0$, but with runtime exponential in $1/\varepsilon$. In other cases, approximations are possible only within a factor of some function of the problem size (e.g., a $\log(n)$ approximation). Even when a problem proves difficult to approximate, it will often be the case that special cases of that problem will prove much easier. If a problem has been proved NP-hard, there is a good chance that its approximability has also been studied. When considering approximation algorithms as an option, it is therefore always useful to start with a literature search to see whether the general problem or any relevant special cases are approximable.

A particularly elegant example of an approximation algorithm is described by Garey and Johnson [33] for the triangle-TSP problem. Triangle-TSP is the traveling salesman problem when edge weights are required to obey the triangle inequality:

$$w(u, w) \leq w(u, v) + w(v, w).$$

This restriction is valid for many practical TSP instances. For example, if the edge weights are actually physical distances on a map, then they will obey the triangle inequality. The algorithm works as follows:

1. Find a minimum spanning tree on the graph.
2. Take a tour of the spanning tree, starting at any arbitrary node and walking across all nodes in depth-first order.
3. Return the order in which nodes are first reached in the depth-first search as the solution to the traveling salesman problem.

Figure 3.8 illustrates this method. The resulting tour is guaranteed to have a cost no more than twice that of the optimal tour in the graph. Informally, the guarantee comes from the fact that we know that walking the tree along the MST edges and going back when we reach a dead end will cost at most twice the weight of edges in the tree. The triangle property guarantees that a direct tour among these nodes cannot cost more than this walk along the MST. Furthermore, the optimal TSP tour cannot have smaller weight than the MST, since the TSP tour is a tree in the graph plus one edge. Thus, the tour we find cannot have cost more than twice the TSP tour.

This algorithm specifically depends on the triangle property and will not work for general TSP. In fact, general TSP is not approximable to any constant factor. Eucli-

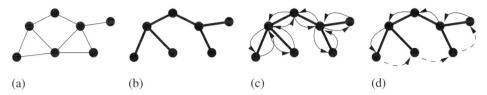

Figure 3.8
Illustration of the 2-approximation algorithm for triangle TSP. (a) An initial graph (edge weights are assumed but not shown in the figure). (b) A possible MST for the input graph. (c) A walk along the graph in depth-first order. (d) Conversion of the depth-first walk into a tour by short-circuiting paths that include nodes already visited.

dean TSP, a special case of triangle TSP in which vertices are treated as points in a Euclidean space and edge weights are linear distances between those points, has an approximation scheme and thus is approximable to any constant factor. Just as a hard problem may be tractable for some special cases of inputs, so that problem may be more accurately approximable for some special cases of inputs even when it is still hard.

It is worth noting that approximation algorithms are popular in pure theoretical computer science circles, but have a bad reputation among computational biologists. The reason is that it is easy to misuse approximation algorithms by developing them without regard to whether an approximation makes sense for the real-world problem being solved. For instance, if we are performing a sequence assembly and our program returns a genome that is twice the size of the actual genome, that will generally be considered a useless result by biologists. Before using an approximation algorithm to solve a real-world problem, one should always consider whether the approximation will actually be useful to someone who cares about that problem.

Branch-and-Bound
Branch-and-bound is a general class of techniques that can be very useful when we genuinely need optimal solutions but are dealing with problems too large for brute force methods. Branch-and-bound methods have exponential runtime in the worst case, but often have reasonable runtimes up to much larger problem sizes than pure brute force methods do. At a high level, the idea behind branch-and-bound is to do a nearly brute force search by building up a solution in pieces, but abort and back up whenever the partial solution provably cannot lead to an optimal solution. This can generally be represented as a strategy for searching a tree of possible solutions where each branch we take from the root to a leaf represents one possible choice in constructing a solution.

We can illustrate the approach with a very simple variant, shown in figure 3.9, for the graph coloring problem. We can color the graph of figure 3.9(a) by performing a

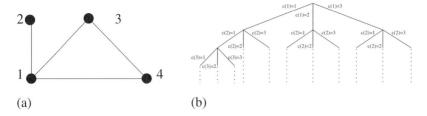

(a)   (b)

Figure 3.9
A highly simplified branch-and-bound solution to the graph coloring problem. (a) A graph we wish to color. (b) A tree of possible solutions. Each edge in the tree represents a choice to color a particular node a particular color.

`vertex-cover-bb` $(G = (V, E)$, K, R, $opt)$
1. If there exists some $u \in R$ such that $(u, v) \in E$ then *reject*
2. If $R + K = V$ and $|K| < opt$ then return $|K|$
3. Let $G' = (V', E')$ where $V' = V - K - R$ and $E' = \{(u, v) \in E | u \in V' \cap v \in V'\}$
4. Find a 2-approximation C to the minimum vertex cover on G'.
5. If $|C|/2 + |K| > opt$ then *reject*
6. Pick any node v in $V - K - R$.
7. $opt \leftarrow$ `vertex-cover-bb` $(G', K, R \cup \{v\}, opt)$
8. $opt \leftarrow$ `vertex-cover-bb` $(G', K \cup \{v\}, R, opt)$
9. return opt

Figure 3.10
Pseudocode for a branch-and-bound algorithm for vertex cover using a vertex cover 2-approximation algorithm. To solve for a graph G, we would invoke the method with a call to `vertex-cover-bb`$(G, \emptyset, \emptyset, C)$, where C is the cost of a vertex cover 2-approximation on G.

depth-first search of the decision tree in figure 3.9(b). If we simply try each possible leaf of the tree and test whether it is a valid coloring and how many colors it uses, we will have a brute force approach. In a branch-and-bound approach, we stop at each internal node of the decision tree and see if the latest decision has created an invalid coloring. Furthermore, we can see if the cost as of that internal node exceeds our best known cost for the whole problem. For example, if we know of a three-color solution, then we will stop searching at any node that adds a fourth color to the tree, even if it is not a leaf node.

A practical branch-and-bound algorithm will generally use a more sophisticated method to try to prove that a solution is nonoptimal. One good way of doing this is through approximation algorithms. For example, suppose we are trying to solve a vertex cover problem on a graph $G = (V, E)$ and we decide that we must have an optimal solution. Figure 3.10 provides pseudocode showing how we can use a 2-approximation algorithm for vertex cover to accelerate a search for an optimal vertex cover solution. The method works through a series of recursive calls to build two sets, K and R, representing the nodes that we propose to be in the vertex cover and those proposed to be removed from it. The approximation algorithm allows us to test at each step whether the number of nodes kept so far plus the minimum possible number needed in the rest of the graph is smaller than our best known solution. If it is not, then the method cannot find an optimum along its current branch of the search tree, allowing it to reject that partial solution and back up. This procedure could potentially allow us to avoid searching a large fraction of the possible solutions without missing any true optima.

What can we do if we do not have an approximation algorithm for our problem? One general and broadly useful approach uses what is called a *linear programming relaxation*. In this approach, we convert our problem into a special NP-complete problem called an *integer linear program*. There are many variations on integer linear programs, but a typical statement (for a version called 0–1 integer programming) is the following:

Input   A set of variables $x_1, \ldots, x_n$, a set of *linear constraints*

$$a_{11}x_1 + a_{12}x_2 + \cdots + a_{1n}x_n \leq y_1$$

$$a_{21}x_1 + a_{22}x_2 + \cdots + a_{2n}x_n \leq y_2$$

$$\vdots$$

$$a_{m1}x_1 + a_{m2}x_2 + \cdots + a_{mn}x_n \leq y_m$$

and an *objective function* $c_1x_1 + c_2x_2 + \cdots c_nx_n$.

Output   An assignment of the value 0 or 1 to the variables $x_1, x_2, \ldots, x_n$ that is consistent with all of our constraints.

Metric $c_1x_1 + c_2x_2 + \cdots c_nx_n$ is maximized (or minimized).

Any NP-complete problem can be cast in this form. We then solve the same problem, except that we allow our variables to be any real number between 0 and 1, and not just the integers 0 or 1. Surprisingly, while the integer variant is NP-hard, the real-valued variant (called a *relaxation* of the integer problem) is solvable in polynomial time. The real-valued variant is guaranteed to give us a metric value at least as good as the 0–1 variant, but often not very much better, making it generally a good bound on the value of the solution of the original problem. Furthermore, it is generally easy to adapt the linear program to reflect a partially solved problem, making it ideal for branch-and-bound formulations. There are several software packages available, some commercial and some free, that have highly optimized solvers for the integer and real-valued versions of these problems, making them often the best choice in practice when one needs to solve a difficult NP-hard problem. We will revisit linear programming in chapter 6.

Heuristic Methods

A heuristic method is essentially any method that seems to work well but cannot be proved to do a good job in any objective sense. Heuristics are often problem-specific, and this book cannot give better advice than just to think about the problem and see if any tricks come to mind. There are a few very general techniques, though, that people often try when they have no idea how to solve a problem of interest. Interestingly, some of the most commonly used techniques were actually inspired by real-world systems in statistical physics and biology.

One such technique is called *simulated annealing*. Annealing is a technique for hardening materials by heating them to a high temperature and gradually cooling them. Simulated annealing is named by analogy to this actual physical process. At a high level, simulated annealing works as follows:

1. Define a series of moves that allow us to transform any solution to our problem into any other. An example of a move would be the conformational changes in our protein lattice model that we saw in chapter 1.
2. Define the energy of each solution to be the negative of its cost.
3. Run the Metropolis method (from chapter 1) on the problem while gradually reducing the temperature to zero.
4. Return the solution that the system "freezes" into when it reaches absolute zero temperature.

Intuitively, a system at a high temperature rapidly jumps between possible solutions. It settles down into a local energy minimum when it cools. If we run this method with a fixed low temperature, we will get stuck in a *local optimum* similar to our initial state but potentially much worse than the *global optimum* among all possible solutions. If we run at a fixed high temperature, we will jump rapidly between very different solutions but will be likely to skip over the optima, local or global. By cooling slowly, it is believed that we can reasonably compromise between the need to explore a large fraction of the space and the need to settle into good solutions as we encounter them. We will see more about this technique when we cover the Metropolis method in greater depth in chapter 9.

Another very general heuristic technique is called a *genetic algorithm*. Genetic algorithms have nothing per se to do with solving problems in genetics, although you can use them for that purpose. They try to find a good solution to a problem by mimicking the process of genetic evolution. In a genetic algorithm, we maintain a set of possible solutions and try to "evolve" them into better solutions. This evolution is accomplished by two operations:

- *Mating*   merge two solutions to construct a hybrid solution.
- *Mutation*   randomly change a solution in some way.

The genetic algorithm typically proceeds by first generating a new set of candidate solutions by choosing some random pairs of existing solutions, mating the members of each pair, and mutating the results. It then evaluates the quality of the new solutions and picks some subset of old and new solutions as its guesses for the next round of the algorithm. For example, we may start with a set of ten solutions, randomly pick ten pairs from those to mate and mutate, and then pick the five best old solutions and the five best new solutions as our solution set for the next round. This process is continued for some number of rounds, and eventually the best solution is returned.

If we want to use a genetic algorithm to solve vertex cover, for example, we can generate a candidate solution by randomly picking nodes until all edges are covered. Repeating this process a few times will give us an initial candidate set. We can then mate two solutions by merging their node sets and randomly deleting nodes until no more can be removed without leaving some edge uncovered. We can mutate a solution by randomly deleting some node, then adding in the other endpoints for any edges left uncovered by that deletion. Putting these operations together will give us a heuristic that may do reasonably well at finding good vertex covers efficiently.

The Kitchen Sink Approach

This is what people generally really do in practice when they need to solve hard instances of a hard problem: throw in whatever tools help. For example, we may use brute force for small problems or subproblems, do branch-and-bound with approximation algorithms for larger problems, and use various heuristics to improve on the solutions at leaves of the tree.

Now Give Up

Sometimes we really do have to recognize that we cannot find a useful solution for a given model of our problem, and go back to the drawing board to find a new and hopefully more tractable model.

References and Further Reading

The classic reference for the concept of NP-completeness is Garey and Johnson [33], which provides an excellent introduction to the theory and to the concepts of hardness proofs and approximation, which receive only perfunctory attention in this chapter. Garey and Johnson also has a thorough compilation of NP-hard problems, including those discussed here and many others from different disciplines. When the hardness of a problem is in doubt, it is often advisable to scan the problem list in Garey and Johnson to see if the problem has already been studied in some form or for suggestions of similar problems with which one may build a reduction. Crescenzi and Kann [34] created a Web repository of NP-hard problems to serve a similar function that is more up-to-date. A version of their compendium is in Ausiello et al. [35]. Their compendium is also a great place to find whether a given problem or anything like it has been studied. All three sources also provide information on hardness and approximation results for various specialized versions of many of the more studied problems.

The primary literature offers many references for the general topic and for the specific problems covered here. This text cannot offer an exhaustive list, but can provide primary references for key topics and problems. The theory of NP-completeness in general derives from a paper by Cook [36], which developed a famous NP-complete problem we have not covered, called circuit satisfiability. Primary literature references for NP-completeness proofs are available for many of the specific problems we have covered. Many were originally established as NP-complete in a classic book chapter by Karp [37], including Hamiltonian path, maximum cut, minimum vertex cover, maximum independent set, k-clique, graph k-coloring, Steiner trees, minimum set cover, and integer programming. Other primary proofs are scattered throughout the literature: the traveling salesman problem [38], maximum/minimum k-cut [39], maximum subgraph/induced subgraph with property Π [40], [41], [42], longest common subsequence/shortest common supersequence [43], shortest common superstring [44], and minimum test set [33].

We also discussed several approximation algorithms for which primary citations are available. The 2-approximation for triangle TSP that we presented is due to Rosenkrantz et al. [45], and was derived for this text from Garey and Johnson [33], but the problem has a more complicated $\frac{3}{2}$ approximation algorithm due to Christofides [46]. The Euclidean TSP approximation scheme is due to Arora [47]. We also briefly discussed a 2-approximation algorithm for the minimum vertex cover, a method attributed to Gavril and Yannakakis but for which our direct source was Cormen et al. [14]. Two other algorithms are available that offer somewhat better bounds for this problem: $2 - \frac{\log \log |V|}{2 \log |V|}$ [48], [49] and $2 - \frac{2 \ln \ln |V|}{\ln |V|} (1 - o(1))$ [50]. There are many other hardness, approximability, and nonapproximability results available for numerous variations on the problems we have examined. We again refer the reader to Ausiello et al. [35], Crescenzi and Kann [34], and Garey and Johnson [33] as starting points for learning more.

Some of the suggestions for practical solution of NP-hard problems above are too obvious to have specific citations, but others can be traced to the primary literature. The simulated annealing method was independently proposed by Kirkpatrick et al.

[51] and Cerny [52]. The earliest known description of a genetic algorithm is due to Barricelli [53].

Readers looking for more background on the general topic of complexity theory would do well to check out Sipser [54] for an introduction to complexity theory in general or Papadimitriou [55] for more advanced theory. Guidance on solving NP-hard problems in practice can be found in many sources. Garey and Johnson [33] provides many useful general suggestions. Hochbaum [56] is a good source for advanced material on the design and use of approximation algorithms.

4 Case Study: Sequence Assembly

This is the first of several full chapters devoted to a single biological topic. Though the point of this text is to learn broadly useful computational methods and how to apply them to new problems, case studies can be a good way to learn about some of the issues that come up in applying the theory we study to the real world. Our topic here is sequence assembly, the problem of figuring out the sequence of a genome using technologies for sequencing small numbers of bases at a time. To begin, we first need to know a bit about the sequencing technologies that produce the data used to assemble a sequence. We can then see some of the ways computational biologists have modeled the assembly problem and how they have worked with these models in practice.

4.1 Sequencing Technologies

There are two classic methods for DNA sequencing: the Maxam–Gilbert (chemical cleavage) and Sanger dideoxy (terminator base) methods. Modern methods we will cover below are primarily based on the Sanger dideoxy method. Both were originally more or less comparable, though, and it is useful to understand how each works. We will then look at how these methods evolved into the standard techniques of today.

4.1.1 Maxam–Gilbert

The Maxam–Gilbert method works by selectively breaking DNA strands at instances of a single base. By looking at the resulting strand lengths, which we can infer through a technique called *polyacrylamide gel electrophoresis*, we can infer where that base occurs in the sequence. Repeating for all four bases then gives us a complete DNA sequence. Assume that we have many copies of our strand to be sequenced, typically because we have used PCR to amplify a region of interest. The Maxam–Gilbert method can be summarized as follows:

1. Attach radioactive phosphorus (P^{32}) to one end of one DNA strand.

2. Break the strand preferentially at a single kind of base (using a specialized chemistry developed for each of the four bases).

3. Run the resulting broken strands on a polyacrylamide gel and look for radioactive bands.

4. Determine positions of the cleavage base from the lengths of the strands on the gel.

To illustrate this method, suppose we start with the strand

CTACGCCT-P^{32}.

If we cut this strand immediately before each C base, then our resulting radioactive strands would be

CTACGCCT-P^{32}

CGCCT-P^{32}

CCT-P^{32}

CT-P^{32}

We ignore the pieces that do not contain the radioactive phosphorus because they will not show up as radioactive bands on our gel. If we run these strands out on a gel, we will see something like the first column in figure 4.1. Repeating the experiment for the other bases will result in something like the other columns of the gel in

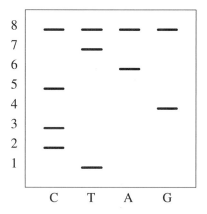

Figure 4.1
A polyacrylamide gel that might be produced by Maxam–Gilbert sequencing from the sequence CTACGCCT. Vertical band positions correspond to different strand lengths, labeled at the left edge of the gel. Each column of the gel shows the lengths of strands produced by breaking preferentially after the base at the bottom.

figure 4.1. Then it should be easy to see how to read off the sequence of the strand simply by looking at which column yields which length of fragment. For example, we see only a single-base strand when we break at T, so the last base must be T. We see only a two-base strand when we break at C, so the next-to-last base must be C. And so forth. Proceeding in this fashion allows us to read off the entire sequence with the exception of the base farthest from the P^{32}.

This seems like a very good general sequencing method, but it presents some practical problems. First of all, working with radioactive material is somewhat inconvenient. And the process of running and reading gels is labor-intensive. The most important problem, though, is that the method does not scale well. If we tried to sequence a very long strand in this way, the bands of the gel would start to run together and would become hard to resolve with certainty. Furthermore, it would simply be very hard to produce strands with very long lengths. If we want a strand that terminates at the 100th T base in a sequence, that means it must be cut at that base and simultaneously not be cut at any of the 99 T bases before that. It will be impossible to ensure a good distribution of cut sites if there are too many possible sites. The method can sequence up to about 100 bases in a row, but not beyond that. Ultimately, though, the Maxam–Gilbert technique has been largely abandoned because it is difficult to automate for reasons we will understand better when we see the other classic technique.

4.1.2 Sanger Dideoxy

The Sanger dideoxy method is in some sense the opposite of the Maxam–Gilbert method. Instead of breaking the strand selectively at a single base, we build up a strand but stop it when it reaches any copy of a single base. This is accomplished by using a dideoxy, or terminator, base. Dideoxy means that the base is lacking the $3'$ hydroxyl group of DNA in addition to the $2'$ hydroxyl lacking in DNA but found in RNA. DNA polymerizes by attaching a phosphate group at the $5'$ hydroxyl of a newly added base to the $3'$ hydroxyl of the last base in the chain. This process is illustrated in figure 4.2. If we incorporate a base lacking the $3'$ hydroxyl into the chain, then we cannot perform this chemistry and therefore cannot incorporate any additional bases into the chain. Thus the alternate name "terminator base."

In the classic Sanger dideoxy method, we introduce the elements needed for polymerization (primers, dATP, dGTP, dCTP, dTTP, and a DNA polymerase) into a solution of the DNA we want to sequence. We then add a single kind of terminator base, say ddGTP, which we have labeled with a fluorescent group. The polymerase will then attempt to copy the DNA strand. Every so often, though, it will incorporate a fluorescent terminator base into the strand and get stuck. For example, if we label our fluorescent terminator ddGTP as *G, then our template strand from above, CTACGCCT, would yield the following set of sequences:

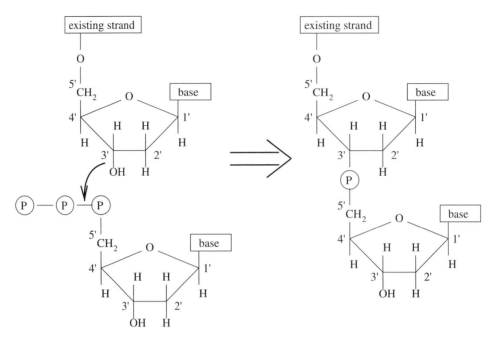

Figure 4.2
Chemical mechanism of DNA polymerization. A newly added DNA nucleotide (bottom) attaches the 3′
end of the existing strand (top) by linking its own 5′ carbon to the 3′ hydroxyl group of the last base of
the existing strand through a phosphate group. The remaining two phosphate groups are cleaved off, providing the energy that drives the reaction.

```
CTACGCCT
─────────
     *GA
    *GGA
   *GCGGA
*GATGCGGA
 GATGCGGA
```

Repeating this process for the other three terminator bases and running them out
on a polyacrylamide gel would give something like the illustration of a gel shown in
figure 4.3. This gel can be read out very much like the Maxam–Gilbert gel, except
that we must keep in mind that the column from each terminator base tells us where
the complementary bases are in the sequence. For example, the column run with
ddGTP tells us where to find C bases in the original sequence.

This method initially appeared to have strengths and weaknesses similar to those
of Maxam–Gilbert. Though the chemistry is somewhat simpler and the lack of radioactive materials is convenient, Sanger dideoxy has similar limitations on sequence
lengths that can be processed. The same issues of bands running together are prob-

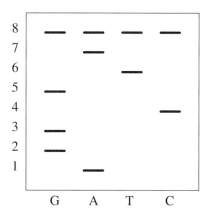

Figure 4.3
A polyacrylamide gel that might be produced by Sanger dideoxy sequencing from the sequence CTACGCCT. Vertical band positions correspond to different strand lengths, labeled at the left edge of the gel. Each column of the gel shows the lengths of strands produced by polymerizing up to the base at the bottom.

lematic for Sanger dideoxy. Furthermore, in order to produce a long strand by Sanger dideoxy, we must incorporate the terminator base at the final position but not at any earlier position, which becomes extremely unlikely for long sequences. Like Maxam–Gilbert, then, the classic Sanger dideoxy method was limited to sequencing approximately 100 bases at a time.

4.1.3 Automated Sequencing
In order for large-scale sequencing to become a reality, it was necessary that some less labor-intensive technology be developed. Two major advances have made that a reality. One of these is the development of fluorescent dyes that could be easily distinguished from each other, allowing reactions for all four bases to be run together. By running a Sanger dideoxy reaction with the polymerization buffer (primers, dNTPs, and polymerase) and all four terminator bases (ddATP, ddCTP, ddGTP, and ddTTP), we will get strands of all lengths incorporating all four kinds of terminator bases. If the four kinds of terminator bases have different colors of fluorescent dye, then we can allow all of them to polymerize together and we can read out the sequence by reading the sequence of dye colors in a single column. Figure 4.4 illustrates how such a four-color gel might appear. Note that this strategy works for the Sanger method, but will not work for the Maxam–Gilbert method, in which the radioactive base is far from the selectively cut base.

Though these four color bases reduce the work of sequencing by roughly a factor of 4, their real advantage is in enabling even further automation through a technology called capillary sequencing. In capillary sequencing, we replace the slab of

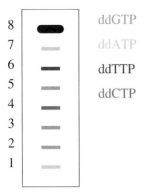

Figure 4.4
A four-color gel produced by running all four terminator bases with distinct fluorescent dyes in a single sample. Distinct shades of gray in the image would be distinct colors in a real four-color gel.

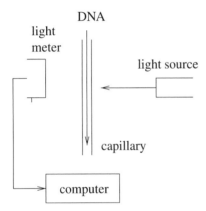

Figure 4.5
A capillary sequencer design. DNA strands move through the polyacrylamide gel in the capillary. A stationary laser excites the fluorescent dyes as they pass by, producing fluorescence that is measured by a light meter and recorded by a computer.

polyacrylamide gel with a thin tube, or capillary, filled with gel. An electric current forces DNA through the capillary, where smaller pieces of DNA move faster than larger ones, as in a standard polyacrylamide gel. A light source excites the DNA as it passes through the strand, and a detector measures fluorescence at all four wavelengths corresponding to the probes on the terminator bases. A computer can then record intensity in each wavelength as the bases pass through the capillary. Figure 4.5 conceptually illustrates the apparatus.

A plot of the intensities of the fluorescence in each wavelength over time as DNA passes through the capillary, called a "sequence trace," can be interpreted by a com-

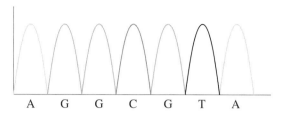

A   G   G   C   G   T   A

Figure 4.6
An idealized model of a sequence trace. Peaks of different shades identify regions of the strand corresponding to particular bases. Although the figure is in gray scale, an actual trace would use four different colors
for the four bases.

puter to directly determine the DNA sequence. Figure 4.6 shows an image of what
one of these traces looks like. An actual trace will be somewhat messier than this,
and bases cannot always be determined with perfect accuracy. This automated approach does greatly improve quality beyond prior methods, though. With a modern
capillary sequencer, one can read sequences of about 500–1000 bases in a row with
99 percent or better accuracy.

4.1.4 What About Bigger Sequences?

Even with a modern sequencing machine reading 1000 bases at a time, it is still not
clear how we will read the thousands of bases of a viral sequence, potentially millions
of a bacterial sequence, or even billions of a large eukaryotic sequence. The classic
method for this problem is a labor-intensive practice called "chromosome walking."
Chromosome walking works approximately as follows:

1. Find a probe sequence anywhere in the DNA strand.
2. Amplify a piece of DNA starting from the probe.
3. Sequence the first few hundred bases of the amplified sequence.
4. Choose a new probe near the end of the newly sequenced region.
5. Start again with the new probe and return to step 2.

Though this may sound tedious, given enough effort it could sequence very long
pieces of chromosome. It might eventually get stuck because some sequences of
DNA are difficult to sequence for various reasons. For example, if the method runs
into a highly repeated piece of DNA, it may prove impossible to find a good probe.
But eventually this approach can be used to get hundreds of thousands or even millions of bases in a row.

Even this will not be enough to sequence large eukaryotic genomes, though. When
the sequencing of the approximately 3-billion-base human genome was first proposed, it was planned that it would be accomplished with a hierarchical "clone-by-
clone" strategy. The idea behind this approach was that scientists would break the

genome into large pieces (on the order of 100 kilobases) and insert them into micro-organisms. Eventually, a construct called a bacterial artificial chromosome (BAC) was chosen, which would allow these large DNA pieces to be inserted into bacteria and replicated. Scientists could then take these colonies of bacteria with various human DNA strands, copy out the strands, and sequence them by chromosome walking. Finally, they could use a long-distance mapping technique to figure out where the BACs were positioned relative to each other on the genome, allowing them to put the whole thing together. There was still criticism that this approach was too slow and labor-intensive, though, which is where computational biology enters the picture.

4.2 Computational Approaches

The basic issue, from a computational modeling point of view, is to try to take the basic capability of sequencing a few hundred bases at a time and combine that with computer science to sequence larger pieces of DNA more efficiently than can be done with the prevailing laboratory methods. We will first examine one approach that has mostly turned out to be a dead end for sequencing large genomes. We do this in part because it illustrates some of the pitfalls we can run into in modeling biological prob-lems, and in part because it did eventually turn out to have important applications. We will then discuss what has actually worked in practice for sequencing large genomes.

4.2.1 Sequencing by Hybridization

One proposed strategy of harnessing computational methods for sequencing involved using a very different technology, called a microarray. A microarray is typically a small glass plate covered with tiny spots of DNA. Each spot has a uniform DNA se-quence, but different spots have different sequences. When we wash a solution of DNA over a microarray, DNA strands stick to spots that have sequences comple-mentary to them. By seeing which spots have DNA bound to them, we can deter-mine which sequences are in our sample. A typical microarray may have tens of thousands of spots, so we can have a separate spot for every possible sequence of length up to about eight bases (called 8-mers, or k-mers for arbitrary length k). Thus, a microarray can in principle tell us exactly which 8-mers occur in some DNA strand of interest. In practice, microarrays are noisy and our data may contain some errors. But for the moment, we can suppose that we really can determine ex-actly which k-mers are in a sequence of interest.

Sequencing by hybridization (SBH) was a method proposed to sequence DNA computationally, using this information about the k-mers a given sequence contains. To illustrate this approach, let us suppose we have a sequence ACGCCATCA and

$$ACGC \longrightarrow CGCC \longrightarrow GCAA \longrightarrow CCAT \longrightarrow CATC \longrightarrow ATCA$$

Figure 4.7
k-mer graph corresponding to the sequence ACGCCATCA for k = 4. Each node corresponds to a 4-mer observed in the sequence. Each edge marks a pair of k-mers for which the (k − 1) suffix of the source matches the (k − 1) prefix of the destination.

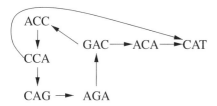

Figure 4.8
k-mer graph corresponding to the sequence ACCAGACAT with k = 3.

we run it on a microarray with all length-4 strands. We should then see hybridization for the spots complementary to ACGC, CGCC, GCCA, CCAT, CATC, and ATCA.

It is easy to go from the full strand to the k-mers, but not obvious how to go from the k-mers to the full strand. The SBH approach starts by modeling the data as a graph. Our nodes are the k-mers present in our input. Our edges correspond to pairs of k-mers that may be consecutive in the sequence. For example, the 4-mer GCCA may be followed by CCAT (if they come from the longer sequence GCCAT), or it may be followed by CCAA, CCAC, or CCAG. In our example above, only CCAT occurs from among these four choices, so we know that CCAT follows GCCA unless GCCA is the last k-mer in the sequence. In general, two k-mers may occur sequentially if the (k − 1) suffix of the first is the (k − 1) prefix of the second. If we construct a graph using this relationship, then a path that passes through all nodes in the graph represents a possible sequence of the DNA. Figure 4.7 shows the graph for our sample sequence. Our sequence can be read out directly from the graph by following the unique path through the k-mers.

The above approach is likely to work fine for short sequences and long k-mers, but it will not work well in general. In particular, a problem appears when we have (k − 1)-mers repeated in our sequence. To illustrate the problem, imagine that we start with the sequence ACCAGACAT, with k = 3. Then we will get the graph shown in figure 4.8. In this case, it is not so obvious what the sequence should be. First of all, the graph has a cycle, and we can go around that cycle an arbitrary number of times. For example, the sequence ACCAGACCAGACCAGACAT will give us the same set of k-mers. The second problem, though, is that trying to find a path in a graph that visits every node is a well-known NP-hard problem we saw in chapter 3: the Hamiltonian path problem. Computer scientists who worked on this SBH

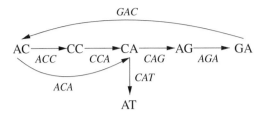

Figure 4.9
Eulerian path graph for the sequence ACCAGACAT with k = 3. Nodes of the graph are 2-mers, and edges correspond to 3-mers.

problem cast it as a Hamiltonian path problem and then tried to use various inexact methods to solve it, but ultimately these approaches were not very useful for solving any real-world instances of the problem. It would seem, then, that the problem, at least as it is formulated here, will not be solvable. It turns out, though, that this intuition is an example of one of the fallacies mentioned in chapter 3—not making sure one's "hard" problem is really hard—as we will see in the next section.

4.2.2 Eulerian Path Method

A different formulation of the SBH problem ultimately led to very efficient algorithms. To understand how this is possible, we need to recognize that although finding a path through a fragment graph is a kind of Hamiltonian path problem and a Hamiltonian path is a hard problem, that does not mean that finding a path through a fragment graph is a hard problem. An alternate graph reduction, proposed by Pevzner [71], shows how we can find paths very quickly.

In the Eulerian path method, we construct a graph from our k-mers by a different reduction. Now, we create a node for each (k − 1)-mer in our graph and create a directed edge between two nodes if there is a k-mer whose prefix is the first node and whose suffix is the second node. Suppose, for example, we use the sequence ACCAGACAT with k = 3. The result will be the graph in figure 4.9, where each edge is labeled with the k-mer that created it.

Now, instead of looking for a path that passes through every node, we need a path that passes through every edge in our graph. The problem of finding a path using every edge in a graph is called an Eulerian path problem. Whereas the Hamiltonian path problem is hard, the Eulerian path problem is tractable, and in fact solvable in linear time. Though proving that a problem of interest is a special case of a hard problem does not prove it hard, proving it is a special case of an easy problem does prove it easy. This is a simple case where having the right model makes a huge difference in the tractability of the problem.

Does this mean sequencing is a solved problem? Unfortunately, no. One problem is that DNA arrays are noisy and often have false positive and false negative values.

The algorithms we have discussed do not deal well with errors in the data, and even tractable problems tend to become hard when one starts including an error model. A more insidious problem, though, is that repeated sequences still defeat the approach. In the Eulerian path SBH example above, we have one cycle in our graph and could endlessly go around that cycle. In general, if we have a k-mer repeated anywhere in our sequence, it will become ambiguous to determine how to put the sequence together. And microarrays pretty much will not allow us to make k-mers longer than about 8. In other words, this approach really seems to have no hope of sequencing large pieces of DNA.

SBH was an extremely popular topic among computational biologists for a number of years because it leads to very elegant theoretical problems. But from the biological point of view, it ultimately seemed to have been a dead end as a means of sequencing large genomes, even in the Eulerian path version. This is an unfortunate but useful example of how computational biologists have sometimes been led astray by putting too much emphasis on doing good computer science and not enough on doing good biology. As we will see later on, though, this work did ultimately prove useful for some kinds of sequencing, although not exactly as originally intended.

4.2.3   Shotgun Sequencing

Though SBH may not have worked out for the problem of sequencing large eukaryotic genomes, computational biology was ultimately crucial to the field through a very different avenue called shotgun sequencing. The basic idea behind shotgun sequencing is to break the DNA into lots of random pieces, sequence them by a sequencing machine to yield many roughly 500 bp fragments, and then put the fragments together computationally by trying to find the shortest possible sequence containing all of them. You should recognize this as an example of the shortest common superstring problem we saw in chapter 3. In contrast to the SBH case, this really is a hard problem, and we cannot solve it exactly. It will turn out, though, that with a lot of tricks we can do a very good job in practice.

There are approximation algorithms for this problem, but their bounds are not good enough for practical use. Traditionally, the way people really solved this problem in practice was through a greedy algorithm that would build up a sequence by repeatedly merging whichever two fragments had the greatest overlap between them. For example, if we had the fragment sequences

1. ACAGGAC

2. AGGTTGA

3. GACTA

4. TGTTCA

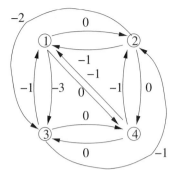

Figure 4.10
Traveling salesman instance created from the fragment set {ACAGGAC, AGGTTGA, GACTA, TGTTCA}. Nodes correspond to fragments, and edges, to the negative of the amount of overlap between fragments.

our sequence of operations would be the following:

1. Merge fragments ACAGGAC and GACTA, using the overlap sequence GAC to get ACAGGACTA.
2. Merge TGTTCA and AGGTTGA to get TGTTCAGGTTGA.
3. Merge TGTTCAGGTTGA and ACAGGACTA to give TGTTCAGGTTGA-CAGGACTA.

This greedy merging is conjectured to be a 2-approximation algorithm, but generally does much better than that in practice.

This process of fragment merging can be visualized by representing the problem as a special case of the traveling salesman problem (TSP). To perform the reduction to TSP, we create one node for each fragment. We then place a directed edge between each pair of nodes v_i and v_j whose value is the negative of the maximum overlap of a suffix of sequence i and a prefix of sequence j. For the fragment set above, this reduction would give the graph of figure 4.10. TSP is a hard problem even compared with other NP-complete problems and is harder to approximate than the shortest common superstring problem, so this reduction would not at first seem to be a very useful thing to do. TSP is, however, a very well studied problem, so there are a lot of heuristics one can bring to bear on the problem by representing it as TSP. This reduction is also a good way to understand the structure of the problem.

The above method really does yield a practical means of performing sequence assembly, although it requires some small modifications. For example, because sequencing technologies are not error-free, we would not use exact overlaps for the edge scores, but rather something analogous to a sequence alignment score that allows for imprecise overlaps. But the TSP reduction is basically the core of how real sequence assemblers for microbial genomes work.

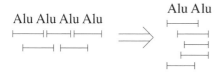

Figure 4.11
Illustration of how shotgun assembly can collapse tandem repeats. A set of fragments (straight lines) gathered from four consecutive Alu repeats could explained by only two Alu repeats.

The method does, however, run into problems when moving from microbial to eukaryotic genomes. In particular, though this shotgun method is not sensitive to the short random repeats that make SBH impractical, it does have difficulty with the longer repeat regions one finds in eukaryotes. For example, if we have a region of *tandem repeats* (the same sequence or something close to it repeated side by side) that is longer than the fragment length, then shotgun assembly has a tendency to "collapse" the region by reducing the number of repeats. Figure 4.11 illustrates the problem by showing how a region of four Alu repeats might be misinterpreted as only two repeats in the assembly.

Repeats that are distant in the genome also create problems; once the assembly enters a repeat region, an assembly algorithm cannot tell which of the copies it is in. For example, if we had the following sequence with three Alus in it

ACCCATG ... Alu ... TTGCTA ... Alu ... GTAGCA ... Alu ... TACTCA

the assembler might misassemble it as follows:

ACCCATG ... Alu ... GTAGCA ... Alu ... TTGCTA ... Alu ... TACTCA

It can infer that the two middle regions start and end in Alus but cannot tell which ones.

4.2.4   Double-Barreled Shotgun
A solution to the problem of repeats was developed using a technique called double-barreled shotgun sequencing. Double-barreled shotgun sequencing exploits a slight inaccuracy in a statement above about the limits of the available sequencing technologies. Earlier, we claimed that modern sequencing machines can sequence strands of up to about 500 bp. In fact, sequencing machines can sequence *the first* 500 bp of a strand even if it is much longer than 500 bp. The importance of this distinction is that we can also sequence the first 500 bp of the complementary strand. As a result, we can really sequence 500 bp from each end of a single strand. Figure 4.12(a) illustrates

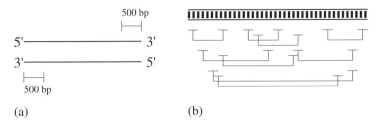

Figure 4.12
Illustration of the concept of double-barreled shotgun sequencing. (a) Sequences derived from both ends of a DNA molecule by sequencing the first 500 bp of the forward and reverse strands. (b) Linked mate-paired fragments providing long-distance connectivity along a piece of sequenced genome.

this concept. We will not know the sequence of the strands between those 500 bp ends, but we can easily determine its approximate length.

In double-barreled shotgun sequencing, we will use these paired fragments (called *mate pairs*) to constrain our assembly. If we have a pair of fragments on both sides of a repeat region and we know the distance between the fragments, then we can determine the size of the repeat region. We can also assemble the genome correctly on both sides of the repeat region. However, we may still make mistakes in determining the correct sequence of the region. Maybe we can determine that it consists of 30 copies of an Alu repeat, but because Alus are not 100 percent identical to each other, we may get individual bases wrong within them. From the point of view of generating a biologically useful sequence, though, this is generally a tolerable problem. In practice, it is helpful to have a range of fragment sizes to help span both large and small problem areas in the genome, as illustrated in figure 4.12(b).

The development of this double-barreled shotgun approach is a great example of how valuable it can be to have an understanding of both the limits of the experimental technology and the needs of the computational problem. In the real world, though, the picture gets somewhat more complicated. Once we reach the point of incorporating mate pairs, it is no longer possible to come up with elegant, practically useful theoretical representations of the problem. Furthermore, there is a strong incentive to use every kind of potentially useful data available. For example, we may have other mapping information giving approximate locations of markers on the genome, as well as other partial sequence information we can use to improve the assembly. Because of these complications, the methods used in practice are essentially a variant on the "kitchen sink" approach described in chapter 3. For those interested in learning more, the References and Further Study section provides citations to some real implementations of shotgun sequence assemblers used for difficult eukaryotic genomes.

Even when we have a good assembly method, though, we are not yet done. The assembly will almost certainly have gaps in it that need to be systematically filled in.

Furthermore, there may be regions of the genome that do not sequence for various technical reasons, and we may need to go back into the laboratory and individually fill these in with specialized methods. These later steps make up a process called "finishing" that is used to polish and complete the genome produced by an assembler. Even then, there are regions of the genome likely to be left undone, particularly the highly repetitive sequences close to the centromeres and telomeres.

4.3   The Future?

What we have covered in the preceding sections more or less brings us to the current state of the art in how computational biology applies to sequencing today. But it is always interesting to consider how things may change in the future and what this will mean to computational modelers.

4.3.1   SBH Revisited

One intriguing possibility raised recently is that SBH may not be quite the dead end indicated earlier. In particular, it may be possible to salvage some of the nice computational properties of the SBH method while eliminating the problems of the experimental technology it is based on by using a technique called "shotgun SBH." Shotgun SBH uses the SBH computational abstraction but applies it to the long sequence fragments produced by sequencing machines rather than to the short k-mers one gets from a microarray. The basic approach is to take fragment sequences from a sequencing machine and artificially "shred" them into k-mers, as illustrated in figure 4.13.

Suppose, for example, we have the following fragments:

AGACTTAC

ACGGTAG

GATTCATA

We can shred them into the following set of k-mers, assuming k = 5:

AGACT, GACTT, ACTTA, CTTAC,

ACGGT, CGGTA, GGTAG

GATTC, ATTCA, TTCAT, TCATA

We can then solve the problem exactly as if we have a microarray that checks every k-mer and tells us that precisely those above are observed. The advantage of this approach is that we can use k-mers much longer than those actually feasible on a microarray, say 30 bp. As long as we have sufficient overlap between our fragments,

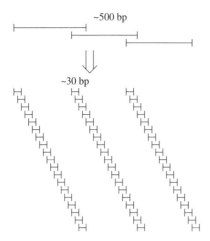

Figure 4.13
Illustration of the concept of shotgun SBH. Shotgun sequence fragments of several hundred base pairs (top
of figure) are computationally "shredded" to produce k-mers approximately 30 bp long (bottom of figure).
These k-mers are then used to simulate the output of a hypothetical 30 bp sequencing array.

we can learn exactly which 30-mers are present in our data. We can then process the
data by using the very fast Eulerian path methods developed for the standard SBH
method, but with a k-mer length long enough to skip over the random repeats we
would expect just by chance even in a 1 Mb microbial sequence.

This method does not solve all of our problems, however. It would still fail to deal
with the long repeat regions that confound standard shotgun assembly, and therefore
may never be appropriate for large genomes. It will also have trouble if we do not
have enough density of fragments. If consecutive fragments do not have an overlap
at least as long as the k-mer length, then we may end up missing k-mers that span the
overlap region, and thus be unable to complete the sequence. Nonetheless, shotgun
SBH, too, appears to be another good example of how a simultaneous appreciation
of what we can do experimentally and what we can do computationally can lead to
very good approaches in practice.

4.3.2 New Sequencing Technologies
One last point to note is that sequencing is a very active area of research. Some
groups are working on incremental improvements to the current technologies that
can be quite helpful to computational methods. For example, increasing fragment
sizes may make assembly computations much easier. Other approaches are lowering
the time and cost of generating fragments, potentially making far greater depth of
coverage available for assembly algorithms. The topic that has people in the field
most excited, though, is called "single-molecule sequencing." The principle behind

single-molecule sequencing is to get rid of fragments entirely and simply sequence an entire DNA strand, possibly hundreds of megabases, all at once. So far, this is just hypothetical, but many scientists feel it may not be far in the future. The most promising approaches to this problem are based on a strategy called "nanopore sequencing," which works by passing DNA through a tiny pore (typically a ring of proteins) and measuring how the electrical field across the pore changes over time. In principle, the voltage gradient across the pore depends on which base is passing through it. Current instruments are sensitive enough to distinguish long runs of one base from another, but no one has managed to get bases to move slowly enough through the nanopores to read out individual bases. But it may be only a matter of time before the whole field of computational sequence assembly is rendered obsolete.

References and Further Reading

For background on the basic sequencing technologies, see Stryer [57], a comprehensive and widely used text on general biochemistry. Those interested in more depth on the development of sequencing technologies can look to the primary literature. Primary references are available for Maxam–Gilbert sequencing [58], Sanger dideoxy sequencing [59], the use of four-color fluorescence to automate DNA sequencing [60], capillary electrophoresis sequencing [61], and chromosome walking [62]. It is difficult to identify a primary reference for the strategy ultimately used for the clone-by-clone sequencing of the human genome since it was hashed out in many individual meetings and position papers. The International Human Genome Sequencing Consortium's first publication announcing the complete genome provides perhaps the most complete primary description of the strategy [63].

The topic of sequence assembly algorithms is covered in depth in a text by Pevzner [64]. Jones and Pevzner [65] provide a more introductory coverage of the topic in the general context of graph algorithms in computational biology. The prospect of sequencing by hybridization was independently proposed by four different groups: Bains and Smith [66], Southern [67], Lysov et al. [68], and Drmanac et al. [69]. The first practical report of its use came from Drmanac et al. [70]. The Eulerian path formulation is due to Pevzner [71]. The shotgun SBH method was proposed by Idury and Waterman [72]. There are many other computational works arising from variations of SBH that we did not cover here. The interested reader can refer to Pevzner's textbook [64] for references to much of this work.

Shotgun sequence assembly was first developed to sequence the genome of *Haemophilus influenzae* [73]. The double-barrel technique was first developed by Edwards et al. [74] to sequence a single gene locus. An influential, and controversial, paper by Weber and Myers [75] first proposed that the double-barrel technique would make whole-genome sequencing possible for complex eukaryotes. Whole-genome double-barrel shotgun sequencing was first demonstrated by Myers et al. [76] to assemble the *Drosophila melanogaster* genome. Readers interested in more detail on what goes into designing a real-world sequence assembler may refer to the Myers et al. paper on *Drosophila*, as well as to the more complex assembler required for the shotgun assembly of the human genome [77]. Batzoglou et al. [78] provide another example of a real-world shotgun assembler designed for general genome assembly problems.

The concept of nanopore sequencing was first suggested by Akeson et al. [79]. For a general review of nanopore sequencing ideas, see Deamer and Akeson [80]. A more recent review of that and other new prospects for rapid sequencing can be found in Shendure et al. [81]. Prospects for single-molecule sequencing are, however, a matter of great speculation as of this writing, and it is uncertain which, if any, of these methods will ultimately prove successful.

5 General Continuous Optimization

We have seen many ways of solving discrete problems, but biological modeling often involves continuous systems. A continuous system is one in which we have one or more real-valued independent variables and one or more real-valued dependent variables. Optimization in continuous systems is often not covered in computing curricula, but from my point of view it is as important as in discrete systems. The material we are going to cover would ordinarily be grouped under the topic of numerical methods. We cannot present an exhaustive treatment of numerical methods here, but will provide an introduction to a few basic and broadly useful techniques.

Continuous optimization comes up in many contexts in biological modeling. For instance, working with molecular models often involves solving continuous optimization problems. Suppose, for example, we want to find how best to fit a small molecule into a binding pocket of a protein. We can represent this as a continuous optimization problem by treating the protein as fixed in space and using six continuous variables to represent the six degrees of freedom (three translational and three rotational) of the small molecule. If we imagine that we have some black box that computes the energy of binding, given the position of the small molecule, $E(x, y, z, \theta, \phi, \psi)$, then our problem is equivalent to finding the values of the six continuous variables that minimize E. In a more sophisticated variant, we may have three degrees of freedom for each atom in our system and seek to minimize $E(x_1, y_1, z_1, x_2, y_2, z_2, \ldots, x_n, y_n, z_n)$ in terms of all the variables. These are both examples of general continuous optimization. Continuous optimization will also be important later in the text when we examine fitting parameters to data sets, a problem that is often a form of continuous optimization. For example, we might have a set of differential equations describing progress of a reaction and a data set of time points from the reaction. Finding the reaction rate constants that minimize the difference between the model and the observations is also a form of continuous optimization.

When working with numerical algorithms, we often have to make some assumptions about the systems we are examining. What we need to assume varies from method to method, but some common assumptions are the following:

1. Continuity (often we assume C_∞ continuity, which means that the function and all of its derivatives are continuous)
2. Bounded derivatives: $\frac{d^k y}{dx^k} < B$ for some (possible unknown) constant B.

It is also important to understand that when talking about general continuous optimization, we are rarely talking about truly optimizing for our systems. Usually, what we will find are *local optima*, which means that they are sets of variables that cannot be improved by infinitesimal changes in any variables. They are distinct from *global optima*, which are the absolute best solutions that exist over the entire space of allowable solutions. It is impossible to develop any method that will find global optima of any arbitrary continuous function, although there are special cases, some of which we will see in the next chapter, for which it is possible to globally optimize continuous systems. For the remainder of this chapter, though, we will generally assume we are looking for local optima.

Suppose we have a continuous function $F(x)$ and we want to find a maximum (or minimum). Those who remember their introductory calculus may recall that extrema of $F(x)$ are found at points where $\frac{dF}{dx}$ (which we will also call $F'(x)$ or $f(x)$) is zero. If we are maximizing over some finite region—for example, finding the maximum of $F(x)$ for $x \in [-1, 1]$—then the maxima may also be at the boundaries of the interval. But maximization and minimization are essentially problems of finding zeros of a function. So how do we find zeros?

There are a few special cases for which zero-finding has a simple analytical solution. For example, with polynomials, there is the well-known quadratic formula for $ax^2 + bx + c$, $\frac{-b \pm \sqrt{b^2 - 4ac}}{2a}$. There is also a less well-known cubic formula that is sometimes useful in practice. There is even a formula for quartic (degree 4) polynomials that is so complicated no one uses it in the real world. There is no general formula beyond quartic. For general functions, though, there is no analytical solution, and we therefore need to use some kind of numerical algorithm. In this chapter, we will cover a few of the most important zero-finding or continuous optimization algorithms.

5.1 Bisection Method

The bisection method is the simplest zero-finding method. The basic idea behind the bisection method is to start with some region that we know includes a zero, then repeatedly cut the region in half until we have zeroed in on our zero. Suppose we are finding a zero of a function f and we know that $f(x_{min}) < 0$ and $f(x_{max}) > 0$ for some $x_{min} < x_{max}$. Then we know there must be a zero somewhere between x_{min} and x_{max}. Figure 5.1 provides pseudocode for the bisection method to find a single

1. $f_{min} \leftarrow f(x_{min})$
2. $f_{max} \leftarrow f(x_{max})$
3. repeat
   A. $x_{mid} \leftarrow (x_{min} + x_{max})/2$
   B. $f_{mid} \leftarrow f(x_{mid})$
   C. if $(f_{mid} > 0)$
      i. $x_{max} \leftarrow x_{mid}$
      ii. $f_{max} \leftarrow f_{mid}$
   D. else
      i. $x_{min} \leftarrow x_{mid}$
      ii. $f_{min} \leftarrow f_{mid}$
4. until $(x_{max} - x_{min} < \epsilon)$
5. return$((x_{min} + x_{max})/2)$

Figure 5.1
Pseudocode for the bisection method, assuming that $f(x_{min}) < 0$ and $f(x_{max}) > 0$. ε is an estimated backward error tolerance for the search.

zero of a function. The code will run until the size of the x interval we are examining is less than some user-specified threshold ε. This interval provides a bound on the final error in x, which is known as a *backward error*. We can also use as our stopping condition an estimate of the error in $f(x)$, which is called the *forward error*. To do that, we might replace line 4 with

4. until $(|f_{mid}| < \varepsilon)$.

For now, however, we can assume we are using backward errors.

To see how the method works, we can look at an example. Suppose we are looking for a zero of $x^2 - 2$. We can easily determine analytically that the answer is $\pm\sqrt{2}$, but let us suppose we do not know how to find that. We will further suppose that we know a zero lies somewhere on the interval $[0, 2]$. If we apply the bisection method, it will repeatedly cut this interval in half until it converges on the zero. Table 5.1 shows how the variables evolve over time. Figure 5.2 shows the progress of the first few steps of the algorithm by illustrating the region of the curve under consideration at each step. The initial region, $[0, 2]$, spans the entire width of the plot. The subsequent region, $[1, 2]$, and the next region after that, $[1, 1.5]$, each covers a smaller fraction of the plot bounding the zero of the function. At the end of four steps of the algorithm, we have established that the zero lies somewhere in the interval $x = [1.25, 1.5]$. Our best guess is that it lies in the middle of the region (1.375). We can also estimate that the error in this guess is about half the size of the interval we are guessing from, or 0.125.

Table 5.1
Changes in variables of the bisection algorithm over successive steps on the curve $f(x) = x^2 - 2$

step	1	2	3	4
x_{min}	0	1	1	1.25
x_{max}	2	2	1.5	1.5
f_{min}	-2	-1	-1	-0.4375
f_{max}	2	2	0.25	0.25
x_{mid}	1	1.5	1.25	1.375
f_{mid}	-1	0.25	-0.4375	-0.1094

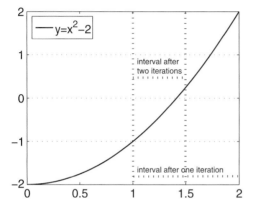

Figure 5.2
Plot of the curve $f(x) = x^2 - 2$ over two successive steps of the bisection algorithm. The full width of the plot shows the initial interval $[0, 2]$. Vertical dashed lines show the intervals after one step ($[1, 2]$) and after two steps ($[1, 1.5]$).

The bisection method is very general, and as long as we can find a good starting region, it will eventually get us arbitrarily close to a zero. It can converge somewhat slowly, though. Each step cuts our interval in half, which may take a while if we are trying to get from a large region to a small error tolerance. It is therefore worth considering faster methods.

5.2 Secant Method

To understand how we can improve on the bisection method, consider the situation in which we are trying to shrink our boundaries but one endpoint of our boundary has an f value much closer to zero than the other. With the bisection method, we shrink our interval by choosing the middle of the region as our next guess for an endpoint of the region. Suppose we pick another point instead of the midpoint, though.

What would be a good point to use? We might guess that our method would converge fastest if we replaced the midpoint with some point very close to the zero.

The secant method uses this intuition by trying to guess where the zero of f lies and using that, rather than the midpoint of the interval, as the new boundary. It does this by linear interpolation between the endpoints we already know. That means we project a line between the two endpoints of our interval and ask where that line crosses $f(x) = 0$. The line passing through our two known points (x_{min}, f_{min}) and (x_{max}, f_{max}) has the form

$$(f(x) - f_{min}) = \frac{f_{max} - f_{min}}{x_{max} - x_{min}} (x - x_{min}).$$

If we solve this equation for x, we get the following:

$$f(x) = f_{min} + \frac{f_{max} - f_{min}}{x_{max} - x_{min}} (x - x_{min}) = 0$$

$$x = x_{min} - \frac{x_{max} - x_{min}}{f_{max} - f_{min}} f_{min}.$$

Thus, to implement the secant method, we use the same algorithm as for the bisection method, except that we replace line 3.A in figure 5.1,

$$x_{mid} \leftarrow \frac{x_{min} + x_{max}}{2},$$

with

$$x_{mid} \leftarrow x_{min} - \frac{x_{max} - x_{min}}{f_{max} - f_{min}} f_{min}.$$

Using the secant method with our example function $f(x) = x^2 - 2$, we begin by projecting a line between the endpoints of our curve in our starting region, $x = [0, 2]$. Figure 5.3 shows the resulting line, $y = 2x - 2$. We then pick as our new midpoint the point where that line crosses the x axis. In this case, it is 1, just as with the bisection method. We next move to the region $x = [1, 2]$ and draw a new line between the endpoints of the curve on that region, $y = 3x - 4$. The new line now crosses the x-axis noticeably to the left of the center of the interval. The secant method will thus choose a new endpoint where f is much closer to zero than it would be with the bisection method, producing an answer slightly closer to the true zero than we would get if we chose the midpoint of the region. Executing one more iteration would give us the line $y = \frac{10}{3}x - \frac{14}{3}$. As we can see, the accuracy of our guess is now far better than we saw with the bisection method.

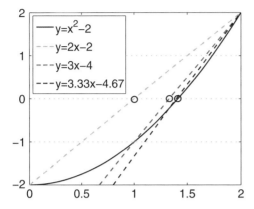

Figure 5.3
Demonstration of three successive steps of the secant method on $f(x) = x^2 - 2$, starting from interval $[0, 2]$. Dashed lines show the secants; and circles, the intercepts. The line $y = 2x - 2$ is the secant for interval $[0, 2]$ with intercept 1. The line $y = 3x - 4$ is the secant for the interval $[1, 2]$ with intercept 1.33. The line $y = \frac{10}{3}x - \frac{14}{3}$ is the secant for the interval $[1.33, 2]$ with intercept 1.4.

Table 5.2
Changes in variables of the secant algorithm over successive steps on the curve $f(x) = x^2 - 2$

step	1	2	3	4
x_{min}	0	1	1.33333	1.4
x_{max}	2	2	2	2
f_{min}	-2	-1	-0.22222	-0.04
f_{max}	2	2	2	2
x_{mid}	1	1.3333	1.4	1.412
f_{mid}	-1	-0.2222	-0.04	-0.00692

Table 5.2 shows the progress of the variables on successive steps of the algorithm. Based on these values, our best guess after the final step will be 1.412 with an estimated error about half the size of the final region, or 0.3. The actual error is much lower, about 0.0022. Often a more accurate measure of error is the amount by which the answer changes between the last two steps, 0.0787, which is still high but closer. In this case, the secant method works much better than the bisection method, although there are no guarantees that it will do so on all curves.

5.3   Newton–Raphson

Although the secant method is generally an improvement over the bisection method, it is typically only a slight improvement. First of all, we need a bounding interval

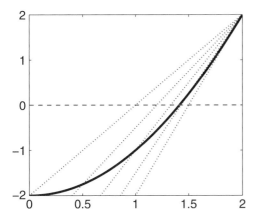

Figure 5.4
Illustration of the intuition behind the Newton–Raphson method using $f(x) = x^2 - 2$. By progressively tightening the interval between two points on a secant line, we converge on a tangent to our curve. This tangent can then be used to estimate the location of a zero of the function.

around a zero to get started, and we may have difficulty finding one. Second, it is still quite slow for some cases. There is another method that can help us with both of these issues: the Newton–Raphson method (often called Newton's method).

To understand Newton–Raphson, imagine that we are applying the secant method but using tighter and tighter intervals to get our secant. Ignore for the moment the fact that our interval may not contain a zero if we make it too small. We will see something like the image in figure 5.4, where the secant line keeps moving to the right as we interpolate between closer and closer points. If we simply keep making the interval tighter and tighter, we eventually end up with a tangent to our curve at a single point. In figure 5.4, the rightmost line is the tangent to $x^2 - 2$ at the point $(2, 2)$. The idea behind Newton–Raphson is to try to project a zero of the curve using linear interpolation, just as with the secant method, but to do it using a tangent to a single point rather than a secant to two endpoints of an interval. We start with a single point, representing an initial guess as to the location of our zero; find the tangent to the curve at that point; and use the tangent to find a point likely to be closer to the zero. This is an example of a very general technique called "locally linearizing," which means pretending that a curve of interest is linear in a local region in order to create an easier computational problem.

To use the Newton–Raphson method, we therefore need to be able to figure out what the tangent is at any point on our curve and where that tangent crosses $f(x) = 0$. The first part is fairly simple. Suppose we start from some initial guess x_0. Then the tangent at x_0 is the line passing through $(x_0, f(x_0))$ whose slope is the derivative of $f(x)$ at x_0, which we can denote by $f'(x_0)$. The equation for this line is

$$f(x) - f(x_0) = f'(x_0)(x - x_0)$$

or, equivalently,

$$f(x) = f'(x_0)(x - x_0) + f(x_0).$$

If we solve for where this line crosses $f(x) = 0$, we get the following:

$$f'(x_0)(x - x_0) + f(x_0) = 0$$

$$f'(x_0)(x - x_0) = -f(x_0)$$

$$x - x_0 = -\frac{f(x_0)}{f'(x_0)}$$

$$x = x_0 - \frac{f(x_0)}{f'(x_0)}.$$

This final equation gives us an iterative formula we can use to successively refine a guess as to the zero of our system.

We will illustrate the method by again using the curve $f(x) = x^2 - 2$. For this $f(x)$ we get $f'(x) = 2x$. If we start from the initial guess $x = 2$, then the values on successive steps of the algorithm are those given in table 5.3. By the fourth step, our actual error is down to 2.5×10^{-6}.

In general, once Newton–Raphson gets close to the answer, it converges very quickly on the correct answer. To understand why this is the case, we can use a Taylor series approximation. We may recall from calculus that any curve $f(x)$ can be approximated by a Taylor series about some point x_i:

$$f(x) = f(x_i) + f'(x_i)(x - x_i) + \frac{f''(\xi)}{2}(x - x_i)^2,$$

where ξ is some unknown number between x and x_i. Therefore, where $f(x) = 0$, we have

Table 5.3
Changes in variables of the Newton–Raphson algorithm over successive steps on the curve $f(x) = x^2 - 2$

step	1	2	3	4
x	2	1.5	1.417	1.414216
$f(x)$	2	0.25	0.00694	8.8×10^{-6}
$f'(x)$	4	3	2.834	2.828

$$f(x_i) + f'(x_i)(x - x_i) + \frac{f''(\xi)}{2}(x - x_i)^2 = 0.$$

If $f'(x_i) \neq 0$, then we can divide through by $f'(x_i)$ to get

$$\frac{f(x_i)}{f'(x_i)} + x - x_i + \frac{f''(\xi)}{2f'(x_i)}(x - x_i)^2 = 0.$$

Rearranging a bit, we get

$$x - x_i + \frac{f(x_i)}{f'(x_i)} = -\frac{f''(\xi)}{2f'(x_i)}(x - x_i)^2.$$

Note that if x_i is our ith guess as to the zero, then $x_i - \frac{f(x_i)}{f'(x_i)}$ is the improved value x_{i+1} that we would get from one more round of Newton–Raphson. Therefore, we can make the following substitution:

$$x - x_{i+1} = -\frac{f''(\xi)}{2f'(x_i)}(x - x_i)^2.$$

Since we assumed that x is a zero of f, then $(x - x_i)$ is the error in our approximation before applying Newton–Raphson, and $(x - x_{i+1})$ is our error afterward. If we can bound $\left|\frac{f''(x)}{2f'(x)}\right|$ by some constant C in the region between our initial guess and our zero, then we can assert that

$$|x - x_{i+1}| \leq C|x - x_i|^2.$$

If we let ε_i be the error at the ith iteration, then the preceding statement is equivalent to saying

$$\varepsilon_{i+1} \leq C\varepsilon_i^2.$$

In other words, each successive iteration of Newton–Raphson approximately squares the error. Once the error gets close to zero, then, it will drop very rapidly.

Note, though, that we did make an important assumption above: the derivative is not zero at our guess, x_i. We can see from our example curve that the method would fail if we picked the initial guess $x = 0$, which is a zero of the derivative. If we happen to pick a guess or land on one that has a derivative of zero, we can always perturb it slightly to move away from zero. A more subtle problem is if the zero we are looking for is also a zero of the derivative. In these cases, Newton–Raphson does generally work, but more slowly than normal. In fact, the accuracy generally will improve by about a constant multiple at each step (as with the bisection method) rather than by squaring the error. Newton–Raphson also may not work at all if we do not have a good guess to start from.

Nonetheless, Newton–Raphson is often the best option for finding zeros, and therefore maxima and minima, quickly. Given a reasonable guess, it is usually much faster than the secant or bisection method and can get errors to within typical machine precision in a few steps. In fact, a variant of Newton–Raphson is often used in modern computer processors for doing division, square roots, and other operations because it is a fast and simple way of doing what would otherwise be costly operations to perform in hardware.

5.4 Newton–Raphson with Black-Box Functions

I have suggested that Newton–Raphson is a very good choice for general optimization problems, but one might wonder what to do if we do not have an expression for the derivative of our function. For example, the function we are optimizing might be the output of a complex simulation, and we do not have any way of analytically determining the derivative of the function. In fact, we can still use Newton–Raphson by using approximate derivatives. To see how this works, we can apply Taylor series again to some arbitrary function $f(x)$:

$$f(x + \Delta x) = f(x) + \Delta x f'(x) + \frac{\Delta x^2}{2} f''(\xi)$$

$$f(x + \Delta x) - f(x) = \Delta x f'(x) + \frac{\Delta x^2}{2} f''(\xi)$$

$$\frac{f(x + \Delta x) - f(x)}{\Delta x} = f'(x) + \frac{\Delta x}{2} f''(\xi).$$

The final line shows us that the term on the left-hand side is an approximation to $f'(x)$ with error $\frac{\Delta x}{2} f''(\xi)$. We can derive this approximation by knowing $f(x)$ and $f(x + \Delta x)$ for some Δx. This is called a *forward difference* approximation and is considered *first-order-accurate* because its error varies linearly with Δx. If we need $f'(x_0)$ and cannot calculate it, we can get a good approximation by picking some small Δx and using $\frac{f(x+\Delta x)-f(x)}{\Delta x}$ in place of $f'(x)$ in our Newton–Raphson formula.

We can also consider other ways to approximate $f'(x)$. For example, suppose we take the following two Taylor expansions,

$$f(x + \Delta x) = f(x) + \Delta x f'(x) + \frac{\Delta x^2}{2} f''(x) + \frac{\Delta x^3}{3} f'''(\xi)$$

and

$$f(x - \Delta x) = f(x) - \Delta x f'(x) + \frac{\Delta x^2}{2} f''(x) - \frac{\Delta x^3}{3} f'''(\xi),$$

and subtract the second from the first. We then get

$$f(x + \Delta x) - f(x - \Delta x) = 2\Delta x f'(x) + 2\frac{\Delta x^3}{3} f'''(\xi),$$

which we can rearrange to

$$\frac{f(x + \Delta x) - f(x - \Delta x)}{2\Delta x} = f'(x) + \frac{\Delta x^2}{3} f'''(\xi).$$

The left-hand side of the equation is therefore also an approximation to $f'(x)$, in this case a *second-order-accurate* approximation known as a *centered difference*.

We can derive formulas for derivatives of arbitrarily high orders of accuracy by combining many Taylor approximations—such as $f(x)$, $f(x + \Delta x)$, $f(x - \Delta x)$, and $f(x + 2\Delta x)$—and solving for the coefficients needed to cancel out as many low-order error terms as possible. The actual calculation of the formulas for the different approximations is an example of a linear system problem, which we will read about in chapter 21. For most applications of Newton–Raphson, we are unlikely to need better than a second-order approximation, though. We will see later where higher-order accuracy derivatives may be needed when doing numerical integration. It is also worth noting that we need the assumption of bounded derivatives from earlier in the chapter in order to establish the accuracy of these approximations. If any derivatives of the function are unbounded on the region of interest to us, we may be unable to generate high-order approximations to the first derivative.

5.5 Multivariate Functions

One other advantage of the Newton–Raphson method over the bisection and secant methods is that it provides a much better way of dealing with systems of several variables. In a multidimensional space, it can be difficult to find the positive and negative points we need in order to start searching for a zero. The Newton–Raphson method still works when we have more than one variable, although it gets a bit more complicated to apply. Suppose that instead of maximizing or minimizing a function $F(x)$, we want to maximize or minimize a multivariate function $F(x_1, x_2, \ldots, x_n)$. To understand how to use Newton–Raphson for this job, we need to consider how some of our concepts from earlier in the chapter generalize to functions of more than one variable.

First, recall that we were using a theorem of calculus which states that maxima or minima of a function occur at zeros of the function's derivative. The multivariate equivalent of this statement is that the maxima or minima occur at places where all of the partial first derivatives are zero. That is,

$$\frac{\partial F}{\partial x_1} = 0$$

$$\frac{\partial F}{\partial x_2} = 0$$

$$\vdots$$

$$\frac{\partial F}{\partial x_n} = 0.$$

An equivalent statement is that the vector of the first derivatives is the all-zero vector. That is:

$$\begin{bmatrix} \dfrac{\partial F}{\partial x_1} & \dfrac{\partial F}{\partial x_2} & \cdots & \dfrac{\partial F}{\partial x_n} \end{bmatrix} = \begin{bmatrix} 0 & 0 & \cdots & 0 \end{bmatrix}.$$

This vector of derivatives is known as the *gradient* of F and is abbreviated ∇F.

For example, suppose we have the function $F(x, y, z) = (x^2 + 2y + z)\sin(x)\cos(y)$.

$$\nabla F = \begin{bmatrix} 2x\sin(x)\cos(y) + (x^2 + 2y + z)\cos(x)\cos(y) \\ 2\sin(x)\cos(y) - (x^2 + 2y + z)\sin(x)\sin(y) \\ \sin(x)\cos(y), \end{bmatrix}$$

which we can abbreviate

$$\nabla F = \begin{bmatrix} f_1(\vec{v}) \\ f_2(\vec{v}) \\ f_3(\vec{v}) \end{bmatrix} = \vec{f}(\vec{v}),$$

where f_i is $\frac{\partial F}{\partial x_i}$ and $\vec{v}$ is the vector of all of our variables, (x, y, z).

Thus, instead of finding a point where our scalar function has derivative zero, we want a point where our function has gradient equal to the all-zero vector. Now how do we find that? The idea behind the scalar Newton–Raphson method was to find a zero of $f(x)$ by locally linearizing about some guess, using the derivative of the derivative ($F''(x) = f'(x)$) at that point. The multidimensional equivalent of the derivative of the derivative is called the *Jacobian* of ∇f or the *hessian* of F. It is a matrix containing all second derivatives of F:

$$J(\vec{v}) = \begin{bmatrix} \frac{\partial^2 F}{\partial x_1^2} & \frac{\partial^2 F}{\partial x_1 \partial x_2} & \cdots & \frac{\partial^2 F}{\partial x_1 \partial x_n} \\ \frac{\partial^2 F}{\partial x_2 \partial x_1} & \frac{\partial^2 F}{\partial x_2^2} & \cdots & \\ \vdots & & \ddots & \\ \frac{\partial^2 F}{\partial x_n \partial x_1} & \cdots & & \frac{\partial^2 F}{\partial x_n^2} \end{bmatrix}.$$

For example, for the function $F(x, y) = (x^2 y + xy - y^2 + y)$, the gradient would be

$$\nabla F(x, y) = [2xy + y \quad x^2 + x - 2y + 1],$$

and the hessian would be

$$J(x, y) = \begin{bmatrix} 2y & 2x + 1 \\ 2x + 1 & -2 \end{bmatrix}.$$

Now that we know the equivalent of a derivative and a second derivative in multiple dimensions, we can figure out how to do a multidimensional Newton–Raphson iteration. Instead of

$$x_{i+1} \leftarrow x_i - \frac{f(x_i)}{f'(x_i)},$$

we would use

$$\vec{v}_{i+1} \leftarrow \vec{v}_i - J(\vec{v}_i)^{-1} \nabla F.$$

Although the formula includes a matrix inverse, it is almost never a good idea to invert a matrix in practice. Almost anything one would want to compute with a matrix inverse can be computed more quickly than the matrix inverse itself. This is no exception. Instead of calculating $J(\vec{v}_i)^{-1} \nabla F$, what we really want to do is define the vector y_i as follows:

$$y_i = J(\vec{v}_i)^{-1} \nabla F,$$

then rearrange to get the linear system

$$J(\vec{v}_i) y_i = \nabla F.$$

We can solve this linear system much faster in practice than we could invert J. We will see methods for this problem in chapter 21. We can then plug y_i into the Newton–Raphson formula.

To see an example of multivariate Newton–Raphson, we can use our function $F(x, y) = (x^2y + xy - y^2 + y)$ from above. If we guess $(0, 0)$ as the maximum of our function, then we will get the following:

$$\nabla F(0, 0) = \begin{bmatrix} 0 & 1 \end{bmatrix}$$

$$J(0, 0) = \begin{bmatrix} 0 & 1 \\ 1 & -2 \end{bmatrix}.$$

Therefore, our Newton–Raphson iteration would be

$$\begin{bmatrix} x_1 \\ y_1 \end{bmatrix} \leftarrow \begin{bmatrix} 0 \\ 0 \end{bmatrix} - \begin{bmatrix} 0 & 1 \\ 1 & -2 \end{bmatrix}^{-1} \begin{bmatrix} 0 \\ 1 \end{bmatrix}.$$

We then solve

$$\begin{bmatrix} 0 & 1 \\ 1 & -2 \end{bmatrix} \begin{bmatrix} u_1 \\ u_2 \end{bmatrix} = \begin{bmatrix} 0 \\ 1 \end{bmatrix}$$

for u_1 and u_2. The linear system is solved by

$$\begin{bmatrix} u_1 \\ u_2 \end{bmatrix} = \begin{bmatrix} 1 \\ 0 \end{bmatrix}.$$

Plugging that into the iteration formula, we get

$$\begin{bmatrix} x_1 \\ y_1 \end{bmatrix} \leftarrow \begin{bmatrix} 0 \\ 0 \end{bmatrix} - \begin{bmatrix} 1 \\ 0 \end{bmatrix} = \begin{bmatrix} -1 \\ 0 \end{bmatrix}.$$

Thus we will use $x = -1$, $y = 0$ as the next guess for the extremum of our function. We still have the same problem as with the one-dimensional case: we need a good initial guess. If we get a good start, though, then essentially all of the nice properties of Newton–Raphson work just as well in multiple dimensions.

The multidimensional Newton–Raphson method is often the best tool for solving difficult continuous optimization problems, but it can be very tricky to use in practice. If our initial guess is not close enough to a solution, the method can end up making the guess worse rather than better. Sometimes it will seem to do exactly the opposite of what we would expect, moving farther and farther away from a seemingly close solution. There is a huge body of literature on how to make Newton–Raphson more robust, and one can take entire courses on that topic. We will not cover these more advanced methods here, but those who plan to make a lot of use of continuous optimization methods would do well to study more advanced coursework on the topic. The References and Further Study section provides some starting points for learning more.

5.6   Direct Methods for Optimization

Although we earlier saw that optimization is equivalent to zero-finding, it is sometimes useful to work directly on the optimization problem rather than through a zero-finding method. In this section we will see a few other optimization approaches that do not explicitly rely on attempts to find zeros of the gradient vector.

5.6.1   Steepest Descent

Though Newton–Raphson is fast but sometimes lacking in robustness, we can use similar ideas to derive a method that tends to be slower but more robust. In using Newton–Raphson to maximize or minimize a function F, we try to find zeros of the gradient ∇F by finding the hessian of F at a point and projecting where the locally linearized function will go to zero. We can similarly work directly with a locally linear model of F by finding F's gradient. ∇F points in the direction of most rapid increase of F, and $-\nabla F$ points in the direction of most rapid decrease. We may therefore propose that we can make the best possible local progress in maximizing or minimizing F by moving in the direction of the gradient (or its negative). Thus, we can propose to get closer to our maximum or minimum by taking a small step along the direction of the gradient. We can then find a new gradient at the updated point, move in the direction of the new gradient, and so on until we get to our extremum. Let us assume here that we are trying to minimize F, and thus moving in the direction of $-\nabla F$. The method resulting from the description above is called *steepest descent* or *gradient descent*.

To perform steepest descent, we first find the gradient of F at our initial starting point $\vec{v}$, $\nabla F(\vec{v}) = \begin{bmatrix} \frac{\partial F}{\partial v_1}(\vec{v}) & \frac{\partial F}{\partial v_2}(\vec{v}) & \cdots & \frac{\partial F}{\partial v_n}(\vec{v}) \end{bmatrix}$. Once we have decided that we want to move in the direction of $-\nabla F$ to improve our guess, we need to decide how far to move in that direction. We want to move along $-\nabla F$ by an amount that minimizes F as much as possible. That is, we ideally want to find the minimum of F along the line passing through $\vec{v}$, our current guess, with direction $-\nabla F(\vec{v})$. We can formulate that as an optimization problem in itself. We want to solve

$$\min_r \; F(\vec{v} - r\nabla F(\vec{v})).$$

We have now transformed our multidimensional optimization problem into a problem of optimizing for a function of a single variable, r. We already know some fine options for minimizing a function of one variable. Once we have r, we replace our past guess $\vec{v}$ with $\vec{v} - r\nabla F(\vec{v})$. We can then keep repeating steepest descent steps until we are satisfied with the quality of our solution.

Figure 5.5 shows pseudocode for the steepest descent method. As with our other methods, we can decide to stop based on an estimate of the backward error (i.e.,

1. Find the gradient $\nabla F(\vec{v}_i)$
2. Solve for r minimizing $F(\vec{v}_i - r\nabla F(\vec{v}_i))$
3. $\vec{v}_{i+1} \leftarrow \vec{v}_i - r\nabla F(\vec{v}_i)$
4. $i \leftarrow i + 1$
5. Go to step 1 unless the solution is sufficiently good.

Figure 5.5
Pseudocode for the steepest descent method, assuming an initial guess $\vec{v}_0$.

looking for when $\|\vec{v}_{i+1} - \vec{v}_i\|$ becomes sufficiently small) or the forward error (i.e., looking for when $|F(\vec{v}_{i+1}) - F(\vec{v}_i)|$ becomes sufficiently small).

Steepest descent tends to be slow in practice compared with Newton–Raphson, but is simpler and often more robust. It is also possible to combine the two, for example, by using steepest descent to get close to a local minimum, then switching to Newton–Raphson to converge very rapidly to the minimum with high precision. This heuristic can be formalized to give rise to the next method we will consider.

5.6.2 The Levenberg–Marquardt Method

We saw above two simple optimization methods, each of which has some very nice properties. Newton–Raphson tends to have very rapid convergence close to zeros, and is therefore an excellent choice if we start with a good initial guess. Steepest descent is a slower but more robust method that can get us to a local minimum pretty reliably but may take a long time to do it. One might reasonably ask if there is some way to get the best of both worlds. It turns out that there very often is, through a hybrid of Newton–Raphson and steepest descent called the Levenberg–Marquardt method.

To understand how Levenberg–Marquardt works, it will be helpful to review the update rules for Newton–Raphson and steepest descent. For Newton–Raphson, our updates take the form

$$\vec{v}_{i+1} \leftarrow \vec{v}_i - J(\vec{v}_i)^{-1}\nabla F.$$

For steepest descent, they take the form

$$\vec{v}_{i+1} \leftarrow \vec{v}_i - r\nabla F(\vec{v}_i)$$

for some scalar r. Levenberg proposed that we can do better by creating a hybrid algorithm whose steps are a mixture of Newton–Raphson and steepest descent steps, switching from primarily steepest descent far from zeros to primarily Newton–Raphson close to zeros. We can do this by using the interpolation formula

$$\vec{v}_{i+1} \leftarrow \vec{v}_i - (J(\vec{v}_i) + \lambda I)^{-1}\nabla F,$$

> 1. Find the gradient $\nabla F(\vec{v}_i)$ and hessian $J(\vec{v}_i)$
> 2. $\vec{v}_{i+1} \leftarrow \vec{v}_i - (J(\vec{v}_i) + \lambda \mathrm{Diag}(J(\vec{v}_i)))^{-1} \nabla F$
> 3. if $F(\vec{v}_{i+1}) < F(\vec{v}_i)$
>   A. $\lambda \leftarrow \lambda/S$
>   B. $i \leftarrow i + 1$
> 4. else
>   A. $\lambda \leftarrow \lambda \times S$
> 5. Go to step 1 unless the solution is sufficiently good.

Figure 5.6
Pseudocode for the Levenberg–Marquardt method, assuming an initial guess $\vec{v}_0$ and scaling factor S.

where I is the identity matrix and λ is a parameter we will adjust based on estimates of how close we are to a solution. Marquardt suggested a minor modification of this formula that turns out to make a big difference in practice:

$$\vec{v}_{i+1} \leftarrow \vec{v}_i - (J(\vec{v}_i) + \lambda \, \mathrm{Diag}(J(\vec{v}_i)))^{-1} \nabla F,$$

where $\mathrm{Diag}(J(\vec{v}_i))$ is the matrix containing the diagonal of the Jacobian and the zeros in all off-diagonal entries. This change has the effect of weighting the gradient so as to take smaller steps in directions of high curvature of the function.

Figure 5.6 provides pseudocode for the full method. The method adjusts λ by either multiplying or dividing it by a small constant scaling factor S on each step. A typical S might be between 2 and 10. When a step leads to a worse solution, then we assume that λ is too small and that we must increase it, shifting the method more toward steepest descent and also reducing the step size. When a step leads to a better solution, then we assume that λ can be reduced, shifting the method more toward Newton–Raphson. By the time it gets near a zero, it should convert to almost pure Newton–Raphson and converge rapidly on the true zero.

The Levenberg–Marquardt method has become a de facto standard for generic nonlinear optimization problems. It will often turn out to be our best option for real-world problems.

5.6.3 Conjugate Gradient

There is another method worth mentioning here, although we will not cover the theory behind it until chapter 21. This is a method called *conjugate gradient*. Conjugate gradient is actually a method for solving linear systems, the most famous of a broadly useful class called *Krylov subspace methods*. By treating our function as locally linear, though, we can transform conjugate gradient into a very effective method for general continuous optimization. Conjugate gradient is very similar to

1. $\vec{p}_0 = [00 \dots 0]$
2. $\vec{\beta}_0 = [00 \dots 0]$
3. for $i \leftarrow 1$ to n
   A. if $||\nabla F(\vec{v}_i)|| < \epsilon$
      i. stop
   B. else
      i. $\vec{\beta}_i = \frac{\nabla F(\vec{v}_i)^T \nabla F(\vec{v}_i)}{\nabla F(\vec{v}_{i-1})^T \nabla F(\vec{v}_{i-1})}$
      ii. $\vec{p}_i = -\nabla F(\vec{v}_i) + \vec{\beta}_i \vec{p}_{i-1}$
      iii. find α_i minimizing $F(\vec{v}_i + \alpha_i \vec{p}_i)$
      iv. $\vec{v}_{i+1} = \vec{v}_i + \alpha_i \vec{p}_i$

Figure 5.7
Pseudocode for the conjugate gradient method, assuming an initial guess $\vec{v}_0$. For a linear system, n is the dimension of the matrix. It can be replaced with an arbitrary maximum number of steps for nonlinear system-solving.

steepest descent, but uses a subtly different minimization criterion that has the effect of avoiding a lot of wasted work the steepest descent method performs.

Figure 5.7 presents pseudocode for the conjugate gradient algorithm. For the most part, the method should remind us of steepest descent. The major difference is that we are moving in the direction of the vector $\vec{p}_i$ instead of the gradient in lines 8 and 9. $\vec{p}_i$ is essentially a "corrected" version of the gradient that we create by subtracting off components of the previous $\vec{p}$ vector to prevent $\vec{p}_i$ from undoing work done with $\vec{p}_{i-1}$. Lines B.i and B.ii identify a $\vec{p}_i$ orthogonal to the previous ones. If we were optimizing in a linear system, this would have the effect of choosing movement vectors that are all at right angles to each other, guaranteeing that work that is done in one step is not undone in subsequent steps. When we apply conjugate gradient to nonlinear systems, some of the theory of optimality breaks down, but in practice it is likely to work extremely well if the system is generally well behaved. The method also depends on the fact that the matrix is positive semidefinite, a concept we will discuss in chapter 21, although it is needed only for efficiency, not for correctness. There are related methods that work on general matrices, which we will also see in that chapter.

References and Further Study

The Numerical Recipes series by Press et al. [82] is an excellent general reference to commonly used numerical algorithms, and was perhaps the single most important source for this text. It covers pretty much everything covered in this chapter, as well as many other methods I have omitted. The series includes the basic *Numerical Recipes*, as well as *Numerical Recipes in Java*, *Numerical Recipes in C*, *Numerical*

Recipes in FORTRAN, and others that provide code samples in a given language for the algorithms covered. The series is sometimes criticized for a "cookbook" approach to presenting algorithms without much regard to the theory behind them and how one might need to adapt them in practice. In that regard, it is similar to this text, and one would be well advised to look deeper into the theory than one will find here or in Press et al. for any method one will be using extensively for particularly demanding applications. Regardless of the merits of this criticism, though, a Numerical Recipes book is a very handy reference for anyone planning to work in any kind of scientific computing.

As with many topics we will cover, we have only scratched the surface of some of the key classic methods for nonlinear optimization. The coverage here is likely to serve one well with reasonably small, well-behaved systems, and at least get one started on the harder problems. There is, however, an extensive body of theory and practical tricks of the trade available for tackling the harder problems. Those interested specifically in a deeper coverage of optimization might look to Dennis and Schnabel [83] or Ruszczyński [84] for more theoretical treatments, or Fletcher [85] for a more practice-oriented treatment. The methods covered here are for the most part quite old as computational methods go, and there are many other texts in which one may look for deeper coverage. Trefethen and Bau [86] is often considered the standard text for numerical linear algebra, and provides information on the conjugate gradient method. It is also useful for learning more about some of the linear systems-solving issues we glossed over in this chapter and will return to in chapter 21. There is also a great deal of information available online on these methods, and many others covered in this text. Those looking for detailed pseudocode, links to related methods, and descriptions of their history can refer, for example, to *Wikipedia* [87] as a great starting point.

The methods covered in this chapter are for the most part quite old, and getting copies of the primary sources will not be easy. Most are simply considered part of the basic knowledge of the numerical methods field, and no source is generally cited for their use. Nonetheless, I can provide some references here. The Newton–Raphson method was first mentioned by Newton [88] as a means of deriving a polynomial approximation to the root of a polynomial, and was first published by Wallis [89] in that form. It was later formalized as a general iterative method by Raphson [90]. The method underwent several refinements, though, and others can lay claim to having developed what we today call the Newton–Raphson method. The conjugate gradient method was first proposed by Hestenes and Stiefel [91]. The Levenberg–Marquardt method derives from a paper by Levenberg [92], proposing the original method, and one by Marquardt [93], proposing the modification of using the hessian to improve the steepest descent component of the steps. I have never seen a primary reference to the bisection or secant methods, but similar methods have been known for thousands of years, and their true origin appears to have been lost.

6 Constrained Optimization

In the last chapter, we examined some general methods for optimization. In this chapter, we will look at a special kind of optimization problem called a *constraint satisfaction* or *constrained optimization* problem. A constraint satisfaction problem is an optimization problem in which we have some rules restricting the allowable solutions and we want to find the optimal value of some objective function consistent with the restrictions.

For example, suppose we have an organism that must produce several proteins—which we can call protein 1, protein 2, and protein 3—for its growth. Its growth rate is limited to the amount of protein 1 it can produce plus twice the amount of protein 2 plus three times the amount of protein 3. We then restrict the amount of an essential amino acid needed by all three proteins. Protein 1 requires two copies of the missing amino acid per molecule, protein 2 requires three copies per molecule, and protein 3 requires four copies per molecule. We then ask how quickly the organism can grow, given these restrictions, if we provide it with the missing amino acid at some fixed rate K. How would we formulate this problem?

We would first assign a variable to the rate of production of each protein, p_1, p_2, and p_3. Our optimization criterion can be expressed in terms of these three rates by saying that we want to maximize $f(p_1, p_2, p_3) = p_1 + 2p_2 + 3p_3$, and that this will provide the maximum growth rate possible for the organism. Our solution, however, must satisfy our other constraints. First, since we cannot have a negative amount of any protein, we have the following three constraints:

$p_1 \geq 0$

$p_2 \geq 0$

$p_3 \geq 0.$

Second, our proteins cannot be produced at a rate that would consume more than K units of amino acid. That gives us the constraint

$2p_1 + 3p_2 + 4p_3 \leq K.$

The problem of maximizing f subject to the constraints on p_1, p_2, and p_3 is an example of a constraint satisfaction problem. In particular, it is something called a *linear programming* problem, an important special case of constraint satisfaction with which we will begin.

6.1   Linear Programming

In the general case, a linear programming problem is defined by three features:

1. A set of variables, $\vec{x} = [x_1, x_2, \ldots, x_n]$
2. A set of linear constraints, $\sum_{i=1}^{n} a_{ij} x_j \leq b_i$ (which can be abbreviated as $A\vec{x} \leq \vec{b}$)
3. A linear function to be optimized, also called an *objective function*, min (or max) $\sum_{i=1}^{n} c_i x_i$ (which can be abbreviated min (or max) $\vec{c}^T \vec{x}$).

For example, we can consider the following system:

maximize $x_1 + x_2$ subject to

$x_1 \geq 0$

$x_2 \geq 0$

$2x_1 + x_2 \leq 6$

$-x_1 + 2x_2 \leq 8.$

One way to understand the problem is to plot the constraints graphically, as in figure 6.1. Each constraint forms a line in a 2-D plot, with solutions constrained to lie

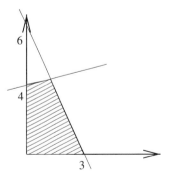

Figure 6.1
Constraints for a sample linear programming problem.

on one side of that line. If we are looking at a problem with more variables, then the constraints will generally form *hyperplanes* (high-dimensional flat surfaces) with solutions constrained to lie on one side of each hyperplane. The shape defined by the set of constraints is called a *polytope*, and in the general case it can be a complicated high-dimensional solid. Because our constraints are linear (depending only on first-order terms in our variables), the faces of the polytope are flat. As we look for solutions, we will consider points in the space of our variables. We require points lying inside the polytope, which are known as *feasible* points. A point outside the polytope, known as an *infeasible* point, will not be a valid solution because it will violate at least one of our constraints. Our goal, then, is to find a feasible point maximizing the value of our function.

6.1.1   The Simplex Method

There is a classic method for this problem, variations of which are still used today, called the simplex method. The simplex method depends on a theorem stating that a constraint satisfaction problem with linear constraints and a linear objective function has its maximum and its minimum at vertices of the polytope. We cannot have an extremum inside the polytope or in the middle of one of its faces, unless there happens to be an equally good solution at a vertex. We therefore only have to look at the vertices to find an optimum.

Given this theorem, one might propose that we can solve this problem simply by testing the value of the objective function at every vertex. While this will work eventually, it is not a practical solution because the number of vertices can be exponential in the size of the problem. The simplex method instead searches only a small fraction of vertices by relying on local movements over the edges of the polytope. For example, imagine that we are trying to maximize the function $f(x_1, x_2) = x_1 + x_2$ on the polytope of figure 6.2. Using the simplex method, we may start at

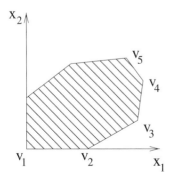

Figure 6.2
A polytope in two variables. The vertices v_1, v_2, v_3, v_4, v_5 mark a path we might follow to maximize $f(x_1, x_2) = x_1 + x_2$ by the simplex method.

vertex v_1 and then move through v_2, v_3, and so on, up to v_5, increasing the value of f at each step. When we get to v_5 and realize that there is no move we can make to increase the objective further, we know we are done and have reached the global optimum.

In order to see how to choose these points, it will help to simplify our problem a bit by converting it into something called *standard form*. A linear programming problem in standard form is expressed as follows:

minimize $c_1x_1 + c_2x_2 + \cdots c_nx_n$ subject to

$$a_{11}x_1 + a_{12}x_2 + \cdots + a_{1n}x_n = b_1$$

$$a_{21}x_1 + a_{22}x_2 + \cdots + a_{2n}x_n = b_2$$

$$\vdots$$

$$a_{m1}x_1 + a_{m2}x_2 + \cdots + a_{mn}x_n = b_m$$

$$x_1 \geq 0$$

$$x_2 \geq 0$$

$$\vdots$$

$$x_n \geq 0.$$

This form can be abbreviated in matrix notation as

minimize $\vec{c}^T\vec{x}$ subject to $A\vec{x} = \vec{b}$, $\vec{x} \geq \vec{0}$,

where $A\vec{x} = \vec{b}$ will typically be an underdetermined system of equations and will therefore have an infinite number of possible solutions.

Given any linear programming problem, there are a few simple steps we can take to put the problem in standard form:

1. Anywhere we have a constraint of the form $\vec{a}_j\vec{x} \geq b_j$ other than $x_i \geq 0$, convert it to $-\vec{a}_j\vec{x} \leq b_j$.

2. Anywhere we have a constraint $\vec{a}_jx \leq b_j$, add a new *slack variable* $\bar{x}_j$ and replace the constraint with the two constraints

$$\vec{a}_j\vec{x} + \bar{x}_j = b_j$$

$$\bar{x}_j \geq 0.$$

3. If we have a variable x_i that may become negative, replace x_i everywhere it occurs with $(x_{i+} - x_{i-})$, and add the constraints

$$x_{i+} \geq 0$$

$$x_{i-} \geq 0.$$

4. If our problem is a maximization problem, max $\vec{c}^T \vec{x}$, convert it to the minimization problem min $-\vec{c}^T \vec{x}$.

The first step turns any of our "greater than or equal to" constraints into "less than or equal to" constraints. The second allows us to convert each "less than or equal to" constraint into an "equality" constraint and a constraint forcing a single variable to be positive. The third converts our problem to one in which all variables are constrained to be positive. The fourth ensures that we have a minimization problem. Putting all of these together, we convert any possible linear program into a program in standard form. We will go through an example of this procedure after we have covered the main portion of the simplex algorithm.

Once our problem is in standard form, we are guaranteed to have at least one variable for each linearly independent constraint. Suppose we have m constraints and n variables. If m is exactly n, then there is only one solution to the $Ax = b$ portion of our constraint set, so our problem reduces to linear system-solving. In general, though, $n > m$, so we have an underdetermined $Ax = b$ and need to look among solutions that satisfy it. In the simplex method, we will search these solutions by looking specifically at points for which $n - m$ of the variables are zero and the remaining m are chosen to satisfy the constraints of $Ax = b$. By fixing all but m of our constraints, we ensure that the remaining m now constitute a full-rank linear system and thus are uniquely determined. Since $n - m$ of the variables are forced to collide with nonnegative constraints, we ensure that the full set of values will fall on a corner of our polytope. The full simplex algorithm for a system in standard form is provided in figure 6.3.

---

1. Pick an initial feasible point at a corner of the polytope by setting $n - m$ of the variables to zero. It will make things easiest if you can pose your problem so that setting all non-slack variables to zero produces a corner of the polytope.
2. Rewrite the constraints and objectives so they are all expressed in terms of the variables that are zero and solve for the non-zero variables.
3. Find some variable x_i that is zero for which increasing x_i reduces the objective function. If no such variable can be found, then terminate.
4. Increase x_i until some other variable becomes zero.
5. Return to step 2.

Figure 6.3
Pseudocode for the simplex method for a linear system in standard form.

Simplex Example We can see how the simplex method works by looking at an example:

maximize $f(x_1, x_2) = x_1 + 3x_2$ subject to

$-x_1 + x_2 \leq 4$

$x_1 + x_2 \leq 10$

$x_1 \geq 0$

$x_2 \geq 0.$

The first thing we need to do is put the problem in standard form. This one is not too far from standard form to begin with, so we do not have very much to do. We need to introduce two slack variables, which we can call x_3 and x_4, and flip the maximization to a minimization. The result will be the following:

minimize $g(x_1, x_2, x_3, x_4) = -x_1 - 3x_2$ subject to

$-x_1 + x_2 + x_3 = 4$

$x_1 + x_2 + x_4 = 10$

$x_1 \geq 0, \quad x_2 \geq 0, \quad x_3 \geq 0, \quad x_4 \geq 0.$

Now we are ready to apply the simplex method.

First, we make an initial guess by forcing the nonslack variables to zero:

$x_1 = x_2 = 0.$

Next, we write the equations in terms of the zero variables, x_1 and x_2. Our minimization function is already expressed in terms of x_1 and x_2, so that is fine. We can rewrite our two other constraints as follows:

$-x_1 + x_2 + x_3 = 4$

$\Downarrow$

$x_3 = x_1 - x_2 + 4$

and

$x_1 + x_2 + x_4 = 10$

$\Downarrow$

$x_4 = -x_1 - x_2 + 10.$

If we then solve for x_3 and x_4, we get

$$x_3 = 0 - 0 + 4 = 4$$

$$x_4 = -0 - 0 + 10 = 10.$$

We are therefore starting at the point $\vec{x} = [x_1 \quad x_2 \quad x_3 \quad x_4] = [0 \quad 0 \quad 4 \quad 10]$.

Third, we need to find a zero variable we can increase. We are minimizing the function $g(x_1, x_2, x_3, x_4) = -x_1 - 3x_2$. We can immediately see that increasing x_1 or x_2 will decrease g, since both have negative coefficients in g. We can therefore use either one. But let us try x_2, since g decreases more quickly with x_2 than with x_1, as x_2's coefficient has a larger absolute value than x_1's.

Fourth, we have to see how much we can increase x_2 before some other variable hits zero. Our first constraint is

$$x_3 = x_1 - x_2 + 4,$$

and since x_1 is currently 0, this is equivalent to

$$x_3 = -x_2 + 4.$$

Therefore, we can increase x_2 up to 4 before it forces x_3 to be 0. The second constraint is

$$x_4 = -x_1 - x_2 + 10,$$

which is currently equivalent to

$$x_4 = -x_2 + 10.$$

We can therefore increase x_2 up to 10 before x_4 becomes 0. Taking the minimum of these two limits on x_2, we will decide to increase x_2 to 4, forcing x_3 to 0 and leaving x_4 positive.

We have now moved x_3 into our set of zero variables and x_2 out of it, so we need to rewrite our objective and constraints in terms of x_3 instead of x_2. We can first do the following rewrite to express x_2 in terms of x_3:

$$x_3 = x_1 - x_2 + 4$$

$$\Downarrow$$

$$x_2 = x_1 - x_3 + 4.$$

We can then plug this new expression for x_2 into the second constraint to get the following:

$$x_4 = -x_1 - x_2 + 10$$

$\Downarrow$

$$x_4 = -x_1 - (x_1 - x_3 + 4) + 10 = -2x_1 + x_3 + 6.$$

We also need to rewrite our objective function:

$$\min -x_1 - 3x_2$$

$\Downarrow$

$$\min -x_1 - 3(x_1 - x_3 + 4) = -4x_1 + 3x_3 - 12.$$

If we now solve again for our nonzero variables, we get the following:

$$x_2 = x_1 - x_3 + 4 = 0 - 0 + 4 = 4$$

$$x_4 = -2x_1 + x_3 + 6 = -2(0) + 0 + 6 = 6.$$

Therefore, we are now at the point $\begin{bmatrix} 0 & 4 & 0 & 6 \end{bmatrix}$, which corresponds to the point $(x_1, x_2) = (0, 4)$ in our original problem.

Now we need to find another zero variable to increase. We have two zero variables, x_1 and x_3. Our objective function is now $-4x_1 + 3x_3 - 12$, so increasing x_1 will reduce the objective but increasing x_3 will increase the objective. We therefore must use x_1 as our variable to increase.

Now we need to figure out how much we can increase x_1. Our constraints are

$$x_2 = x_1 - x_3 + 4$$

$$x_4 = -2x_1 + x_3 + 6.$$

Increasing x_1 will only increase x_2, so the first constraint does not place any limits on how much we can increase x_1. The second constraint limits x_1 to 3 before x_4 is forced to 0. Therefore, we can increase x_1 to 3 and convert x_4 to 0 in the process.

Now we have removed x_1 from our zero set and put x_4 into it. We therefore need to rewrite our objective and constraints again, to put them in terms of x_4 instead of x_1. We first use the second constraint to get an expression for x_1 in terms of x_4:

$$x_4 = -2x_1 + x_3 + 6$$

$\Downarrow$

$$x_1 = \frac{1}{2}x_3 - \frac{1}{2}x_4 + 3.$$

We then substitute this new expression for x_1 into the other constraint:

$$x_2 = x_1 - x_3 + 4$$

$$\Downarrow$$

$$x_2 = \left(\frac{1}{2}x_3 - \frac{1}{2}x_4 + 3\right) - x_3 + 4 = -\frac{1}{2}x_3 - \frac{1}{2}x_4 + 7.$$

We also substitute it into the objective function:

$$\min -4x_1 + 3x_3 - 12$$

$$\Downarrow$$

$$\min -4\left(\frac{1}{2}x_3 - \frac{1}{2}x_4 + 3\right) + 3x_3 - 12 = x_3 + 2x_4 - 9.$$

We now look for another variable to increase. Our objective function now has only positive coefficients, though, so we cannot increase any of our zero variables without increasing the objective function. This tells us that we cannot improve our solution, and have hit the global optimum. Our constraints tell us the values of our nonslack variables:

$$x_1 = \frac{1}{2}x_3 - \frac{1}{2}x_4 + 3 = 3$$

$$x_2 = -\frac{1}{2}x_3 - \frac{1}{2}x_4 + 7 = 7.$$

Therefore, we are at the point $[3 \quad 7 \quad 0 \quad 0]$, which corresponds to the point $(x_1, x_2) = (3, 7)$ in our original problem. With our transformed objective function, $g(x_1, x_2, x_3, x_4) = -x_1 - 3x_2$, we therefore have an objective value of -24. If we flip back to our original maximization objective, $f(x_1, x_2) = x_1 + 3x_2$, we see that $f(3, 7) = 24$. This, then, is the global optimum of our original problem.

There are a few details worth noting about the simplex method. Note that while the simplex method is guaranteed to find the global optimum, there is no guarantee it will not visit an exponentially large number of vertices before it does so. In practice, though, the method tends to work very well, typically requiring a number of steps that is approximately linear in the number of constraints. It is also not always trivial to find an initial vertex as a starting point, although that is often something we can control through choice of model. Finally, there are other ways to do the computation of the model that some may find easier. For example, the step of rewriting the equations in terms of the zero variables after each change of values is equivalent to solving a full-rank linear system for the nonzero variables. We can automate this step through the process of Gaussian elimination, which we will see in chapter 21.

6.1.2 Interior Point Methods

Although the simplex method generally works well in practice, it can in the worst case require exponential time. It was long an open question whether there was any provably tractable method to solve linear programming problems. It was eventually shown that linear programs can in fact be solved in polynomial time in input size through a class of methods called *interior point methods*. As the name implies, interior point methods work by exploring the interior of the simplex, rather than just the surface. Although the solution is guaranteed to lie on a vertex, it turns out that it can nonetheless be useful to perform a search within the interior of the simplex.

Linear programming was first proved to be a tractable problem through an interior point method called the *ellipsoid method*. The ellipsoid method works by defining an ellipsoid (a higher-dimension equivalent of an ellipse) that contains part of the feasible region. It then checks if the center of the ellipsoid is feasible, and if it is, the method can guarantee that the center contains an optimal solution to the problem. If the center of the ellipsoid is not feasible, the method finds another, slightly smaller ellipsoid that is guaranteed to overlap with the feasible region unless the feasible region is empty. This continues until either a solution is found or the ellipsoid becomes so small that one can prove there is no solution. The method guarantees that the ellipsoid shrinks by at least a constant factor at each step, which makes it possible to put a polynomial bound on the runtime. Nonetheless, the method ultimately proved impractical, and for purposes of solving real linear programming problems, it is essentially of purely historical interest now. We therefore mention it only in passing.

Interior point methods did ultimately prove practical for real-world linear programming problems, though, starting with a different kind of method called *Karmarkar's method*, which is an example of a broader class of interior point methods called *barrier methods*. We will look at a simplified version of Karmarkar's method called an *affine method*. Pseudocode for the basic affine method for solving a system in standard format is presented in figure 6.4.

To see how this works in practice, we can examine some of the steps in isolation. Assume we are solving min $c^T x$ subject to $Ax = b$, $x_i \geq 0$. Suppose we have a feasible initial guess $\vec{x}_0$. For the moment, let us ignore step 2 and pretend that our initial guess is far from the boundaries of the polytope. We can return to the issue of scaling at the end. Our method, then, is the following:

Find the steepest descent direction. This is the vector $-\vec{c}$. So if we are minimizing $-x_1 - 3x_2$, the steepest descent direction will be the vector $\begin{bmatrix} 1 & 3 \end{bmatrix}$.

Project into the null-space of A. Null-space is a concept one would ordinarily learn about in a full linear algebra course, but it can be understood with a minimal understanding of linear algebra. Essentially, if we find any set of linearly independent vec-

---

1. Start with some feasible guess $\vec{x}_0$.
2. Transform the coordinate system by scaling variables so as to move the feasible guess far from any boundaries of the polytope.
3. Find the steepest descent direction $-\nabla g = -\vec{c}$ of the objective.
4. Project ∇g into the *nullspace* of A. This means that you find the component of ∇g that gives you a zero vector when multiplied by A (see below). Call this component $\vec{h}$.
5. Move from the current guess $\vec{x}_i$ along $\vec{h}$ until it almost reaches a boundary of the polytope. Call the resulting point $\vec{x}_{i+1}$.
6. If we are not at a solution, return to step 2.
7. Transform the solution back into normal space.

Figure 6.4
Pseudocode for an affine method for linear programming.

tors $\vec{v}_1, \vec{v}_2, \ldots, \vec{v}_k$ for which $A\vec{v}_i = \vec{0}$ for all i, then any linear combination of these vectors, $\vec{v} = d_1\vec{v}_1 + d_2\vec{v}_2 + \cdots + d_k\vec{v}_k$, will also have the property that $A\vec{v} = \vec{0}$. If we imagine that our vectors $\vec{v}_1, \ldots, \vec{v}_k$ are the axes of a coordinate system, then the space they define will consist entirely of points having the property $A\vec{v} = \vec{0}$. The space of all possible vectors of this form is called the null-space of A because A transforms any vector in the space into the zero vector. We can find the component of $-\nabla g$ lying in the null-space of A by the following formula:

$$\vec{h} = (I - A^T(AA^T)^{-1}A)(-\nabla g).$$

Now we need to move along $\vec{h}$ so as to almost hit a boundary of the polytope. In other words, we need to find a scalar α for which $\vec{x}_{i+1} = \vec{x}_i + \alpha\vec{h}$ is nearly at a boundary. So how do we find a good α? First, we can figure out the value α_{min} for which we first hit a boundary. To do so, we separately examine each constraint and determine how large α can be before that constraint is violated. The smallest of all of those is α_{min}. We can then simply pick something slightly smaller than α_{min}, say $0.99\alpha_{min}$, and that gives us our answer.

Next, we can reconsider the issue we neglected above about transforming our coordinate system to move the point far from any boundaries of the polytope. Because we assume our system is in standard format, our boundaries occur where variables go to zero. A good way to ensure we are not near a boundary is to scale all variables so none of them are close to zero. We can accomplish this by dividing each variable x_i by its current value, thus transforming every x_i into a new variable $\bar{x}_i$, which has a value of 1. We also have to transform the rest of the problem, though, into an equivalent problem with these new scaled variables. To understand how to do that, we can express what we are doing as matrix operations.

Suppose we create a diagonal matrix X for which entry (i, i) is x_i. That is,

$$X = \begin{bmatrix} x_1 & 0 & \cdots & 0 \\ 0 & x_2 & & \vdots \\ \vdots & & \ddots & \vdots \\ 0 & \cdots & \cdots & x_n \end{bmatrix}.$$

Transforming each x_i to $\bar{x}_i$ is equivalent to multiplying $\vec{x}$ by X^{-1}. That is,

$$\bar{x} = X^{-1}\vec{x}.$$

If we want to create an equivalent problem using $\bar{x}$ instead of $\vec{x}$ as our set of variables, then we need to transform our constraints and objective function as follows:

$$\min \vec{c}^T \vec{x}$$

$$\Downarrow$$

$$\min (X\vec{c})^T \bar{x} (= (X\vec{c})^T X^{-1}\vec{x} = \vec{c}^T XX^{-1}\vec{x} = \vec{c}^T \vec{x})$$

and

$$A\vec{x} = \vec{b}$$

$$\Downarrow$$

$$AX\bar{x} = \vec{b} (= AXX^{-1}\vec{x} = A\vec{x}).$$

Note that

$$\vec{x} \geq \vec{0}$$

$$\Downarrow$$

$$\bar{x} \geq \vec{0},$$

which is essentially unchanged since we have simply scaled the x_is by positive constants.

Now that we are through specifying the method, it is helpful to consider intuitively what it is doing at each step, and why. We begin by finding the gradient of the objective. This is the direction of greatest increase of the objective, so moving approximately in the opposite direction will reduce the value of the objective, improving the solution quality. We cannot just move in the direction of the gradient, though, because that may cause us to violate our constraints. By moving only along the component of $-\nabla g$ that lies in the null-space of A, we guarantee that we are not changing the value of $A\vec{x}$, and therefore that we still satisfy all of our equality constraints, $A\vec{x} = \vec{b}$. We also need to make sure we do not violate our inequality constraints, $\vec{x} \geq 0$. Since our variables go to zero exactly at boundaries of the polytope, we guarantee we are not violating our inequality constraints by stopping just short of the boundary. Putting all of this together, we get a new solution vector that has improved cost and still satisfies all of the constraints. This affine method will give a provably polynomial time solution to the linear programming problem, although in practice it might be slower or faster than the simplex method on any given problem.

6.2   Primals and Duals

There are some important variants on the methods we have seen so far, called *primal-dual methods*, that we will mention briefly. For every linear programming problem for which we are seeking to minimize $\vec{c}^T\vec{x}$ there is a related problem for which we maximize a function $\vec{b}^T\vec{y}$. The original problem is called the *primal*, and the paired problem is the *dual*. Suppose we are solving the primal problem

$\min z = \vec{c}^T\vec{x}$ subject to $A\vec{x} \geq \vec{b}$, $\vec{x} \geq \vec{0}$.

Then the dual problem has the form

$\max w = \vec{b}^T\vec{y}$ subject to $A^T\vec{y} \leq \vec{c}$, $\vec{y} \geq \vec{0}$.

Note that A, $\vec{b}$, and $\vec{c}$ are the same for the primal and the dual. The primal and the dual are related by two important properties:

- For all feasible $\vec{x}$ and $\vec{y}$, $\vec{c}^T\vec{x} \geq \vec{b}^T\vec{y}$
- For optimal $\vec{x}$ and $\vec{y}$, $\vec{c}^T\vec{x} = \vec{b}^T\vec{y}$.

   In other words, solutions to the two problems approach the same optimum from opposite directions. There are variants of both the simplex and the interior point methods that exploit these properties to try to solve linear programs more efficiently by switching back and forth between the primal and dual variants of the problems, trying to close in on a solution from both above and below. We will not look into these methods any further, but it is useful to be familiar with the terminology of primals and duals because they are terms one will often hear when looking further into the topic of linear programming.

6.3   Solving Linear Programs in Practice

For practical linear programming problems, there is no clear best method. The simplex method was long the standard, and the ellipsoid method, although theoretically superior, proved inferior in practice. Karmarkar's method and its derivatives quickly proved competitive with, and even superior to, the simplex method, and for a time were the preferred way of solving hard linear systems. Both simplex and interior point methods have continued to advance, though, and one can make a case for either being the best approach for hard linear programming problems.

Although it is not too difficult to write correct linear programming solvers, there is a huge bag of tricks known to experts in the field that leads to much better performance than we are likely to get by coding the methods as we have seen them here. If we need to solve a reasonably hard linear programming problem in the real world, we are much better off using code written by people who know the field well. We can find code for the methods discussed above on the Internet. Unfortunately for scientists, though, linear programming has many important applications in the business world, and the best codes for it are therefore prohibitively expensive. Nonetheless, good codes are freely available for scientific work.

6.4 Nonlinear Programming

So far, we have only been talking about linear programming problems, those for which our constraints and objective are linear functions of our variables. Linear functions are important, but they are also a fairly restrictive class. It turns out, though, that the interior point methods will work for a somewhat broader class of problems. Specifically, we can solve efficiently for any such constraint satisfaction problem, provided our constraints and objective function are *convex*.

A set S is formally defined as convex if for any two points x and y in S and any $\alpha \in [0, 1]$, $\alpha x + (1 - \alpha) y$ is in S. A function f on $\mathscr{R}$ is defined to be convex if

$$f(\alpha x + (1 - \alpha) y) \leq f(x) + (1 - \alpha) f(y) \quad \forall x \in \mathscr{R}, \ y \in \mathscr{R}, \ \alpha \in [0, 1].$$

Informally, what the first definition means is that a set is convex if when we pick any two points in the set, the entire line between them is contained in the set. The second definition informally means that if we pick two points on a convex curve, the line between them is entirely above the curve. Figure 6.5 illustrates the concepts of convex and nonconvex sets of functions. Linear objectives are always convex functions and linear constraints always define convex sets. The opposite of a convex function is a *concave* function, defined by

$$f(\alpha x + (1 - \alpha) y) \geq f(x) + (1 - \alpha) f(y) \quad \forall x \in \mathscr{R}, \ y \in \mathscr{R}, \ \alpha \in [0, 1].$$

Note that a function is not necessarily convex or concave. Figure 6.5(b) shows a function that does not satisfy either condition. The problem of minimizing a convex

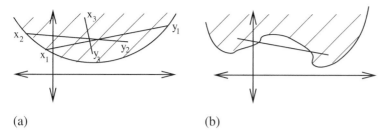

Figure 6.5
Illustration of the concept of convexity. (a) A convex function, which we can observe because, for each pair of points it bounds, all points on the line between the points are in the set. The space above the function is thus a convex set. (b) A function that is neither convex nor concave.

function over a convex set is solvable by the interior point methods we have discussed. Likewise, maximizing a concave function over a convex set is solvable. This fact follows from a property of such problems that any local optimum is also a global optimum.

How can we tell if a function is convex? The exact condition that tells us whether a function is convex over a space is if its hessian is *positive semidefinite* over that space. Positive semidefinite and positive definite matrices come up in many contexts in numerical methods for linear algebra and have several equivalent definitions. Two particularly useful definitions of positive definite are

- For all x, $x^T A x > 0$ (or $x^T A x \geq 0$ for a positive semidefinite matrix).
- All eigenvalues of A are real.

Intuitively, what these definitions mean is that the function has positive curvature at every point, similar to the usual English meaning of the term "convex." The formal definitions can be difficult to work with in the general case, though, without a lot more linear algebra than we can cover here. In particular, it can be difficult to show that a hessian whose entries are functions of several variables is in fact positive definite across some space. We can show it in some special cases, though.

One special case worth considering is *quadratic programming*. In quadratic programming, we have a linear constraint set, just as in a linear program, but an objective of the form

$$x^T B x + c^T x,$$

where B is a constant matrix and c is a constant vector. The hessian for this objective is $B + B^T$, which is positive definite exactly when B is. Thus, quadratic programming can be solved optimally when the constraint matrix B is positive definite. One place where such matrices sometimes show up is in objectives of the form $(Ax)^T(Ax) = x^T(A^TA)x$. Any matrix that can be expressed in the the form A^TA where A has all real entries will be positive semidefinite. We can see this because x^TA^TAx is the dot product of Ax with itself and thus must be nonnegative.

It is worth noting that semidefinite programming has recently been appearing quite a bit for optimization problems in biology. Science has its fads in which a technique becomes popular and suddenly gets a lot of exposure, and this may be an example. But semidefinite programming is a very broadly useful tool, and it may also be that as more in the community have become aware of what it can do, it has taken on a role appropriate to its potential. In any event, it is a useful method of which to be aware, and it is a valuable skill to be able to recognize when the mathematical programming problems we encounter are solvable. If we are solving for a nonconvex function or constraint set, then we generally will not be able to find a global optimum. It is likely that our interior point methods or the general continuous methods we saw in the last chapter will be able to find local optima, though.

References and Further Study

We have only scratched the surface of the field of constraint satisfaction and examined a couple of very basic tools for the problem. This elementary coverage will hopefully serve the reader well in recognizing constraint optimization problems, formulating them in practice, and understanding the basic principles behind their solution. For our purposes as modelers, there is usually little practical value in knowing more about how to solve linear programs, for the simple reason that it would almost never be a good idea to write one's own linear or nonlinear program solver. There is, however, a vast literature on methods for solving these problems to which one can refer for more depth. This chapter was prepared using a text by Nash and Sofer [94], and relied heavily on their presentations of the simplex and affine methods. That text provides an excellent introduction to these topics as well as much greater depth than we can cover in this one chapter. As with many of our other topics, *Wikipedia* [87] has become an excellent resource for the background, history, and methods of constrained optimization.

The simplex method was developed by George Dantzig in 1947 and first described in a book chapter from 1951 [95]. The ellipsoid method, and thus the first proof that linear programming was efficiently solvable, was developed by Khachiyan [96], based on an earlier method for convex optimization of Nemirovskii and Iudin [97]. Karmarkar's method [98] was the first that was efficient in both theory and practice. It formed the basis for the somewhat simpler affine method covered here, which comes from Nash and Sofer [94].

Those needing to solve linear programs in practice can choose from many available packages. The COIN-OR [99] linear program solver, the lpsolve [100] library, and the Gnu Linear Programming Kit (GLPK) [101] are all fine alternatives for free, open-source solution of linear programs. Many others are available commercially, in some cases free or at a reduced rate for students, academics, or other researchers. This is far from an exhaustive list, and a Web search may turn up better options for some situations.

II SIMULATION AND SAMPLING

7 Sampling from Probability Distributions

At the most basic level, most of what we are doing in simulation is sampling from probability distributions. That is, we can treat a simulation as set of coupled random variables (the final outputs of the simulation), and then we wish to pick one possible set of outcomes according to the joint distribution of all of the variables. For example, we will see in chapter 12 how we can simulate molecular evolution in a population, generating a set of k individuals, each with its individual DNA sequence. We may treat each base in each sequence as a random variable, with all of the variables related to each other through some complicated joint distribution. We may not be able to explicitly determine the joint distribution function, but we can create a model that implies the desired distribution and then run simulations to sample from the implied distribution.

In some cases, though, we will be interested in sampling from some simpler distributions we can explicitly express and analyze. These may be self-contained problems in themselves, such as determining the probability of success of an experiment, or they may be tiny modules of more complicated simulations. In either case, it is helpful to know when and how we can easily choose a random variable according to some arbitrary distribution. We will start out by considering continuous distributions and then see, toward the end, what to do about discrete distributions.

7.1 Uniform Random Variables

For now, we will assume we have a way to generate random numbers from a uniform discrete distribution. That is, if we have k elements, we can pick one of the k with equal probability. In practice, pretty much any programming language will have some routine for choosing approximately uniformly among a large set of integers (e.g., rand( ) and random( ) in C and C++). Computers are not actually capable of generating random numbers, so they instead use *pseudorandom numbers* that are generated by a deterministic procedure but "look" random. If we know how the generator works, we can generally devise tests that will show the pseudorandom numbers to be nonrandom, and in some applications this nonrandomness can create problems for scientific computing. Most built-in random number generators belong to a class called *linear congruential generators* whose numbers fail some tests of randomness when treated as points in high-dimensional spaces. For serious applications of random numbers, it is useful to understand how these methods work and how one can test whether a number is sufficiently good for a given application. In the interests of space, we will not cover random number generation and tests of randomness here. Pointers to information on those topics will be provided under References and Further Study.

Given the ability to generate uniform discrete numbers over a large set, we can generate almost any kind of random number we want. We can sample uniformly from any small discrete set almost perfectly by taking the original variable modulo the number of elements. So, for example,

```
x=random()%k
```

in C or C++ would set x to be a random number chosen almost perfectly uniformly from the integers $0, 1, \ldots, k - 1$, provided k is much smaller than the range of numbers produced by `random()`.

We can also use the capability of generating random integers uniformly from a large discrete space to generate approximately uniform continuous random variables. We can accomplish this by dividing the discrete random variable by the size of the set. Thus, if we can sample uniformly from the integers $0, 1, 2, \ldots, m - 1$, then dividing by m will give us approximately a $U[0, 1]$ random variable (i.e., a uniform random variable on the interval $[0, 1]$). We can convert this $U[0, 1]$ variable into a $U[a, b]$ random variable by the transformation

$$U[a, b] = a + (b - a)U[0, 1].$$

The precision of the real-valued variable will be $\frac{b-a}{m}$.

Since we can assume that we can generate uniform random numbers, our main concern now will be how we translate this capability of generating uniform random numbers into a capability of generating other distributions. We will see that we can accurately sample from pretty much any distribution so long as we can calculate its density function. We will accomplish this through two basic methods: the transformation method and the rejection method.

7.2 The Transformation Method

Suppose we know how to sample from some continuous density function $f(x)$ and we want to sample from some other continuous density function $g(y)$. There is a basic theorem called the *fundamental transformation law of probabilities* that provides a solution in many cases. The law states that if we sample a random variable from some density $f(x)$, then apply a function $y(x)$ to x, the density $g(y)$ of y will be related to that of x by the following rule:

$$g(y) = f(x) \left| \frac{dx}{dy} \right|.$$

We can show why this is so by using a proof from the Numerical Recipes series [82]. Suppose $x = U[0, 1]$. Then

$$f(x)\,dx = \begin{cases} dx & 0 < x < 1 \\ 0 & \textit{otherwise} \end{cases}.$$

We can therefore derive a method to convert x into some y obeying some desired density $g(y)$ as follows:

$$\frac{dx}{dy} = g(y)$$

$$x = \int_{-\infty}^{y} g(u)\,du = G(y)$$

$$\Rightarrow y = G^{-1}(x).$$

That is, we find the distribution $G(y)$ by integrating the density $g(y)$, invert the distribution to get G^{-1}, and apply this inverse distribution to x to get y distributed according to $g(y)$. This is called the *transformation method*. Figure 7.1(a) provides pseudocode for the transformation method.

1. Integrate $g(y)$ to get distribution $G(y)$
2. Invert $x = G(y)$ to get $y = G^{-1}(x)$
3. Sample $x = U[0, 1]$
4. Return $G^{-1}(x)$

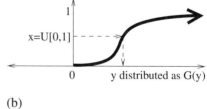

(a)   (b)

Figure 7.1
The transformation method for sampling from an arbitray density $g(y)$, using a $U[0, 1]$ uniform random number generator. (a) Pseudocode for the transformation method. (b) Graphical interpretation of the method. We invert $G(y)$, which normally maps from the real numbers to $[0, 1]$, by picking a number uniformly on $[0, 1]$ on the vertical axis and mapping it into $\mathscr{R}$ on the horizontal axis through $G(y)$.

The transformation method has a simple graphical interpretation, illustrated in figure 7.1(b). The probability of picking y in some range $[y_0, y_1]$ from density $g(y)$ is $G(y_1) - G(y_0)$. Therefore, the probability of landing within any range on the vertical axis is proportional to the amount of space on the horizontal axis covered by the corresponding region of the curve. We can use this intuition to sample from the desired density $g(y)$ by choosing a point uniformly between 0 and 1 on the vertical axis, and using the distribution $G(y)$ to map that point to its corresponding value on the horizontal axis. This point on the horizontal axis will then be distributed according to $G(y)$. That is exactly what we are doing above in the transformation method: we sample a $U[0, 1]$ variable representing a value on the vertical axis, then see what value it corresponds to on the horizontal axis mapped through $G(y)$.

We are assuming we know $g(y)$, but how do we then get $G(y)$ and $G^{-1}(x)$? Sometimes it is easy: we integrate analytically to get $G(y)$ and invert analytically to get $G^{-1}(x)$. Other times, we can integrate analytically but may not know how to invert G. In those cases, we still can numerically invert, that is, find $y = G^{-1}(x)$ by finding a zero of $h(y) = G(y) - x$. In other cases, we will not be able to integrate analytically. We can still generally solve those problems by integrating numerically, a topic we will cover in chapter 14.

Example Sampling from an exponential distribution. Suppose we want to sample an exponential random variable with parameter λ. Exponential random variables have the following density function:

$$g(y) = \begin{cases} 0 & y < 0 \\ \lambda e^{-\lambda y} & y \geq 0 \end{cases}.$$

To sample from $g(y)$, we first need to find its distribution $G(y)$:

$$G(y) = \int_{-\infty}^{y} g(u)\, du = \int_{0}^{y} g(u)\, du = -e^{-\lambda u}\big|_0^y = 1 - e^{-\lambda y}.$$

We then need to invert $G(y)$:

$$x = 1 - e^{-\lambda y}$$

$$e^{-\lambda y} = 1 - x$$

$$-\lambda y = \ln|1 - x|.$$

$$y = -\frac{1}{\lambda} \ln|1 - x|.$$

Thus, if $x = U[0, 1]$, then $-\frac{1}{\lambda} \ln|1 - x|$ is exponential with parameter λ. Note that we can actually simplify a little in this case, since $x = U[0, 1]$ implies that $|1 - x|$ is also $U[0, 1]$. We can thus use $y = -\frac{1}{\lambda} \ln x$. Also, note that if we do this on a computer, there is a small possibility that x will be exactly zero. This will produce an underflow when we try to take a logarithm of zero, so we should check for that possibility before taking the logarithm.

7.2.1 Transformation Method for Joint Distributions

It is also possible to use the transformation method on joint distributions, sampling from one distribution using random variables sampled from another. The method is similar in theory, although much harder to use in practice. If we want to transform joint density function $f(x_1, \ldots, x_k)$ into some desired $g(y_1, \ldots, y_k)$, then we use a modified version of the fundamental transformation law of probabilities:

$$g(y_1, \ldots, y_k) = f(x_1, \ldots, x_k) \begin{vmatrix} \frac{\partial x_1}{\partial y_1} & \cdots & \frac{\partial x_1}{\partial y_n} \\ \vdots & \ddots & \vdots \\ \frac{\partial x_k}{\partial y_1} & \cdots & \frac{\partial x_k}{\partial y_k} \end{vmatrix},$$

where $|A|$ refers to the *determinant* of matrix A.

So if we have a way of sampling variables $x_1, \ldots, x_k$ according to joint density $f(x_1, \ldots, x_k)$, then we can sample $y_1, \ldots, y_k$ from joint density $g(y_1, \ldots, y_k)$. In particular, if we assume that each x_i is an independent $U[0, 1]$ variable, then we can sample for the y_is if we can find a set of functions

$y_1(x_1, \ldots, x_k)$

$y_2(x_1, \ldots, x_k)$

$\vdots$

$y_2(x_1, \ldots, x_k)$

such that we can invert these functions to get

$x_1(y_1, \ldots, y_k)$

$x_2(y_1, \ldots, y_k)$

$\vdots$

$x_2(y_1, \ldots, y_k).$

We can then verify that

$$
\begin{vmatrix}
\frac{\partial x_1}{\partial y_1} & \cdots & \frac{\partial x_1}{\partial y_n} \\
\vdots & \ddots & \vdots \\
\frac{\partial x_k}{\partial y_1} & \cdots & \frac{\partial x_k}{\partial y_k}
\end{vmatrix} = g(y_1, \ldots, y_k).
$$

Unfortunately, this knowledge is not very helpful in figuring out what the transformation functions $y_1(x_1, \ldots, x_k), \ldots, y_k(x_1, \ldots, x_k)$ should be. If we have a reasonable guess for what might work, though, it does give us a way to prove whether it will or will not work.

The transformation method for joint distributions does have one important practical application, though, in that it gives us a way to sample from normal distributions. The normal distribution is at first glance hard to use for the transformation method because we cannot analytically integrate the density, $g(y) = \frac{1}{\sqrt{2\pi}} e^{-y^2/2}$. There is, however, a set of transformations that lets us generate two independent $N(0, 1)$ normals simultaneously by the transformation method. This technique for sampling normal variables is called the *Box–Müller* method.

In performing Box–Müller, we seek to sample two variables according to the joint distribution

$$
g(y_1, y_2) = \frac{1}{\sqrt{2\pi}} e^{-y_1^2/2} \frac{1}{\sqrt{2\pi}} e^{-y_2^2/2}.
$$

We can separate this function into a product of two functions $g(y_1, y_2) = g_1(y_1) g_2(y_2)$ that each refer to only one variable. As a result, y_1 and y_2 are independent. Furthermore, they have the densities

$$
g_1(y_1) = \frac{1}{\sqrt{2\pi}} e^{-y_1^2/2}, \quad g_2(y_2) = \frac{1}{\sqrt{2\pi}} e^{-y_2^2/2},
$$

so they are both $N(0, 1)$ normals. I will assert that we can take two $U[0, 1]$ independent uniform variables x_1 and x_2 and determine y_1 and y_2 from them as follows:

$$
y_1(x_1, x_2) = \sqrt{-2 \ln x_1} \, \cos(2\pi x_2)
$$

$$
y_2(x_1, x_2) = \sqrt{-2 \ln x_1} \, \sin(2\pi x_2).
$$

As noted above, the theory we have seen is not particularly helpful in figuring out what these formulas should be. I do not know how Box and Müller came up with them, and cannot recommend any general technique for coming up with a similar method for other distributions. Nonetheless, the transformation method does let us verify that the formulas are correct, as follows:

$$y_1^2 + y_2^2 = -2 \ln x_1 (\cos^2(2\pi x_2) + \sin^2(2\pi x_2)) = -2 \ln x_1 \Rightarrow x_1 = e^{-(y_1^2 + y_2^2)/2}$$

$$\frac{y_2}{y_1} = \tan(2\pi x_2) \Rightarrow x_2 = \frac{1}{2\pi} \arctan \frac{y_2}{y_1}.$$

Therefore

$$\begin{vmatrix} \frac{\partial x_1}{\partial y_1} & \frac{\partial x_1}{\partial y_2} \\ \frac{\partial x_2}{\partial y_1} & \frac{\partial x_2}{\partial y_2} \end{vmatrix} = \begin{vmatrix} -2y_1 e^{-(y_1^2 + y_2^2)/2} & -2y_1 e^{-(y_1^2 + y_2^2)/2} \\ -\frac{y_2}{y_1^2} \frac{1}{2\pi} \frac{y_1^2}{y_1^2 + y_2^2} & \frac{1}{y_1} \frac{1}{2\pi} \frac{y_1^2}{y_1^2 + y_2^2} \end{vmatrix}$$

$$= -\frac{1}{2\pi} e^{-(y_1^2 + y_2^2)/2} \left[1 + \frac{y_2^2}{y_1^2} \right] \left[\frac{y_1^2}{y_1^2 + y_2^2} \right] = -\frac{1}{2\pi} e^{-(y_1^2 + y_2^2)/2}$$

$$= \frac{1}{\sqrt{2\pi}} e^{-y_1^2/2} \frac{1}{\sqrt{2\pi}} e^{-y_2^2/2}.$$

7.3 The Rejection Method

The rejection method is an alternative to the transformation method that we can use to generate random variables from a distribution $f(x)$. In order to perform the rejection method, we need to find a function $g(x)$ that strictly upper bounds $f(x)$. That is:

$$g(x) \geq f(x) \quad \forall x.$$

Furthermore, we require that we can figure out the area under $g(x)$, $A = \int_{-\infty}^{\infty} g(u) \, du$. Note that $g(x)$ is not a probability density because its integral from $-\infty$ to ∞ is not 1, but $\frac{1}{A} g(x)$ is a probability density. The rejection method works by sampling points uniformly under $g(x)$, using the fact that $\frac{1}{A} g(x)$ is a probability density, and then throwing away those points that are not also under $f(x)$. The method is described by the pseudocode of figure 7.2(a) and is is depicted graphically in figure 7.2(b).

Of course, this assumes that we know how to sample from $\frac{1}{A} g(x)$. But we get to pick $g(x)$, so that is not too restrictive. We can pick any $g(x)$ that upper bounds $f(x)$ and that has a finite integral we can calculate and invert; then we can use the transformation method to sample from $g(x)$. The more tightly $g(x)$ bounds $f(x)$, the more efficient the method will be. The ratio of the area under the curves is the average number of points under $g(x)$ we will need to generate before we succeed in picking a point under both curves.

> 1. Sample x from density $\frac{1}{A}g(x)$ by the transformation method.
> 2. Sample y from $U[0, g(x)]$.
> 3. If $y < f(x)$ then return(x) else return to step 1.

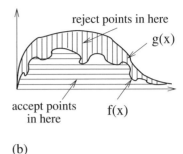

reject points in here

$g(x)$

accept points
in here   f(x)

(a)   (b)

Figure 7.2
The rejection method for sampling from a distribution $f(x)$ using an upper-bounding function $g(x)$. (a) Pseudocode for the rejection method. (b) Graphical interpretation of the rejection method. We want to pick points uniformly from the space under $f(x)$, so we instead pick them uniformly under $g(x)$ and reject any choice that is not also under $f(x)$.

Example   We can use the rejection method to find another way to sample from normal distributions. Suppose we want to sample from

$$f(x) = \frac{1}{\sqrt{2\pi}} e^{-x^2/2},$$

that is, from an $N(0, 1)$ normal. First, we can note that the $N(0, 1)$ normal is symmetric about $x = 0$, so we can sample from positive values of x and then pick a single random bit to decide whether to use $+x$ or $-x$. So we really just need a function that bounds $f(x)$ for positive x.

We will use a function of the form

$$g(x) = Ce^{-x/2}$$

on the logic that this will decay more slowly than $e^{-x^2/2}$, and therefore is guaranteed to bound the normal density for sufficiently large C. So how do we pick C? We require a C satisfying

$$Ce^{-x/2} \geq \frac{1}{\sqrt{2\pi}} e^{-x^2/2}$$

$$C \geq \frac{1}{\sqrt{2\pi}} e^{(x-x^2)/2}.$$

The right-hand side is maximized where $x - x^2$ is maximized, which is where

$$\frac{d}{dx}(x - x^2) = 1 - 2x = 0 \Rightarrow x = \frac{1}{2}.$$

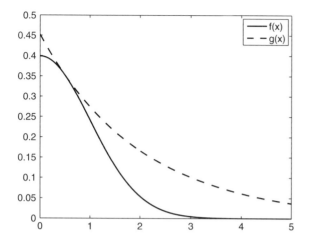

Figure 7.3
Plot of $f(x) = \frac{1}{\sqrt{2\pi}} e^{-x^2/2}$ for the unit normal and bounding function $g(x) = \frac{e^{1/8}}{\sqrt{2\pi}} e^{-x/2}$ for use with the rejection method.

So we require

$$C \geq \frac{1}{\sqrt{2\pi}} e^{(1/2-1/4)/2} = \frac{e^{1/8}}{\sqrt{2\pi}}.$$

Thus we will choose

$$g(x) = \frac{e^{1/8}}{\sqrt{2\pi}} e^{-x/2}.$$

Figure 7.3 plots these choices of $f(x)$ and $g(x)$.

Now must figure out how to sample under $g(x)$. We first need to know the area under $g(x)$:

$$A = \int_0^\infty \frac{e^{1/8}}{\sqrt{2\pi}} e^{-u/2} \, du = \frac{2e^{1/8}}{\sqrt{2\pi}}.$$

If we scale $g(x)$ by $\frac{1}{A}$, we will get a probability distribution:

$$\frac{1}{A} g(x) = \frac{\sqrt{2\pi}}{2e^{1/8}} \frac{e^{1/8}}{\sqrt{2\pi}} e^{-x/2} = \frac{1}{2} e^{-x/2}.$$

Thus the scaled version of $g(x)$ is in fact an exponential with parameter $\frac{1}{2}$. We already know how to sample from an exponential by the transformation method, so to sample from the unit normal distribution, we have to do the following:

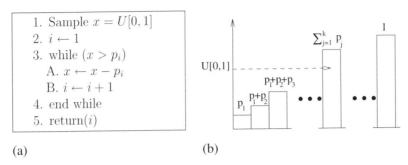

(a)   (b)

Figure 7.4
Discrete transformation method for sampling from a distribution $p_1, \ldots, p_k$. (a) Pseudocode for the discrete transformation method. (b) Graphical interpretation of the discrete transformation method.

1. Sample $x = Exp\left(\frac{1}{2}\right)$.
2. Sample $y = U\left[0, \frac{e^{1/8}}{\sqrt{2\pi}} e^{-x/2}\right]$.
3. If $y \geq \frac{1}{\sqrt{2\pi}} e^{-x^2/2}$, return to step 1.
4. If $I\left(\frac{1}{2}\right) = 1$, return x, else return $-x$.

The probability of accepting any point will be the ratio of the areas under the two curves. We know the area under $f(x)$ is $\frac{1}{2}$ since it covers the right half of the normal distribution function. The area under $g(x)$ is $\frac{2e^{1/8}}{\sqrt{2\pi}}$. Thus the probability of accepting a point should be $\frac{\sqrt{2\pi}}{4e^{1/8}} \approx .553$. We therefore need to try about two points each time we want to sample from the normal distribution.

7.4 Sampling from Discrete Distributions

Sampling from discrete distributions is generally much easier than sampling from continuous distributions. If we have a small finite set of outcomes $1, \ldots, k$ with probabilities $p_1, \ldots, p_k$, then we can sample from the distribution implied by those probabilities, using the pseudocode of figure 7.4(a). We first sample a uniform random number and then move through the possible outcomes, accumulating probabilities until they exceed our uniform random number. The outcome on which we first exceed the uniform number is then selected. This method is basically a discrete version of the transformation method, as illustrated by figure 7.4(b). We sample a point uniformly along the vertical axis, then see to which element on the horizontal axis it corresponds. This discrete transformation method may be feasible even if we have an infinite sample space, so long as the probabilities fall off relatively quickly. The number of iterations it takes on average will be the mean of the distribution, so it can be slow if we are using a distribution with a large expectation.

In those cases where the discrete transformation method is not practical, we can create a discrete version of the rejection method. We can do this by creating a con-

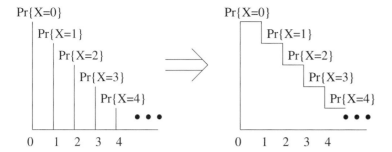

Figure 7.5
Conversion of a density over a discrete set into a density over the real numbers by creation of a step function.

tinuous analog of the discrete density by turning it into a step function, as illustrated in figure 7.5. $Pr\{x = k\}$ in the discrete density is transformed into $Pr\{k \leq x < k + 1\}$ in the continuous density. We can then use the rejection method by bounding our step function above with a continuous curve $g(x)$.

Example   Suppose we want to sample from a $Geom(\frac{1}{2})$ geometric variable. This random variable has the density function

$$Pr\{X = k\} = \left(\frac{1}{2}\right)^{k-1} \left(\frac{1}{2}\right) = \left(\frac{1}{2}\right)^{k}.$$

To use the discrete rejection method, we first convert this discrete probability function into a continuous step function density $f(x)$:

$$f(x) = \begin{cases} 0 & x \leq 0 \\ 1/2 & 0 < x \leq 1 \\ 1/4 & 1 < x \leq 2 \\ 1/8 & 2 < x \leq 3 \\ \vdots & \vdots \end{cases}.$$

Then we bound $f(x)$ with a continuous function $g(x)$. We can use the following function:

$$g(x) = \begin{cases} 0 & x \leq 0 \\ \left(\frac{1}{2}\right)^{x} = e^{-x \ln 2} & x > 0 \end{cases}.$$

Figure 7.6 plots these two functions.

We then need to sample from the area under $g(x)$. To do that, we need the total area under $g(x)$:

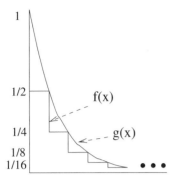

Figure 7.6
A geometric $Geom(\frac{1}{2})$ step function $f(x)$ and an exponential bounding curve $g(x)$.

$$A = \int_{-\infty}^{\infty} g(u)\, du = \int_{0}^{\infty} e^{-u \ln 2}\, du = \frac{1}{\ln 2}.$$

We then apply the main procedure for the rejection method:

1. Sample X from density $\frac{1}{A} g(x) = \ln 2\, e^{-X \ln 2}$ (which is $\mathrm{Exp}(\ln 2)$).
2. Sample from $Y = U[0, e^{-X \ln 2}]$.
3. If $Y \leq \left(\frac{1}{2}\right)^{\lceil X \rceil}$, then return X, else go to step 1.

Since we know the area under $f(x)$ must be 1, the ratio of the areas is $\frac{1}{\ln 2} \approx 1.44$, so we need an average of about 1.44 trials before we pick a valid point.

References and Further Study

Press et al. [82] provides excellent coverage of the methods cited here as well as related methods we have not seen. Texts by Ross on probability models [102] and simulation [103] also offer detailed coverage of these and many related topics. Those interested in learning more about random number generation, a topic referenced only briefly above, can refer to Press et al. [82]. For greater depth, readers can refer to Knuth's *The Art of Computer Programming* [104], particularly volume 2 (*Seminumerical Algorithms*). Knuth's series is among the most useful reference sets for anyone working in computer research. I highly recommend that those planning a career involving large amounts of computing acquire their own copy of that series, as well as Press et al.'s Numerical Recipes.

The methods we have covered here are simple enough that one rarely consults primary references, and the first uses of sampling and rejection on various specific distributions are probably mostly forgotten. Nonetheless, we can acknowledge a few of the most important results. Both the transformation method and the rejection method are due to John von Neumann. The earliest written statement of either appears to be in a letter von Neumann sent to Stanislaw Ulam in 1947. Ulam later inspired some of the earliest work in computational biology. Interested readers can find the letter reproduced in an article by Eckhardt on Ulam and von Neumann's contributions to Monte Carlo simulation [105]. The first citable reference of which I am aware is from 1951 [106]. The Box–Müller method is, unsurprisingly, due to Box and Müller [107]. Some of the specific distributions examined in this chapter are covered in Press et al., Ross, or Knuth.

8 Markov Models

We will now begin the first of several chapters on a broadly useful class of discrete models called Markov models. A Markov model is generally represented as a graph containing a set of *states* represented as nodes and a set of *transitions* with probabilities represented by weighted edges. Figure 8.1 shows a Markov model with four states.

We simulate a Markov model by starting at some state and moving to successive neighboring states by choosing randomly among neighbors according to their labeled probabilities. For example, if we start in state q_4, then we would have probability p_1 of moving to q_1, p_2 of moving to q_2, and $1 - p_1 - p_2$ of moving to q_3. If we move to q_2, then we have probability p_3 of moving to q_1 and $1 - p_3$ of moving to q_3, and so on. The result is a walk through the state set (e.g., $q_1, q_2, q_1, q_2, q_3, q_3, \ldots$). The resulting sequence of states is called a *Markov chain*.

Markov models show up in many areas of biology. For example, figure 8.2 shows a pair of Markov models that may be used to simulate random strings of DNA. Figure 8.2(a) is a simple Markov chain for generating DNA bases. We can develop much more sophisticated models, though. Figure 8.2(b) shows how we might organize a model specifically for coding DNA to account for different base frequencies in the three codon positions. We can go further and join the models of figures 8.2(a) and (b) to create a new model capable of representing both coding and noncoding DNA. As we will discuss later, real-world DNA models can get far more complex, merging special modules for many different kinds of DNA. There are limits to what can be modeled with Markov models, though. For example, exon lengths must be geometrically distributed with any such model. Nonetheless, they can be very versatile, as we shall see.

A Markov model is formally defined by the following components:

- A state set $Q = \{q_1, q_2, \ldots, q_n\}$.
- A starting distribution $Pr\{q(0) = q_i\} = p_i$ (which can be represented by a vector $\vec{p}$).

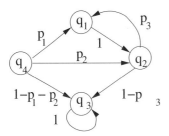

Figure 8.1
A Markov model.

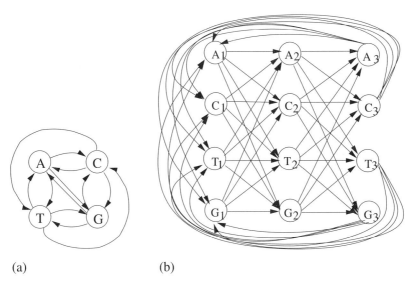

(a)   (b)

Figure 8.2
Examples of Markov models we might use to generate random strings of DNA. (a) A simple model we might use to generate bases with a single set of frequencies for all positions. (b) A more complicated model we might use for simulating coding DNA to account for variations in base frequencies in the three codon positions.

• A set of transition probabilities $Pr\{q(n + 1) = q_j \,|\, q(n) = q_i\} = p_{ij}$ (which can be represented by a matrix P).

To simulate a Markov model, we pick an initial state $q(0)$ from distribution $\vec{p}$, then repeatedly pick the next state $q(i + 1)$ from distribution P, given prior state $q(i)$.

This is actually a definition of what is called a *first-order Markov model* because the transition probabilities are of the form $Pr\{q(n + 1) = q_j \,|\, q(n) = q_i\} = p_{ij}$, with the probability of entering each possible next state dependent only on the current state. In a *kth-order Markov model*, transition probabilities take the form $Pr\{q(n) =$

$q_{i,n} \mid q(n-1) = q_{i,n-1} \wedge q(n-2) = q_{i,n-2} \wedge \cdots q(n-k) = q_{i,n-k}\} = p_{ij}$. That is, the probability of the next state depends on the previous k states. This representation can sometimes be convenient. For example, in a DNA model, the distribution of the third base in a codon will generally depend on the previous two, suggesting that a second-order Markov model would be a good way to generate more exonlike sequences. However, any kth-order Markov model can be transformed into a first-order Markov model by defining a new state set $Q' = Q^k$ (i.e., each state in Q' is a set of k states in Q), with the current state in Q' being the last k states visited in Q. Then a Markov chain in the kth-order model Q—$q_1, q_2, q_3, q_4, \ldots$—becomes the chain $\{q_1, q_2, \ldots, q_k\}, \{q_2, q_3, \ldots, q_{k+1}\}, \{q_3, q_4, \ldots, q_{k+2}\}, \ldots$ in Q'. We can therefore generally ignore higher-order Markov models when talking about the theory behind Markov models, even though they may be conceptually useful in practice.

8.1 Time Evolution of Markov Models

Although the behavior of Markov models is random, it is also in some ways predictable. One way to understand how Markov models behave is to work through a few steps of a Markov model simulation. Suppose we have a two-state model, $Q = \{q_1, q_2\}$, with initial probabilities p_1 and p_2 and transition probabilities p_{11}, p_{12}, p_{21}, and p_{22}. We will then ask how likely we are to be in any given state at each point in time.

At step zero, the distribution of states is exactly described by the initial probability vector $\vec{p}$:

$$\begin{bmatrix} Pr\{q(0) = q_1\} \\ Pr\{q(0) = q_2\} \end{bmatrix} = \begin{bmatrix} p_1 \\ p_2 \end{bmatrix}.$$

After one step of the Markov model, the probabilities will be the following:

$$\begin{bmatrix} Pr\{q(1) = q_1\} \\ Pr\{q(1) = q_2\} \end{bmatrix} = \begin{bmatrix} p_1 p_{11} + p_2 p_{21} \\ p_1 p_{12} + p_2 p_{22} \end{bmatrix}.$$

That is, the probability of being in state 1 at time 1 is the probability of being in state 1 at time 0 and staying there, plus the probability of being in state 2 at time 0 and moving from state 2 to state 1. Likewise, the probability of being in state 2 at time 1 is the probability of being in state 1 at time 0, then moving from state 1 to state 2 plus the probability of starting in state 2 at time 0 and staying there.

After the next step, the probability distribution will be the following:

$$\begin{bmatrix} Pr\{q(2) = q_1\} \\ Pr\{q(2) = q_2\} \end{bmatrix} = \begin{bmatrix} (p_1 p_{11} + p_2 p_{21})p_{11} + (p_1 p_{12} + p_2 p_{22})p_{21} \\ (p_1 p_{11} + p_2 p_{21})p_{12} + (p_1 p_{12} + p_2 p_{22})p_{22} \end{bmatrix}.$$

In other words, the probability of being in state 1 at time 2 is the probability of being in state 1 at time 1 and staying there plus the probability of being in state 2 at time 1 and moving from state 2 to state 1. Similarly, the probability of being in state 2 at time 2 is the probability of being in state 2 at time 1 and staying there plus the probability of being in state 1 at time 1 and moving from state 1 to state 2.

We can see the pattern here by using the matrix and vector representations of the Markov model probability distributions. If we know the distribution at step i, the probability distribution at step $i+1$ will be the following:

$$\begin{bmatrix} Pr\{q(i+1) = q_1\} \\ Pr\{q(i+1) = q_2\} \end{bmatrix} = \begin{bmatrix} p_{11} & p_{21} \\ p_{12} & p_{22} \end{bmatrix} \begin{bmatrix} Pr\{q(i) = q_1\} \\ Pr\{q(i) = q_2\} \end{bmatrix}.$$

This implies that if we want to know the distribution at state n, we can find it by multiplying the initial distribution by the transition matrix n times:

$$\begin{bmatrix} Pr\{q(n) = q_1\} \\ Pr\{q(n) = q_2\} \end{bmatrix} = \begin{bmatrix} p_{11} & p_{21} \\ p_{12} & p_{22} \end{bmatrix} \times \cdots \times \begin{bmatrix} p_{11} & p_{21} \\ p_{12} & p_{22} \end{bmatrix} \begin{bmatrix} p_1 \\ p_2 \end{bmatrix}.$$

Example   Suppose we design a Markov model to represent our probability of being in an intron, an exon, or an intergenic region in a genetic sequence. For the sake of argument, we can suppose the transition probabilities are as illustrated in figure 8.3. We can say that q_1 is the exon state, q_2 is the intron state, and q_3 is the intergenic state. Then the transition matrix P is

$$P = \begin{bmatrix} .8 & .1 & .1 \\ .1 & .9 & 0 \\ .1 & 0 & .9 \end{bmatrix}.$$

Further assume the initial state vector $\vec{p}$ is the following:

$$\begin{bmatrix} p_1 \\ p_2 \\ p_3 \end{bmatrix} = \begin{bmatrix} 0.2 \\ 0.3 \\ 0.5 \end{bmatrix}.$$

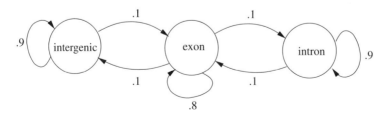

Figure 8.3
Hypothetical Markov model for generating distributions of introns, exons, and intergenic sequences.

Then the probability of being in each state at time 0 is

$$\begin{bmatrix} .2 \\ .3 \\ .5 \end{bmatrix}$$

At time 1, the distribution is

$$P = \begin{bmatrix} .8 & .1 & .1 \\ .1 & .9 & 0 \\ .1 & 0 & .9 \end{bmatrix} \begin{bmatrix} .2 \\ .3 \\ .5 \end{bmatrix} = \begin{bmatrix} .24 \\ .29 \\ .47 \end{bmatrix}.$$

At time 2, the distribution is

$$P = \begin{bmatrix} .8 & .1 & .1 \\ .1 & .9 & 0 \\ .1 & 0 & .9 \end{bmatrix} \begin{bmatrix} .24 \\ .29 \\ .47 \end{bmatrix} = \begin{bmatrix} .268 \\ .285 \\ .447 \end{bmatrix}.$$

In general, for time n the distribution will be

$$P = \begin{bmatrix} .8 & .1 & .1 \\ .1 & .9 & 0 \\ .1 & 0 & .9 \end{bmatrix}^n \begin{bmatrix} .2 \\ .3 \\ .5 \end{bmatrix}.$$

We can more efficiently find the distribution of states for larger n through *successive squaring*, a process by which we can recursively compute $F(n) = P^n$. The successive squaring algorithm is illustrated in figure 8.4. There is an even more efficient way that we will see shortly.

We can generalize this notion of how the distribution of states of a Markov model evolves over time through the *Chapman–Kolmogorov equations*. Suppose we have a Markov model with $|Q|$ states where we define $p_{ij}(n)$ to be the probability of going

function F(n)
  1. if n is zero
    a. return(I)
  2. else if n is odd
    a. return($PF(n-1)$)
  3. else
    a. return($(F(n/2))^2$)

Figure 8.4
Pseudocode for the successive squaring method for computing P^n in O($\log n$) steps.

from state i to state j in exactly n steps in this model. The Chapman–Kolmogorov equations for the model are the following:

$$p_{ij}(n+m) = \sum_{k=1}^{|Q|} p_{ik}(n)p_{kj}(m)$$

for all $n \geq 0$, $m \geq 0$, and any states i and j. That is, the probability of getting to state j from state i in $n+m$ steps is the sum over all possible intermediate states k of the probability of getting from i to k in n steps, then from k to j in the remaining m steps.

8.2   Stationary Distributions and Eigenvectors

Suppose we want to look at the evolution of our Markov model over really long time scales. We can keep multiplying by our transition matrix. Eventually, though, the example above will converge on a single probability distribution that will not change on further multiplication:

$$\begin{bmatrix} .2 \\ .3 \\ .5 \end{bmatrix} \rightarrow \begin{bmatrix} .24 \\ .29 \\ .47 \end{bmatrix} \rightarrow \begin{bmatrix} .268 \\ .285 \\ .447 \end{bmatrix} \rightarrow \cdots \rightarrow \begin{bmatrix} .33333 \\ .33333 \\ .33333 \end{bmatrix} \rightarrow \begin{bmatrix} .33333 \\ .33333 \\ .33333 \end{bmatrix}.$$

Once we have multiplied enough times that we are at a distribution which does not change appreciably, then we know we no longer need to do further multiplications. Furthermore, it does not matter what our initial distribution is for this model; we will always converge to the same final distribution vector, regardless of our starting point. This vector on which the state distribution converges after a large number of steps is called the *stationary distribution*.

One might wonder if this property of convergence on a unique stationary distribution, regardless of starting point, will work for any example. The answer is no. It is possible that the final vector we converge to is not unique. Consider the example of figure 8.5(a). Here, the transition matrix is

$$P = \begin{bmatrix} .8 & 0 & 0 \\ .1 & 1 & 0 \\ .1 & 0 & 1 \end{bmatrix}$$

If we have starting distribution

$$\begin{bmatrix} 0 \\ 1 \\ 0 \end{bmatrix}$$

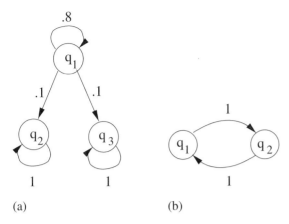

Figure 8.5
Markov models that do not converge on unique stationary vectors.

then it will get stuck at

$$\begin{bmatrix} 0 \\ 1 \\ 0 \end{bmatrix}$$

If we have starting distribution

$$\begin{bmatrix} 0 \\ 0 \\ 1 \end{bmatrix}$$

then it will get stuck at

$$\begin{bmatrix} 0 \\ 0 \\ 1 \end{bmatrix}$$

The distribution of states still eventually converges, but where it converges depends on where we start. The final distribution can be any combination of probabilities for states q_2 and q_3, depending on where we start.

A Markov model is not even guaranteed to converge on any vector. Consider the Markov model of figure 8.5(b) with transition matrix

$$\begin{bmatrix} 0 & 1 \\ 1 & 0 \end{bmatrix}$$

If we initialize this model with the probability vector

$$\begin{bmatrix} p \\ 1-p \end{bmatrix}$$

then the distribution of states on successive steps will be the following:

$$\begin{bmatrix} p \\ 1-p \end{bmatrix} \rightarrow \begin{bmatrix} 1-p \\ p \end{bmatrix} \rightarrow \begin{bmatrix} p \\ 1-p \end{bmatrix} \rightarrow \begin{bmatrix} 1-p \\ p \end{bmatrix} \rightarrow \begin{bmatrix} p \\ 1-p \end{bmatrix} \rightarrow \cdots$$

The model thus never converges on any final state.

We can understand why these examples do not converge on a single distribution by considering what properties our first example has that the second and third do not. What the second model, with

$$P = \begin{bmatrix} .8 & 0 & 0 \\ .1 & 1 & 0 \\ .1 & 0 & 1 \end{bmatrix}$$

is lacking is a property called *ergodicity*. Ergodicity means that for any two states q_i and q_j there is some sequence of transitions with nonzero probability that go from q_i to q_j. An ergodic Markov chain is also sometimes called *irreducible*. In the above example, we cannot move from q_2 to q_1 or q_3, or from q_3 to q_1 or q_2, so the Markov model is not ergodic. If our model is not ergodic, then it is possible for it to converge to different stationary distributions, depending on the starting distribution.

We can understand why this is so by reviewing our linear algebra. A Markov model is in its stationary distribution if multiplying the transition matrix by the distribution vector results in the same distribution vector. This means that the distribution vector must be an eigenvector of the transition matrix. Recall that vector x is an eigenvector of matrix A if $Ax = \lambda x$, where λ is a scalar value called the eigenvalue. The stationary distribution is then an eigenvector of P that has eigenvalue $\lambda = 1$. A Markov model will converge to a unique stationary distribution if its transition matrix has exactly one eigenvector with eigenvalue $\lambda = 1$ and has $|\lambda| < 1$ for every other eigenvector.

To see why this is so, suppose we call our eigenvalues $\lambda_1, \ldots, \lambda_n$ and our eigenvectors $\vec{x}_1, \ldots, \vec{x}_n$, and we assume $\lambda_1 = 1$, $|\lambda_i| < 1$ for $i = 2, \ldots, n$. Then, if we have some starting distribution $\vec{p}$, we can decompose $\vec{p}$ into a linear combination of the eigenvectors as follows:

$$\vec{p} = c_1\vec{x}_1 + c_2\vec{x}_2 + \cdots + c_n\vec{x}_n.$$

Then the distributions at successive steps will look like the following:

$$P\vec{p} = c_1 P\vec{x}_1 + c_2 P\vec{x}_2 + \cdots + c_n P\vec{x}_n$$

$$= c_1\lambda_1\vec{x}_1 + c_2\lambda_2\vec{x}_2 + \cdots + c_n\lambda_n\vec{x}_n$$

$$P^2 \vec{p} = c_1 P^2 \vec{x}_1 + c_2 P^2 \vec{x}_2 + \cdots + c_n P^2 \vec{x}_n$$

$$= c_1 \lambda_1^2 \vec{x}_1 + c_2 \lambda_2^2 \vec{x}_2 + \cdots + c_n \lambda_n^2 \vec{x}_n$$

$$\vdots$$

$$P^k \vec{p} = c_1 P^k \vec{x}_1 + c_2 P^k \vec{x}_2 + \cdots + c_n P^k \vec{x}_n$$

$$= c_1 \lambda_1^k \vec{x}_1 + c_2 \lambda_2^k \vec{x}_2 + \cdots + c_n \lambda_n^k \vec{x}_n.$$

As we move to large k, all of the λ_i^k terms except λ_1^k will converge toward zero. Eventually, we will approach the limit

$$\lim_{k \to \infty} P^k \vec{p} = c_1 \cdot 1 \cdot \vec{x}_1 + c_2 \cdot 0 \cdot \vec{x}_2 + \cdots + c_n \cdot 0 \cdot \vec{x}_n = c_1 \vec{x}_1.$$

In other words, the model will converge on a distribution proportional to $\vec{x}_1$.

If a Markov model is not ergodic, then its state set can be partitioned into discrete graph components unreachable from one another. Each such component will have its own eigenvector with eigenvalue 1. Depending on which component we start in, we may converge on any of them. In our nonergodic example above, both

$$\begin{bmatrix} 0 \\ 1 \\ 0 \end{bmatrix} \quad \text{and} \quad \begin{bmatrix} 0 \\ 0 \\ 1 \end{bmatrix}$$

are eigenvectors of the transition matrix with eigenvalue 1. (The third eigenvector is

$$\begin{bmatrix} .81650 \\ -.40825 \\ -.40825 \end{bmatrix}$$

with eigenvalue 0.8.)

Eigenvectors also help explain why the second model, with

$$P = \begin{bmatrix} 0 & 1 \\ 1 & 0 \end{bmatrix}$$

failed to converge on a stationary distribution. This matrix has two eigenvectors:

$$\vec{x}_1 = \begin{bmatrix} \frac{1}{2} \\ \frac{1}{2} \end{bmatrix}$$

with eigenvalue 1, and

$$\vec{x}_2 = \begin{bmatrix} \frac{1}{2} \\ -\frac{1}{2} \end{bmatrix}$$

with eigenvalue -1. When we decompose the initial distribution into a linear combination of eigenvectors, any component of $\vec{x}_2$ will never die away. The model will exhibit an oscillatory behavior that comes from flipping the sign of the $\vec{x}_2$ component on each successive step of the model. For example, if we had the initial vector

$$\vec{p} = \begin{bmatrix} 1 \\ 0 \end{bmatrix}$$

then that would be decomposed into

$$\vec{p} = \begin{bmatrix} 1 \\ 0 \end{bmatrix} = \begin{bmatrix} \frac{1}{2} \\ \frac{1}{2} \end{bmatrix} + \begin{bmatrix} \frac{1}{2} \\ -\frac{1}{2} \end{bmatrix} = \vec{x}_1 + \vec{x}_2.$$

After k steps, this would become

$$P^k \vec{p} = \lambda_1^k \begin{bmatrix} \frac{1}{2} \\ \frac{1}{2} \end{bmatrix} + \lambda_2^k \begin{bmatrix} \frac{1}{2} \\ -\frac{1}{2} \end{bmatrix} = \begin{bmatrix} \frac{1}{2} \\ \frac{1}{2} \end{bmatrix} + (-1)^k \begin{bmatrix} \frac{1}{2} \\ -\frac{1}{2} \end{bmatrix} = \vec{x}_1 + (-1)^k \vec{x}_2.$$

Each successive step of the Markov model would flip the sign on the $\vec{x}_2$ term, so we would end up cycling between $\vec{x}_1 + \vec{x}_2 = \begin{bmatrix} 1 & 0 \end{bmatrix}^T$ and $\vec{x}_1 - \vec{x}_2 = \begin{bmatrix} 0 & 1 \end{bmatrix}^T$.

There is a test we can do to determine whether a given Markov model converges on a stationary distribution. A Markov model is guaranteed to converge on a stationary distribution $\vec{\pi}$ if there exists some integer $N > 0$ such that

$$\min_{i,j} \; p_{ij}(N) = \delta, \quad \delta > 0.$$

That is, there is some number of steps N such that no matter where we start, we have some bounded nonzero probability of getting to any given ending position in exactly N steps. The preceding example fails this test because there is no single N for which $p_{11}(N)$ and $p_{12}(N)$ are both nonzero.

8.3   Mixing Times

Eigenvalues and eigenvectors can also tell us about one more important property of a Markov model: its *mixing time*. Informally, the mixing time is the time needed for the Markov model to get close to its stationary distribution. We will see a more formal discussion of this concept in chapter 9. For now, though, we should be aware that we can approximately determine the mixing time by knowing the ratio of the two largest eigenvalues, $|\frac{\lambda_2}{\lambda_1}|$. Since the largest eigenvalue, λ_1, is 1, this is really just $|\lambda_2|$. $|\lambda_2|$ is approximately the amount by which the transient behavior decays on each successive step, assuming we have run the model long enough. Therefore, if we want to run the model long enough for the transients to die away by some factor r, then we need to run for a number of rounds k, approximately defined by

$$|\lambda_2|^k = r$$

$$k \log \lambda_2 = \log r$$

$$k = \frac{\log r}{\log \lambda_2}.$$

Note that this is only approximate, since the exact contribution of each eigenvalue will depend on the starting distribution vector. If we start with a stronger component of $\vec{x}_2$, then we may need more steps, and if we start with a stronger component of $\vec{x}_1$, then we may need fewer.

References and Further Study

There are many fine references on the subject of Markov models. Ross's text on probability models [102] provides a strong foundation of the topic in greater depth than we can cover here. Ross's text on simulation [103] also provides some coverage of Markov models in their specific application to simulations. Many other texts on probability or simulation might serve equally well, though. This chapter was prepared in part with guidance from Rozanov [108].

There are several texts one might consider for the topic of Markov models specifically for scientific simulation. Gillespie [109] is written for physical scientists in general, and describes applications particularly relevant, for our purposes, to biochemical and biophysical modeling. One recent addition to the literature that is particularly useful for our purposes is Wilkinson [110], which discusses Markov model theory extensively with application to systems biology models.

The theory of Markov chains was, unsurprisingly, first developed by Markov [111]. Seminal advances from this first formulation are too numerous to cite individually and at this point are generally regarded as part of the core knowledge of anyone working with probability models.

9   Markov Chain Monte Carlo Sampling

One of our major concerns in studying simulation methods is figuring out how to sample efficiently from a complicated distribution implied by some model. As we have seen, there are some general sampling methods that can handle many situations we will encounter in biological domains. There are some situations, though, in which it would be prohibitively inefficient to sample exactly from such a model. For example, we may want accurate sampling of a small portion of the probability space. If we construct a Markov model of particles moving in a diffuse space and we want accurate sampling of the rate of particle collisions, we may have to simulate the model for an extremely long time to observe any collisions. If we develop a structural model of an enzymatic reaction and we want to know how often the model enters a rare transitional state, again we may need to run the model for an extremely long time to get accurate sampling of the small fraction of the probability distribution that is of concern to us.

In this chapter, we will consider a few specialized methods by which Markov models can help us with these hard-to-sample probability densities. We will first examine *Metropolis sampling*, a method that was developed for finding stationary distributions of thermodynamic systems but has much broader application. We will then examine *Gibbs sampling*, a specialized method for sampling from complicated distributions of many variables. Both Metropolis and Gibbs sampling are often used as approximate optimization methods in addition to sampling methods. We will then examine a general strategy, called *importance sampling*, for selectively accelerating estimates of some regions of the stationary distribution of a model. Finally, we will examine *umbrella sampling*, a special case of importance sampling widely used in biophysics contexts for improving estimates of the frequencies of rare events.

9.1 Metropolis Method

We will begin with a model we examined very briefly in chapter 1, called a *Metropolis model*. Metropolis models are very useful for looking at problems in thermodynamics, which include many problems in biophysics. The Metropolis method (also called the Metropolis–Hastings or Hastings–Metropolis method) is a technique for using a Markov model to determine the thermodynamic equilibrium of a system of discrete states for which we know potential energies. The method creates a Markov model whose stationary distribution is the distribution of states at the thermodynamic equilibrium of the system. Simulating the Markov model then samples from the states of the system at thermodynamic equilibrium.

Suppose we have a system of five states. These may represent folds of a single molecule, binding states in a reaction system, or any other set of discrete conditions in which a given system may be found. If we define some possible ways of moving between states (e.g., reaction events or structural rearrangements), then we end up with a system we can represent by a graph in which nodes represent states and edges represent allowed transitions between states. Figure 9.1(a) shows a hypothetical graph model of this form. Further assume that each state q_i has a potential energy E_i. For example, if the states correspond to folds of a protein, then each fold may have some free energy. Thermodynamics tells us that at equilibrium, the stationary probability of being in state q_i, which we can call π_i, is described by a *Boltzmann distribution*:

$$\pi_i = \frac{e^{-E_i/kT}}{\sum_{j=1}^{n} e^{-E_j/kT}},$$

where k is Boltzmann's constant and T is the absolute temperature. If the number of states is small, then we can calculate this distribution directly. The Metropolis

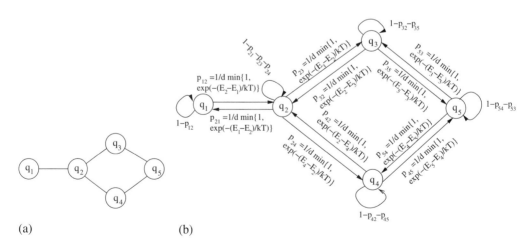

(a)   (b)

Figure 9.1
A state transition graph (a) and a corresponding Metropolis Markov model (b).

method is helpful when the state set is extremely large and we do not have time to explicitly compute the energy of each state. The Metropolis method creates a Markov model whose stationary distribution will be the Boltzmann distribution defined by the state energies. For this system, the Metropolis model will be that shown in figure 9.1(b), where d is the maximum degree of any node in the graph (in this case, 3).

Although the probabilities look complicated, the model is actually fairly easy to simulate. Starting from state q_i, we can find the next state as follows:

1. Pick a random neighbor of q_i, which we will call q_j, with probability $\frac{1}{d}$, or $q_j = q_i$ with probability $1 - \frac{d_i}{d}$ if the degree of node i (d_i) is less than d.
2. If $E_j \leq E_i$, then move to q_j.
3. If $E_j > E_i$, then with probability $e^{-(E_j - E_i)/kT}$ move to q_j, otherwise stay in q_i.

Note that we do not need to know anything about the global structure of the graph or the global energy landscape in order to simulate these transitions. We just need to know the maximum degree of the entire graph, or even an upper bound on it, and to have a way to determine the degree of the current node and the energy of each of its neighbors. If we have a very large state space but spend almost all time at equilibrium in a tiny fraction of the states (as might be the case for a protein-folding simulation, for example), then the Metropolis method provides a way to sample the equilibrium distribution efficiently without needing to explicitly create the full state graph.

The simplicity of the model is a virtue only if it is correct, though. We can in fact prove that this model will accurately sample from the Boltzmann distribution. As long as the energies are finite, all of the edges are reversible; if you can go from q_i to q_j, you can go from q_j to q_i, although generally with different probabilities. That immediately tells us that the model is ergodic as long as the original graph is connected. Furthermore, as long as we wait enough steps, we have a nonzero probability of getting from any node to any other. For example, suppose the largest number of steps necessary to travel between any two states is D (the *diameter* of the graph). Then we can get from any q_i to any q_j with some nonzero probability in exactly N steps for any $N \geq 2D$ by walking to any node q_k that has a self-loop, which must exist unless every node has the same energy; taking a few passes around the self-loop, then continuing from q_k to q_j, provided we linger at the self-loop just long enough to bring the total path length to N. The probability of this transition is bounded below by the minimum edge probability to the Nth power. This tells us that min $p_{ij}(N) > 0$ for some sufficiently large N, proving that the model converges to a unique stationary distribution.

The Markov models created by the Metropolis method not only have a stationary distribution; they also obey a stronger condition called *detailed balance* (also known as *microreversibility*). This property says that given any two states q_i and q_j with transition probabilities p_{ij} and p_{ji} and stationary probabilities π_i and π_j, then

$$\pi_i p_{ij} = \pi_j p_{ji}.$$

We will see in chapter 10 how to prove that a given Markov model obeys detailed balance. Detailed balance implies a stationary distribution, but not every chain having a stationary distribution satisfies detailed balance. If we know we have a chain obeying detailed balance, we often can easily establish the relative stationary probabilities of a set of states using the detailed balance condition.

Suppose we look at two states q_i and q_j for which $E_j < E_i$. Then

$$\pi_i p_{ij} = \pi_i \frac{1}{d}$$

$$\pi_j p_{ji} = \pi_j \frac{1}{d} e^{-(E_i - E_j)/kT}.$$

If we now plug in $\pi_i = \frac{e^{-E_i/kT}}{Z}$ and $\pi_j = \frac{e^{-E_j/kT}}{Z}$, where $Z = \sum_k e^{-E_k/kT}$, then

$$\pi_i p_{ij} = \frac{e^{-E_i/kT}}{Z} \frac{1}{d}$$

$$= \frac{e^{-E_i/kT}}{Zd}$$

$$= e^{-E_j/kT} \frac{e^{-(E_i - E_j)/kT}}{Zd}$$

$$= \frac{e^{-E_j/kT}}{Z} \frac{e^{-(E_i - E_j)/kT}}{Zd}$$

$$= \pi_j p_{ji}.$$

This verifies that these two values of π_i and π_j satisfy the detailed balance condition. Any set of probabilities satisfying detailed balance must be a stationary distribution, since the net change in the state distribution from each step of the Markov model must be zero if all pairs of state probabilities are in detailed balance. That is, the probability lost from p_i to q_j by one application of P when $Pr\{q = q_i\} = \pi_i$ is $\pi_i p_{ij}$, and the probability gained by p_i from p_j when $Pr\{q = q_j\} = \pi_j$ is $\pi_j p_{ji}$. When these two are equal for all i and j, then the model is at its stationary distribution. Since these values are exactly the Boltzmann weights that we should get at the thermodynamic equilibrium, this tells us that the Boltzmann distribution is a stationary distribution of the model, and thus that the model will uniquely converge on that distribution.

Example   Let us see an example of how this method would be applied to a model of protein-folding. Suppose we start with a lattice model of protein-folding like that we saw in chapter 1. We will assume that the protein sits on a regular grid, with consecutive amino acids at adjacent grid points and no two amino acids allowed to occupy the same grid point. The energy of any particular configuration of the chain will be given by a contact potential between pairs of amino acids occupying adjacent grid points but not consecutive in the chain backbone. We want to know how often the protein takes on each possible configuration. If the chain is short, we may be able to try every possible shape, determine the energy of each, and directly calculate the Boltzmann distribution of the energies. This will be impractical for even moderate chain sizes (maybe 30 amino acids), though, because the number of configurations grows exponentially with the chain size.

Once we have decided we will use a Metropolis model, we need to consider how the details of the model will be set. The states of the model are simply the possible configurations of the chain. We do not have an easy way to enumerate all of the states, but it is enough to know they exist and to be able to distinguish the ones we visit.

Next, we need to know the transitions of the Markov model. They are defined by the move set we choose for our lattice protein model. Suppose we decide that our allowed moves are chain changes that take a single set of three consecutive amino acids and bend them to a different position, as in figure 9.2. This move set tells us which pairs of states have transitions: those that are separated by a single chain bend. We can also infer from this an upper bound on the maximum degree of the graph. If we have n amino acids, then on each move we have $n - 2$ positions we can bend and up to two possible positions to which each can be bent, giving a maximum

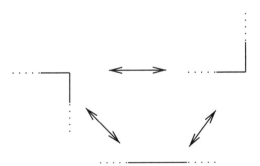

Figure 9.2
Allowed moves for a hypothetical lattice protein-folding model.

degree of $2(n-2)$. Not all of these moves will lead to valid conformations, since some may cause overlaps in the chain, so any particular node may have lower degree. In fact, for the purposes of correctly implementing the Metropolis algorithm, it does not actually matter if any node has exactly this degree, so long as it is an upper bound. If it is a poor upper bound, that will affect runtime, but not correctness.

We can then implement the Metropolis method for this problem as follows:

1. Choose a random amino acid from position 2 through $n-1$.
2. Choose a random bend direction from among the two possibilities (flip 90° left or 90° right if it is currently straight, or flip 90° or 180° if it is currently bent).
3. If the flipped chain is not self-avoiding, reject the move.
4. If the change in energy is less than or equal to zero, accept the move.
5. Otherwise, accept with probability $e^{-\Delta E/kT}$.

Running this model long enough will yield accurately sampled Boltzmann probabilities for all of the states, even if our upper bound on the node degree is not tight.

Caveats About Metropolis   The Metropolis method can be a powerful and easy-to-use means of estimating the thermodynamic equilibrium of a system, but there are two important cautions to consider when using it.

1. The Metropolis method is *thermodynamically correct*, but it is not generally *kinetically correct*. This means that if we have all of the energies right, then the stationary distribution will be the thermodynamic equilibrium distribution, but the pathways between states may not be in any way connected to how the real system will move between states. Metropolis is often used as a simulation method for kinetic processes, but it is not correct to do so.
2. Mixing times can be very long, especially if the transitions are poorly chosen. The Metropolis method can easily get stuck in local minima for long periods of time or have difficulty finding the correct trajectories. And it can be very difficult to judge whether the method is getting a good sample or is temporarily stuck in one part of the state space.

9.1.1 Generalizing the Metropolis Method

Although we described the Metropolis method in terms of thermodynamics, there is nothing in the algorithm that requires the stationary distribution to be a Boltzmann distribution. If we look carefully at the Metropolis algorithm, we can note that we do not really need the energies of the states. For any state q_i, we can find transition probabilities to each neighbor state q_j by knowing only the ratio of their stationary probabilities. When the stationary distribution is a thermodynamic equilibrium, then these ratios are of the form $e^{-\Delta E/k_B T}$. But we can ignore the energies and work directly with the ratios. This will give us the following generalized Metropolis move operation:

1. Pick a random neighbor of q_i, which we will call q_j, with probability $\frac{1}{d}$, or $q_j = q_i$ with probability $1 - \frac{d_i}{d}$ if the degree of node i, d_i, is less than d.
2. If $\frac{\pi_j}{\pi_i} \geq 1$, then move to q_j.
3. If $\frac{\pi_j}{\pi_i} < 1$, then with probability $\frac{\pi_j}{\pi_i}$ move to q_j, otherwise stay in q_i.

The proof of correctness above will follow identically for this general method as when we derive the ratios from energies. We can therefore apply the Metropolis method to estimating any discrete distribution, provided we can establish a connected graph on the states and determine the ratio of stationary probabilities along any edge in the graph.

9.1.2 Metropolis as an Optimization Method

Although the Metropolis method is, strictly speaking, a sampling method, it is also sometimes used as a heuristic method for optimization. Given a discrete optimization problem, we can declare that each possible solution to the problem has an "energy" which is determined by the value of the optimization metric on that solution. If we define "moves" between possible solutions, then a Metropolis simulation will be expected to move toward low-energy (high-quality) solutions. This method can be useful for locally improving on the solution returned by other methods, such as an approximation algorithm or another heuristic.

For example, suppose we want to find solutions to a traveling salesman problem (TSP). We can begin with some cycle in our input graph, such as that in figure 9.3(a). We can then declare that the "energy" of that path is the sum of the weights of its edges. We can then run the Metropolis method for a while, expecting that it will tend to move toward "low-energy" (i.e., short) paths. If we eventually return

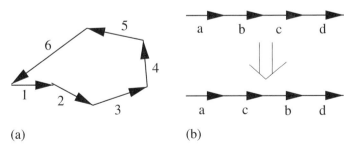

(a)   (b)

Figure 9.3
Illustration of use of Metropolis models for creating a heuristic traveling salesman problem solver. (a) Hypothetical initial tour in a graph. (b) A "move" we can use to define new tours, given existing ones.

the lowest-energy (shortest) path encountered by the Metropolis model, then we may expect this to be a good guess as to the shortest traveling salesman tour in the graph.

It is important when using this approach to make sure the chosen move set defines an ergodic Markov model. For TSP, we may use the move shown in figure 9.3(b), which allows us to flip the order of two nodes in a subpath. This move will define an ergodic Markov model. To show this, we need to show that for any two cycles π and π' in the TSP graph, there is a path of nonzero probability in the Markov model graph connecting them. Suppose we define π^* to be the tour connecting all nodes in some canonical order $q_1, q_2, q_3, \ldots, q_n, q_1$. Then we can get from π to π^* by bubble-sorting the list of nodes, which we can do solely by exchanging consecutive pairs of nodes using our one move. We can also go from π' to π^* by bubble-sorting nodes and, therefore, go from π^* to π' by following the reverse of those moves. Thus, we can get from π to π' by some sequence of these moves. Since any such move will have nonzero probability (assuming finite energies), the model must be ergodic.

One improvement on this Metropolis optimization approach that we have briefly mentioned before is *simulated annealing*. Simulated annealing is exactly the same as simulating a Metropolis model, except that we gradually reduce the temperature in the model. In the limit of high temperature,

$$\lim_{T \to \infty} e^{-(E_2 - E_1)/kT} = e^0 = 1.$$

Therefore, when the temperature is very high, almost all moves will be accepted and the model will choose transitions almost uniformly at random. It should therefore move easily out of local minima and get a good sample of the probability space. In the limit of low temperature, if $E_2 > E_1$, then

$$\lim_{T \to 0} e^{-(E_2 - E_1)/kT} = e^{-\infty} = 0.$$

Therefore, at very low temperatures, the method will almost never choose a move that is energetically unfavorable. It will behave as a pure local optimizer and will settle into some local minimum energy state, for which no one move can lead to an improvement. The hope is that somewhere between very high and very low temperatures, the method will be good at escaping local minima while still settling into low-energy regions of the graph. Then, as the temperature cools below that point, the method will seek the lowest energy state in that low-energy region. The actual performance will depend on exactly how the method goes from high to low temperatures (known as the *cooling schedule*) in ways that are not well understood. It can also be helpful to run the method multiple times, since it may get stuck in a bad region on some runs but not others.

9.2 Gibbs Sampling

Gibbs sampling is another technique for constructing a Markov model whose stationary distribution is some probability distribution from which it would otherwise be hard to sample. Gibbs sampling is particularly useful for sampling from joint distributions on many variables. We accomplish this by allowing states of the Markov model to correspond to possible assignments of the full state vector of the jointly distributed variables. We then allow transitions corresponding to possible changes in a single element of the state vector.

Assume we are given a joint distribution

$$Pr\{X_1 = x_1, X_2 = x_2, \ldots, X_n = x_n\} = p(x_1, x_2, \ldots, x_n),$$

where $R_1, R_2, \ldots, R_n$ are the ranges of the respective random variables. The state set of the Gibbs sampler is the product of the ranges of the variables:

$$Q = R_1 \times R_2 \times \cdots \times R_n.$$

That is, there is one state for each possible assignment of values to the random variables. There is a transition possible for any change of a single random variable. The probability of making any possible transition obeys the following density:

$$Pr\{(x_1, x_2, \ldots, x_i, \ldots, x_n) \rightarrow (x_1, x_2, \ldots, x_i', \ldots, x_n)\}$$

$$= \frac{1}{n} \, Pr\{X_i = x_i' \mid X_1 = x_1, X_2 = x_2, \ldots, X_{i-1} = x_{i-1}, X_{i+1} = x_{i+1} \ldots, X_n = x_n\}.$$

Just as with Metropolis models, Gibbs sampling Markov models lend themselves to a very simple procedure for making transitions:

1. Pick a variable X_i uniformly at random.
2. Sample a new value for that one variable from the conditional distribution of that variable, given the current values of all of the other variables:

$$Pr\{X_i = x_i' \mid X_1 = x_1, X_2 = x_2, \ldots, X_{i-1} = x_{i-1}, X_{i+1} = x_{i+1} \ldots, X_n = x_n\}.$$

Repeatedly applying that transition step will produce a correct sample of the joint distribution of the variables. We can show the correctness of the method by establishing that the desired joint distribution is a stationary distribution satisfying detailed balance for the Gibbs sampler. We can show that as follows:

$$Pr\{(x_1, x_2, \ldots, x_i, \ldots, x_n) \rightarrow (x_1, x_2, \ldots, x_i', \ldots, x_n)\}$$

$$\times Pr\{(x_1, x_2, \ldots, x_i, \ldots, x_n)\}$$

$$= \frac{1}{n} Pr\{X_i = x_i' \mid X_1 = x_1, \dots, X_{i-1} = x_{i-1}, X_{i+1} = x_{i+1}, \dots, X_n = x_n\}$$

$$\times Pr\{(x_1, x_2, \dots, x_i, \dots, x_n)\}$$

$$= \frac{1}{n} \frac{Pr\{X_1 = x_1, \dots, X_i = x_i', \dots, X_n = x_n\}}{Pr\{X_1 = x_1, X_2 = x_2, \dots, X_{i-1} = x_{i-1}, X_{i+1} = x_{i+1}, \dots, X_n = x_n\}}$$

$$\times Pr\{(x_1, x_2, \dots, x_i, \dots, x_n)\}$$

$$= \frac{1}{n} \frac{Pr\{X_1 = x_1, \dots, X_i = x_i, \dots, X_n = x_n\}}{Pr\{X_1 = x_1, X_2 = x_2, \dots, X_{i-1} = x_{i-1}, X_{i+1} = x_{i+1}, \dots, X_n = x_n\}}$$

$$\times Pr\{(x_1, x_2, \dots, x_i', \dots, x_n)\}$$

$$= \frac{1}{n} Pr\{X_i = x_i \mid X_1 = x_1, \dots, X_{i-1} = x_{i-1}, X_{i+1} = x_{i+1}, \dots, X_n = x_n\}$$

$$\times Pr\{(x_1, x_2, \dots, x_i', \dots, x_n)\}$$

$$= Pr\{(x_1, x_2, \dots, x_i', \dots, x_n) \rightarrow (x_1, x_2, \dots, x_i, \dots, x_n)\}$$

$$\times Pr\{(x_1, x_2, \dots, x_i', \dots, x_n)\}.$$

Thus, $Pr\{(x_1, x_2, \dots, x_i, \dots, x_n)\}$ is the unique stationary distribution of the model. Running it for a sufficient number of steps will therefore accurately sample from that joint distribution. As with any Markov model, though, it is important to remember that the time to converge on the stationary distribution may be quite long.

Example   Consider a two-variable example. Suppose we want to sample a pair of DNA bases, which we can call X_1 and X_2, consecutive in a protein-coding sequence. It is likely that these bases are highly correlated. We can study these correlation patterns and establish a set of joint probabilities:

$$Pr\{A_1 \wedge A_2\} = p_{AA}, \quad Pr\{A_1 \wedge C_2\} = p_{AC}, \quad Pr\{A_1 \wedge G_2\} = p_{AG},$$

$$\dots Pr\{T_2 \wedge C_1\} = q_{TC}, \quad Pr\{T_2 \wedge G_1\} = q_{TG}, \quad Pr\{T_2 \wedge T_1\} = q_{TT}.$$

Then we can begin by picking an arbitrary pair of bases, say $C_1 G_2$. We then choose to change either the first base or the second with equal probability. If we choose to change the first, then we will sample a new base from the distribution $Pr\{X_1 \mid X_2 = G\}$. Alternatively, we will sample a new X_2 from the distribution

$Pr\{X_2 \mid X_1 = C\}$. Putting these possibilities together, we will get the following transition probabilities from our current state:

$$Pr\{C_1 G_2 \rightarrow A_1 G_2\} = \frac{1}{2} \ Pr\{A_1 \mid G_2\} = \frac{p_{AG}}{2(p_{AG} + p_{CG} + p_{GG} + p_{TG})}$$

$$Pr\{C_1 G_2 \rightarrow G_1 G_2\} = \frac{1}{2} \ Pr\{G_1 \mid G_2\} = \frac{p_{GG}}{2(p_{AG} + p_{CG} + p_{GG} + p_{TG})}$$

$$Pr\{C_1 G_2 \rightarrow T_1 G_2\} = \frac{1}{2} \ Pr\{T_1 \mid G_2\} = \frac{p_{TG}}{2(p_{AG} + p_{CG} + p_{GG} + p_{TG})}$$

$$Pr\{C_1 G_2 \rightarrow C_1 A_2\} = \frac{1}{2} \ Pr\{A_2 \mid C_1\} = \frac{q_{AC}}{2(q_{AC} + q_{CC} + q_{GC} + q_{TC})}$$

$$Pr\{C_1 G_2 \rightarrow C_1 C_2\} = \frac{1}{2} \ Pr\{C_2 \mid C_1\} = \frac{q_{CC}}{2(q_{AC} + q_{CC} + q_{GC} + q_{TC})}$$

$$Pr\{C_1 G_2 \rightarrow C_1 T_2\} = \frac{1}{2} \ Pr\{T_2 \mid C_1\} = \frac{q_{TC}}{2(q_{AC} + q_{CC} + q_{GC} + q_{TC})}$$

$$Pr\{C_1 G_2 \rightarrow C_1 G_2\} = \frac{1}{2} \ Pr\{C_1 \mid G_2\} + \frac{1}{2} \ Pr\{G_2 \mid C_1\}$$

$$= \frac{p_{CG}}{2(p_{AG} + p_{CG} + p_{GG} + p_{TG})} + \frac{q_{GC}}{2(q_{AC} + q_{CC} + q_{GC} + q_{TC})}.$$

Note that we have to count the self-transition twice, since it can be reached from either initial choice of which variable to modify.

This method can be applied to almost any joint distribution, provided the conditional distributions of isolated variables are easy to sample. Note that this technique can generalize even to continuous distributions. Given some joint continuous density $f_{xy}(x, y)$, we can construct a Gibbs sampler by repeatedly sampling x or y individually from the conditional densities $f_x(x \mid y = y_0)$ and $f_y(y \mid x = x_0)$. These are calculated as follows:

$$f_x(x \mid y = y_0) = \frac{f_{xy}(x, y_0)}{\int_x f_{xy}(u, y_0) \, du}$$

$$f_x(y \mid x = x_0) = \frac{f_{xy}(x_0, y)}{\int_y f_{xy}(x_0, v) \, dv}.$$

These two formulas describe continuous densities on one variable each, and they are therefore likely to be easier to sample. The resulting Gibbs sampler will become a

```
            AGCTAGCTACGCATCA
     AGGTACGACTCTAGCT
           ACTCGATCTACGATCT
     CATCGACTCATATCTC
          ACTGAATCTACATCAT
        GCATCGAGTTACTCTT
        ACTGCATCTATCTACT
```

Figure 9.4
A Gibbs sampling for motif-finding.

continuous-state Markov model, meaning that its state set has an uncountably infinite number of elements.

9.2.1   Gibbs Sampling as an Optimization Method

Gibbs sampling, like Metropolis sampling, is often used as a heuristic optimization method, particularly for maximum likelihood problems. A classic example of this practice in biology is motif-finding. Given a set of DNA sequences, each of which is presumed to contain some motif, our goal is to infer the motifs by aligning the sequences to some window. Suppose we are given a window size k and a set of n sequences each of at least length k. We want to find the alignment of all sequences to the window that best captures the similarity of the sequences. Figure 9.4 illustrates the problem. In this figure, we see several sequences aligned to a common window of four bases. If this alignment correctly identifies a binding motif, then we will conclude that the motif has a strongly conserved A in the first position, preference for either T or G in the second position, a weakly conserved C in the third position, and a strongly conserved T in the fourth position.

If we want to find the best alignment by Gibbs sampling, we first create a probability model expressing the probability of generating any given multiple alignment of the sequences to the window. We may assume that there is some position-specific score matrix identifying the probability of outputting any given base in each position of the sequence:

$$
\begin{bmatrix}
p_{1A} & p_{2A} & p_{3A} & p_{4A} \\
p_{1C} & p_{2C} & p_{3C} & p_{4C} \\
p_{1G} & p_{2G} & p_{3G} & p_{4G} \\
p_{1T} & p_{2T} & p_{3T} & p_{4T}
\end{bmatrix}
$$

where p_{iN} is the probability of emitting base N from position i. We can then say that the probability of emitting the observed sequences in the windows, given the alignment, is the product of the per-base output frequencies over all observed bases. We can write the sequences in the windows for a given alignment as follows:

$$\Sigma_1 = \sigma_{\delta_1+1}\sigma_{\delta_1+2}\sigma_{\delta_1+3}\sigma_{\delta_1+4}$$

$$\Sigma_2 = \sigma_{\delta_2+1}\sigma_{\delta_2+2}\sigma_{\delta_2+3}\sigma_{\delta_2+4}$$

$$\vdots$$

$$\Sigma_n = \sigma_{\delta_n+1}\sigma_{\delta_n+2}\sigma_{\delta_n+3}\sigma_{\delta_n+4},$$

where δ_i is the offset of the window in the alignment of sequence i. Then the likelihood of a full global alignment Δ is

$$Pr\{\Sigma \mid \Delta\} = \prod_{i=1}^{n} p_{1,\sigma_{\delta_i+1}} \times p_{2,\sigma_{\delta_i+2}} \times p_{3,\sigma_{\delta_i+3}} \times p_{4,\sigma_{\delta_i+4}}.$$

We can then construct a Markov model where each state corresponds to a possible alignment $\Delta = \{\delta_1, \delta_2, \ldots, \delta_n\}$ and the stationary probability of any given state Δ is proportional to the likelihood of Δ:

$$\pi_{\Delta_i} = \frac{Pr\{\Sigma \mid \Delta_i\}}{\sum_{\Delta_j} Pr\{\Sigma \mid \Delta_j\}}.$$

We accomplish this by saying that any given transition from some $\Delta_1 = \{\delta_1, \delta_2, \ldots, \delta_j, \ldots, \delta_n\}$ to some $\Delta_2 = \{\delta_1, \delta_2, \ldots, \delta_j', \ldots, \delta_n\}$ has probability

$$Pr\{\delta_j' \mid \delta_1, \delta_2, \ldots, \delta_{j-1}, \delta_{j+1}, \ldots, \delta_n\}.$$

To be strictly correct, we should also consider that there may be a distribution over possible scoring matrices. This is known as a *prior distribution*, and reflects our previous knowledge about likely solutions to the problem. Often in practice this will be accomplished by assuming a uniform prior distribution, in which all scoring matrices are presumed to be equally likely. In that case, we can choose transitions in the Markov model as follows:

1. Pick a sequence k uniformly at random.
2. Compute the base frequency within the window for all sequences other than k as an estimate of the score frequencies p_{iN}.
3. For each possible offset j of sequence k, compute the probability $p_j = p_{1,j+1} \times p_{1,j+2} \times p_{1,j+3} \times p_{1,j+4}$.
4. Choose some new offset j for sequence k with probability $p_j/(\sum_{j'} p_{j'})$.

We can, however, weight the probabilities in step 3 by the probability of the derived scoring matrix to create a nonuniform prior distribution. For example, we may have some prior reason to believe that the motif is pyrimidine-rich, and may therefore give extra weight to offsets that tend to produce pyrimidine-rich motifs.

Regardless of our prior, repeated application of the steps above will eventually converge on a stationary distribution of possible alignments. We can then pick the most commonly occurring alignment from that distribution as our maximum-likelihood estimate.

9.3 Importance Sampling

The Metropolis and Gibbs sampling methods provide ways of sampling from distributions that may be hard to express analytically. Sometimes, though, having a correct sampler is not enough. We also need the sampler to be efficient, in that it gets close to its stationary distribution in a small number of steps. Importance sampling is a technique we can use when we have a sampler for a given distribution but want to accelerate it. The basic idea is fairly simple. Given some model with state set

$$Q = \{q_1, \ldots, q_n\}$$

and corresponding stationary distribution

$$\Pi = \{\pi_1, \ldots, \pi_n\},$$

we construct a new model with the same state set Q but biased stationary distribution,

$$\hat{\Pi} = \{\hat{\pi}_1, \ldots, \hat{\pi}_n\},$$

where $\hat{\pi}_i = w_i \pi_i$ for some set of weights w_i. We then sample from distribution $\hat{\Pi}$ but adjust the estimated frequencies for each state q_i sampled from $\hat{\Pi}$ by a factor of $\frac{1}{w_i}$. The result is an accurate estimator of π_i, but one in which the variance of the estimates has been changed. In particular, we generally want to choose the weights w_i such that the variance of the model is reduced, leading to faster estimation of Π.

One common use of importance sampling is to bias a model toward those states that account for the majority of the probability density, thus accelerating estimation of those states. For example, in a protein-folding model, we may be most interested in compact states of the protein, where it is likely to spend most of its time at equilibrium. If we sample states by a Metropolis model, though, we may need to run the model for a long time before we start to see an accurate sample of these compact states. We can likely accelerate this convergence by biasing the probability based on the radius of gyration (essentially the diameter) of the current state. Suppose we have a Monte Carlo model of protein-folding with the following distribution:

$$\Pi = \{\pi_1, \ldots, \pi_n\}.$$

We can generate a modified distribution by attaching a penalty of e^{-kr_i} to each state, where r_i is the radius of gyration of state q_i and k is a scaling constant. Then we get the new distribution

$$\hat{\Pi} = \{\pi_1 e^{-kr_1}/Z, \ldots, \pi_n e^{-kr_n}/Z\},$$

where Z is a scaling factor used to make the probabilities sum to 1. We can sample from this new distribution by multiplying the ratio π_j/π_i by $e^{-k(r_j-r_i)}$ when picking our Metropolis moves. The result will be a sampler of $\hat{\Pi}$ that is likely to sample compact states more quickly. When we believe we have adequately sampled the distribution, we can then convert $\hat{\Pi}$ into Π by retroactively scaling the states we have sampled. We know each r_i and can therefore find the vector Π', defined as

$$\pi_i' = \hat{\pi}_i \times e^{kr_i} = \pi_i/Z.$$

We can then solve for Z using the fact that since Π is a probability distribution, $\sum_i \pi_i = 1$:

$$Z = \left(\sum_i \pi_i'\right)^{-1}.$$

Finally, we can estimate

$$\pi_i = \pi_i' \times Z.$$

For sufficiently many trials, we will expect this procedure to derive the same estimates as if we had not used importance sampling. Our modified method should, however, perform these estimates more quickly. In particular, it should give us a more accurate sample of the compact states quickly.

9.3.1 Umbrella Sampling

There is a special case of importance sampling, called *umbrella sampling*, commonly used in statistical physics problems to give more accurate estimates of frequencies of rare events in a model. For example, imagine that we have a protein that has been destabilized by a mutation. It spends most of its time correctly folded, but occasionally unfolds. We would like to understand how that protein behaves when it is in an unfolded form. We might, for example, be concerned with whether it maintains its secondary structure when it loses its tertiary structure, or how often some crucial binding pocket is formed when it is mostly unfolded.

Umbrella sampling proceeds in essentially the same way as general importance sampling, but is biased so that our modified chain spends a disproportionate amount of time in the portion of space we want to estimate accurately. For our mutant protein problem, we may simply reverse the weight function we used in the preceding protein-folding example to illustrate importance sampling. That is, we can create a model in which we provide extra weight to states with large radii of gyration by using weight terms of the form e^{+kr_i}/Z. We can then proceed exactly as in the previous section to estimate $\hat{\Pi}$ and convert it into an estimate of Π:

1. Scale each Metropolis ratio $\frac{\pi_j}{\pi_i}$ by $e^{k(r_j-r_i)}$ to get a sampler for distribution $\hat{\Pi}$.
2. Scale each estimated $\hat{\pi}_i$ by e^{-kr_i} to get $\pi'_i = \pi_i/Z$.
3. Solve for $Z = \left(\sum \pi'_i\right)^{-1}$.
4. Find the unbiased estimates $\pi_i = \pi'_i \times Z$.

As before, sufficiently long runtimes will yield identical estimates of Π whether or not we use umbrella sampling. With the protocol we describe above, though, the model will tend to quickly find accurate samples of the rare space we elevated in weight. The overall model, however, will tend to converge more slowly on an accurate stationary distribution because states with high equilibrium frequency will now rarely be sampled.

9.3.2 Generalizing to Other Samplers

Although the examples assumed we were sampling from Metropolis models, that is not a necessary assumption to importance or umbrella sampling. The same basic procedure will work no matter how we choose to estimate the distribution. For example, suppose we want to sample from a distribution of many variables $p(x_1, \ldots, x_n)$ and are interested in the probability space in which some particular x_i is large. We can construct a Gibbs sampler of p and weight each state by e^{x_i} when varying x_i to derive a Gibbs sampler of some $\hat{p}$ biased toward large x_i. We can then adjust the final estimated probabilities by e^{-x_i} to convert the probabilities derived from the $\hat{p}$ sampler into an estimate of the stationary distribution of p. As long as we have a sampling method that allows us to weight particular parts of the distribution, we can use the methods above to perform importance sampling or umbrella sampling.

We can even do this with continuous distributions, just as we could with a Gibbs sampler. For example, suppose we define a continuous model of the mutant protein example we considered above. We may, for instance, define a molecular dynamics energy function for our protein structure $E(\vec{C})$, where $\vec{C}$ is our protein's conformation expressed as a point in a continuous, high-dimensional parameter space (e.g., a vector of ϕ-ψ angles.) Our protein's time evolution would be expressed deterministically by the following second-order differential equation:

$$\frac{d^2\vec{C}}{dt^2} = -M^{-1}\nabla E(\vec{C}),$$

where M is a matrix with atom masses on the diagonal for each degree of freedom. We can convert this expression to a system of first-order equations:

$$\frac{d\vec{V}}{dt} = -M^{-1}\nabla E(\vec{C})$$

$$\frac{d\vec{C}}{dt} = \vec{V}.$$

From there, we can create a probabilistic model of the protein's movement in continuous space using stochastic differential equations, a technique we will see in chapter 16:

$$d\vec{V} = -M^{-1}\nabla E(\vec{C})\,dt + \mu M^{-1}\,d\vec{W}$$

$$d\vec{C} = \vec{V}\,dt,$$

where $d\vec{W}$ is a vector of normal random variables representing Brownian noises acting on each degree of freedom of the model. Stochastic integration of these equations will then provide a continuous sampler of a probability distribution describing the range of motion of the protein in the presence of Brownian noise. We can directly integrate this model to study the protein's dynamics, but it may be very slow if the protein only rarely unfolds. If we want to selectively sample the unfolded space by umbrella sampling, we may change our energy function as follows:

$$E'(\vec{C}) = E(\vec{C}) - kr(\vec{C}),$$

where r now specifies the radius of gyration of the continuous-space model $\vec{C}$. We can then integrate

$$d\vec{V} = -M^{-1}\nabla E'(\vec{C})\,dt + \mu M^{-1}\,d\vec{W}$$

$$d\vec{C} = \vec{V}\,dt$$

to sample the probability space biased toward unfolded chains. If we then want to know, for example, the fraction of time the radius exceeded some r_0, we can take the states at all time points simulated by the biased model, scale each by $e^{kr(\vec{C})/k_bT}$ to cancel the effect of the biasing energy, and rescale the whole set by the sum of these scaling factors to get a corrected distribution from the original probability space. The weighted sum of states for which $r(\vec{C}) > r_0$ will be an estimate of our desired probability that will be expected to provide accurate probabilities for the unfolded states much more quickly than if we have used the uncorrected SDEs. We can also ask questions about properties of the chain when it is unfolded, such as how often $r > r_0$ and a given hydrogen bond is formed, or how often $r > r_0$ and two given amino acids are within 1 nm of one another.

References and Further Study

The topics in this section are commonly covered in probability texts of various levels. Several we have seen previously are fine references for these topics, including Rozanov [108] and Ross [102]. Hochbaum [56] is a good reference for the use of sampling methods as heuristics for optimization. These techniques are also widely used in statistical applications for sampling from complicated densities, and a good text on statistical modeling is likely to have some coverage of them. Wasserman [112], for example, provides coverage of all of the samplers we have seen here.

Primary references are available for several of the topics covered here. The Metropolis method was first described in a seminal paper by Metropolis et al. [7], which remains one of the most widely cited papers in the entire scientific literature. The simulated annealing method is due to Kirkpatrick et al. [51] and Cerny [52]. Although Gibbs sampling was named after the physicist J. W. Gibbs, it was actually invented by Geman and Geman [113]. Importance sampling is considered part of the basic knowledge of the statistical sampling field, and I have never seen a primary citation for the method in general. The umbrella sampling method was developed by Torrie and Valleau [114].

10   Mixing Times of Markov Models

When we first introduced Markov models in chapter 8, we briefly discussed the concept of *mixing time*: the time it takes for a Markov model to approximately reach its stationary distribution. As we saw in chapter 9, we will often design a Markov model so that its stationary distribution will be some probability distribution we care about but that is difficult to state explicitly. It is therefore important to know how long we need to run the model to estimate the stationary distribution accurately. We saw that we could get an approximate idea of the mixing times for some Markov models by looking at the eigenvalues of the transition matrix. When we cannot explicitly state the transition matrix, perhaps because it is too large, or we cannot find its eigenvalues, then we need some other options. In this chapter, we will explore some theoretical methods we can use to put rigorous bounds on the mixing time. Much of the material in this chapter is derived from a chapter on the topic by Sinclair and Jerrum in Hochbaum [56].

For the remainder of this chapter, we will make the following assumptions about our Markov models:

1. ergodicity
2. $p_{ii} \geq \frac{1}{2}$ for all i
3. detailed balance $(\pi_i p_{ij} = \pi_j p_{ji} = Q_{ij})$.

It is not generally meaningful to talk about the stationary distribution for a nonergodic Markov chain, so that is not a difficult criterion on which to insist. If the second condition is not satisfied for a model of interest, it is easy to convert the model to one that has the same stationary distribution and does satisfy the condition by cutting all of the non-self-transition probabilities in half and then adding $\frac{1}{2}$ to all of the self-transition probabilities. This operation will exactly double the mixing time by causing the model to linger in each state chosen for an average of two steps, but it will not change the equilibrium state distribution. The third condition is trickier. If we have a distribution from which we want to sample, then we can easily define a model to satisfy the third condition if we know the ratios between the equilibrium probabilities of neighboring states. We assign a transition probability in one direction for each edge, and the ratio tells us the transition probability in the other direction needed to satisfy detailed balance. In the Metropolis–Hastings method, for example, we can determine the ratios of the transition probabilities from the energy differences between the states. If we do not know these ratios, it is not necessarily easy to design a model satisfying detailed balance. It is, however, possible to determine after the fact if a given model does satisfy detailed balance. This is established by the *Kolmogorov criterion*:

$$p_{12} \times p_{23} \times \cdots \times p_{k-1,k} \times p_{k1} = p_{1k} \times p_{k,k-1} \times \cdots \times p_{3,2} \times p_{21}$$

for all cycles $q_1, \ldots, q_k, q_1$ in the graph. The criterion tests whether the probability of moving around a cycle in one direction is equal to the probability of moving around the cycle in the other direction, for all cycles in the graph. If the model satisfies the Kolmogorov criterion, then it has a stationary distribution obeying detailed balance.

10.1   Formalizing Mixing Time

Before we can talk about formal bounds on the mixing time, we need to be more rigorous about what we mean by mixing time. To do so, we first need to define a concept called the *variation distance*, where we follow Jerrum and Sinclair's terminology:

$$\Delta_q(t) = \max_{S \subseteq Q} \left| Pr\{q(t) \in S \mid q(0) = q\} - \sum_{q_i \in S} \pi_i \right|$$

$$= \frac{1}{2} \sum_{q_i \in Q} \left| Pr\{q(t) = q_i \mid q(0) = q\} - \pi_i \right|,$$

where, as previously, Q is the state set, $q(t)$ is the state at time t, and π_i is the stationary probability of state i. Variation distance is a measure of how much the distribution of states at time t differs from the distribution at equilibrium, given some starting state q. There is an alternative definition of variation distance that is sometimes more intuitive:

$$\Delta = \max_i \frac{Pr\{q(t) = q_i\} - \pi_i}{\pi_i},$$

but we will not use that latter definition here.

We can define the mixing time formally in terms of the variation distance as

$$\tau_q(\varepsilon) = \min\{t \mid \Delta_q(t') \leq \varepsilon \; \forall t' \geq t\}.$$

In other words, the mixing time is the time at which the variation distance first falls below some fixed ε and remains there. Note that we are defining mixing time relative to some fixed starting state q. We often really want to know the mixing time maximized over all possible starting states, but that is an easy generalization to make.

10.2   The Canonical Path Method

We will start looking at formal bounds on mixing time using a proof technique called *the canonical path method*. In the canonical path method, we put a bound on the mixing time by showing that, for most pairs of nodes, there is some path allowing rapid transitions between those nodes. For each pair of nodes q_i and q_j, we will identify one *canonical path* from q_i to q_j, which we will call γ_{ij}. This can be any path in the graph from q_i to q_j, although how we choose the path will affect the tightness of the mixing time bound. We will then define Γ to be the set of all canonical paths, $\Gamma = \{\gamma_{ij} \mid q_i, q_j \in Q\}$.

We then define the *maximum edge loading* of the canonical path set Γ to be

$$\rho(\Gamma) = \max_{e \in E} \frac{1}{Q_e} \sum_{\gamma_{ij} \ni e} \pi_i \pi_j |\gamma_{ij}|,$$

where E is the set of transitions in the Markov graph and Q_e is $\pi_i p_{ij}$ for $e = (q_i, q_j)$. Our goal will be to choose a set of canonical paths which ensures that the edge loading is not too high for any edge in the graph. That will establish that it is in some sense easy to get from any node to any other, and thus that the mixing time is small.

Example   We will now see how we would compute the maximum edge loading of a sample graph. Suppose we want to simulate the Markov model of figure 10.1. The stationary distribution of this graph is $\Pi = \begin{bmatrix} \frac{1}{3} & \frac{1}{6} & \frac{1}{3} & \frac{1}{6} \end{bmatrix}$, a fact we can verify by showing that Π satisfies detailed balance for the graph:

$$\pi_1 p_{12} = \frac{1}{3} \times \frac{1}{8} = \frac{1}{6} \times \frac{1}{4} = \pi_2 p_{21}$$

$$\pi_2 p_{23} = \frac{1}{6} \times \frac{1}{4} = \frac{1}{3} \times \frac{1}{8} = \pi_3 p_{32}$$

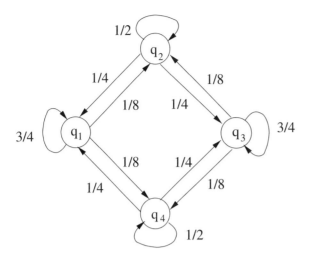

Figure 10.1
A Markov model graph with states and transition probabilities labeled.

$$\pi_3 p_{34} = \frac{1}{3} \times \frac{1}{8} = \frac{1}{6} \times \frac{1}{4} = \pi_4 p_{43}$$

$$\pi_4 p_{41} = \frac{1}{6} \times \frac{1}{4} = \frac{1}{3} \times \frac{1}{8} = \pi_1 p_{14}.$$

We then need to define the canonical paths. We choose one canonical path for each pair of states. Since we want to keep edge loading low, we want to avoid placing too many paths through any edge and to avoid using low-capacity edges altogether. In this case, it is fairly easy since we have only two choices for any pair of nodes: clockwise or counterclockwise. We will declare that γ_{ij} is clockwise for $i < j$ and counterclockwise for $i > j$.

We can then compute the maximum edge loading $\rho(\Gamma)$ by examining the edge loading for each edge (q_i, q_j):

$$\rho_{12} = \frac{1}{\pi_1 p_{12}} \left(\pi_1 \pi_2 |\gamma_{12}| + \pi_1 \pi_3 |\gamma_{13}| + \pi_1 \pi_4 |\gamma_{14}| + \pi_2 \pi_1 |\gamma_{21}| + \pi_3 \pi_1 |\gamma_{31}| + \pi_4 \pi_1 |\gamma_{41}| \right)$$

$$= 24 \left(\frac{1}{3} \cdot \frac{1}{6} \cdot 1 + \frac{1}{3} \cdot \frac{1}{3} \cdot 2 + \frac{1}{3} \cdot \frac{1}{6} \cdot 3 + \frac{1}{6} \cdot \frac{1}{3} \cdot 1 + \frac{1}{3} \cdot \frac{1}{3} \cdot 2 + \frac{1}{6} \cdot \frac{1}{3} \cdot 3 \right)$$

$$= 24 \left(\frac{1}{18} + \frac{4}{18} + \frac{3}{18} + \frac{1}{18} + \frac{4}{18} + \frac{3}{18} \right) = \frac{(24)(16)}{18} = \frac{64}{3}$$

$$\rho_{23} = \frac{1}{\pi_2 p_{23}} \left(\pi_1 \pi_3 |\gamma_{13}| + \pi_1 \pi_4 |\gamma_{14}| + \pi_2 \pi_3 |\gamma_{23}| + \pi_2 \pi_4 |\gamma_{24}| + \pi_3 \pi_1 |\gamma_{31}| + \pi_4 \pi_1 |\gamma_{41}| \right.$$

$$\left. + \pi_3 \pi_2 |\gamma_{32}| + \pi_4 \pi_2 |\gamma_{42}| \right)$$

$$= 24 \left(\frac{1}{3} \frac{1}{3} \cdot 2 + \frac{1}{3} \frac{1}{6} \cdot 3 + \frac{1}{6} \frac{1}{3} \cdot 1 + \frac{1}{6} \frac{1}{6} \cdot 2 + \frac{1}{3} \frac{1}{3} \cdot 2 \right.$$

$$\left. + \frac{1}{6} \frac{1}{3} \cdot 3 + \frac{1}{3} \frac{1}{6} \cdot 1 + \frac{1}{6} \frac{1}{6} \cdot 2 \right)$$

$$= 24 \left(\frac{4}{18} + \frac{3}{18} + \frac{1}{18} + \frac{1}{18} + \frac{4}{18} + \frac{3}{18} + \frac{1}{18} + \frac{1}{18} \right) = \frac{(24)(18)}{18} = 24.$$

The other cases are each equivalent to one of these two, so we can conclude that the maximum edge loading in the graph is 24.

The reason we care about the edge loading is that we can use it to bound the mixing time with the following theorem.

Theorem   Given a finite, reversible, ergodic Markov chain with maximum edge loading ρ, $\tau_{q_i}(\varepsilon) \leq \rho(\ln(\pi_i^{-1}) + \ln(\varepsilon^{-1}))$ for any initial state q_i.

Intuitively, what the theorem means is that if there is a set of paths allowing the model to move quickly between all pairs of nodes, then the chain will mix rapidly. For our example above, the maximum edge loading is 24 and the minimum π_i is $\frac{1}{6}$. Therefore, we can bound $\tau_q(\varepsilon)$ by $24(\ln(6) + \ln(\varepsilon^{-1}))$.

Example   We will now look at a more complicated, and biologically motivated, example: mixing time of a simple model of DNA evolution. Suppose we have a model of the evolution of a strand of DNA. We will assume there is no selective pressure; bases randomly mutate over time according to some Markov process. We want to know how long the process must run before we essentially have a random sequence. This may be useful for inferring, for example, how long the similarity between two common ancestors will be preserved, which we can use to understand when homology approaches are likely to be successful.

Suppose we assume that we have a string of n bases (e.g., AACATGAT if $n = 8$) defining a Markov model with 4^n states. We will assume that these evolve by some random process and that each transition can flip a single DNA base. If mutation is rare enough, then that should be a reasonable assumption. Thus, we will have a transition between any pair of sequences that differ by a single base. Thus, AACAT-GAT's neighbors will be CACATGAT, GACATGAT, TACATGAT, ACCAT-GAT, AGCATGAT, and so on. We may incorporate an estimate of the rate of molecular evolution into the model by controlling the self-transition probabilities, p_{ii}. For example, if bases flip an average of every 1000 generations, we may set

$p_{ii} = 0.999$ for all i. For now, though, we will keep it simple and assume the following transition probabilities:

$$p_{ij} = \frac{1}{6n}, \quad i \neq j$$

$$p_{ii} = \frac{1}{2}.$$

In this case, it is pretty obvious what the stationary distribution will be—all bases will eventually become equally likely in all positions—but we want to know how quickly we approach it. We will use the canonical path method to put a bound on the mixing time.

We first need to verify that this model meets the preconditions of the canonical path method:

• Ergodicity: We can transition from any sequence to any other by converting bases one at a time wherever they differ between the two sequences. Therefore, we have ergodicity.
• Self-transitions have probability of at least $\frac{1}{2}$. This is true by the design of the model.
• Detailed balance: We can verify detailed balance by noting that all non-self-transitions have the same probability $\left(\frac{1}{6n}\right)$, so the product of the transition probabilities along any cycle of length k is $\left(\frac{1}{6}\right)^k$, regardless of whether we take the cycle in the forward or reverse direction. The model therefore satisfies the Kolmogorov criterion and exhibits detailed balance.

Having verified that we can use the canonical path method for this model, we next need to choose a set of canonical paths. Suppose we define the canonical path between any pair of sequences to be the path we will get by changing the bases that differ between them in the order they occur in the sequence. For example, the canonical path from ATCCAG to GACTAC would be

ATCCAG $\rightarrow$ GTCCAG $\rightarrow$ GACCAG $\rightarrow$ GACTAG $\rightarrow$ GACTAC.

This procedure establishes a full canonical path set Γ.

We then need to establish $\rho(\Gamma)$. In this case, we know that each node will have stationary probability $\pi_i = \frac{1}{4^n}$. We can therefore infer that the capacity of each edge, Q_{ij}, will be $\frac{1}{4^n} \times \frac{1}{6n}$ for all i and j. The hard part is figuring out which paths use any given edge. For any particular edge that flips the kth base of its sequence, the paths using that edge are those going from a starting node with the same suffix following base k as the edge's endpoints to an ending node with the same prefix before base k as the edge's endpoints. For example, if we consider the edge

$$A^{k-1}AA^{n-k} \to A^{k-1}TA^{n-k},$$

which moves from a sequence of n As to a sequence with a T in position k, then the canonical paths using that edge will be those of the form

$$N^{k-1}AA^{n-k} \to \cdots \to A^{k-1}TN^{n-k},$$

where an N stands for any base. We know this because we have chosen our paths such that we flip base k after we have set the first $k-1$ bases to their final values, but before we have made any changes to bases $k+1$ to n.

Suppose we now pick some edge (q_1, q_2) and evaluate its edge loading:

$$\rho = \frac{1}{Q_{12}} \sum_{i,j,e \in \gamma_{ij}} \pi_i \pi_j |\gamma_{ij}| = 6n \times 4^n \sum_{i,j,e \in \gamma_{ij}} \frac{1}{4^n} \times \frac{1}{4^n} \times |\gamma_{ij}|$$

$$= 6n \times 4^n \times \frac{1}{4^n} \times \frac{1}{4^n} \sum_{i,j,e \in \gamma_{ij}} |\gamma_{ij}| = \frac{6n}{4^n} \sum_{i,j,e \in \gamma_{ij}} |\gamma_{ij}|.$$

The number of paths of any given length m will correspond to the number of ways of choosing m_1 bases from the $k-1$ prefix bases of the first node in the path and $m - m_1$ bases from the $n - k$ suffix bases of the last node in the path, then choosing among the three possible ways to flip each base chosen. The choice of bases to flip is equivalent to simply picking m bases from a set of size $n - 1$, $\binom{n-1}{m}$, giving us the following sum:

$$\rho = \frac{6n}{4^n}\left(0 \times 1\binom{n-1}{0} + 1 \times 3\binom{n-1}{1} + 2 \times 3^2\binom{n-1}{2}\right.$$

$$\left. + \cdots + (n-1) \times 3^{n-1}\binom{n-1}{n-1}\right)$$

$$= \frac{6n}{4^n}\left(\sum_{i=0}^{n-1} i3^i\binom{n-1}{i}\right).$$

Since we are trying to bound the mixing time, we do not need to solve exactly for ρ. We just need to put a reasonably tight upper bound on it. We can do that by replacing the factor of i in the sum with an upper bound of $(n-1)$, which can then be pulled out of the sum as follows:

$$\rho \le \frac{6n(n-1)}{4^n}\sum_{i=0}^{n-1} 3^i\binom{n-1}{i}.$$

We can then evaluate the sum, using the binomial theorem. The binomial theorem says that $(a + b)^m = \sum_{i=0}^{m} \binom{m}{i} a^i b^{m-i}$ for any a and b and any integer m. Plugging in $a = 1$, $b = 3$, and $m = n - 1$, we then get

$$\rho \leq \frac{6n(n-1)}{4^n}(1+3)^{n-1}$$

$$= \frac{6n(n-1)}{4}.$$

Using this bound on ρ, we can put the following bound on the mixing time:

$$\tau_q(\varepsilon) \leq \frac{6n(n-1)}{4}(\ln(4^{-n}) + \ln(\varepsilon^{-1})) = O(n^3 + n^2 \ln(\varepsilon^{-1})).$$

In other words, we can show that a number of steps cubic in the number of bases is sufficient for this Markov chain to mix effectively. A chain that mixes in a number of steps polynomial in its parameters is called a *rapidly mixing* Markov chain.

10.3 The Conductance Method

If we have trouble proving a good mixing time with the canonical path method, we can try an alternative called the conductance method. Intuitively, the conductance method estimates how prone the model is to getting stuck in some subset of the states. If it is unlikely to be stuck in any subset of states for very long, then its mixing time must be short. More formally, we will define the conductance out of a given state set S to be

$$\phi(S) = \frac{\sum_{q_i \in S, q_j \notin S} Q_{ij}}{\Pi(S)},$$

where $\Pi(S) = \sum_{i, q_i \in S} \pi_i$. The conductance of the full Markov model is then defined to be the minimum conductance over all choices of S for which $\Pi(S)$ is at most $\frac{1}{2}$:

$$\Phi = \min_{S \subset Q, 0 \leq \Pi(S) \leq 0.5} \frac{\sum_{q_i \in S, q_j \notin S} Q_{ij}}{\Pi(S)}.$$

For example, given our Markov model from figure 10.1, the conductances we would get for different possible choices of S are as follows:

$S = \{q_1, q_2\}$:

$$\frac{Q_{14} + Q_{23}}{\pi_1 + \pi_2} = \frac{1/24 + 1/24}{1/3 + 1/6} = \frac{1/12}{1/2} = \frac{1}{6}$$

$S = \{q_2, q_4\}$:

$$\frac{Q_{12} + Q_{14} + Q_{23} + Q_{24}}{\pi_2 + \pi_4} = \frac{1/24 + 1/24 + 1/24 + 1/24}{1/6 + 1/6} = \frac{1/6}{1/3} = \frac{1}{2}$$

$S = \{q_1\}$:

$$\frac{Q_{12} + Q_{14}}{\pi_1} = \frac{1/24 + 1/24}{1/3} = \frac{1/12}{1/3} = \frac{1}{4}$$

$S = \{q_2\}$:

$$\frac{Q_{21} + Q_{23}}{\pi_2} = \frac{1/24 + 1/24}{1/6} = \frac{1/12}{1/6} = \frac{1}{2}.$$

The other cases will all be equivalent to one of these, so we can conclude that the conductance of the model is $\Phi = \frac{1}{6}$.

We can use the conductance to establish a bound on the mixing time by the following theorem:

Theorem   Given a finite, ergodic, reversible Markov chain with $p_{ii} \geq \frac{1}{2}$ for all i, which has conductance Φ, $\tau_{q_i}(\varepsilon) \leq 2\Phi^{-2}(\ln(\pi_i^{-1}) + \ln(\varepsilon^{-1}))$.

Thus, for the example system above, we can say that $\tau_{q_i}(\varepsilon) \leq 2 \times 36 \times (\ln(6) + \ln(\varepsilon^{-1})) = 72(\ln(6) + \ln(\varepsilon^{-1}))$. In this case, we get a slightly worse bound than we do using the canonical path method. In other cases, the conductance method may yield a tighter bound, depending on the specific Markov model and the canonical paths chosen.

Example   We will now examine how the conductance method might apply to a more involved example, a bounded random walk. A random walk is a kind of Markov process in which we assume we have a particle on some grid and we allow it to move randomly to adjacent grid points. Random walks are often used as models of diffusive processes, such as movement of a molecule under Brownian motion. In a bounded random walk, we assume that there are boundaries the particle cannot move beyond. For example, figure 10.2(a) shows a Markov model corresponding to a one-dimensional bounded random walk. We want to know how long we need to run this model in order to get it to mix sufficiently, uniformly randomizing the particle position. We will use the conductance method to establish a bound.

First, we need to see if our model fits the preconditions for these methods:

• *Ergodicity*   We can get from any position in the walk to any other by stepping between consecutive positions, so the model is ergodic.

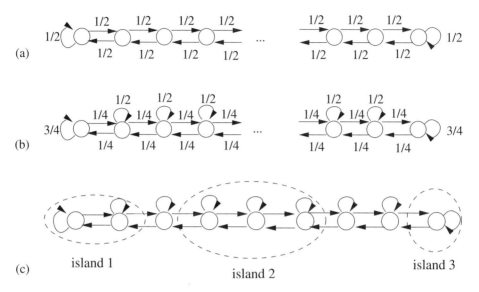

Figure 10.2
Markov model defining a one-dimension random walk. (a) The initial Markov model. (b) A modification of the model with the same stationary distribution but self-transition probabilities of at least $\frac{1}{2}$. (c) A possible choice of state subsets we need to consider in bounding the conductance Φ.

- *Detailed balance* The graph has no cycles, which means that it trivially satisfies the Kolmogorov criterion. In fact, any Markov model whose underlying graph is a tree (neglecting self-loops and directed edges) will satisfy the Kolmogorov criterion, and therefore will satisfy detailed balance.
- *Self-transitions have a probability of at least $\frac{1}{2}$* This condition is not satisfied for our model, so we will need to convert to a nearly equivalent model that does satisfy it. Suppose we cut all non-self-transition probabilities in half and then correct the self-probabilities accordingly. We will then get the model of figure 10.2(b), which has the same stationary distribution and exactly twice the mixing time of the model in figure 10.2(a). If we bound the mixing time of this modified model, then we will have shown the original model has a mixing time with half that bound.

We next need to find S minimizing $\frac{\sum_{q_i \in S, q_j \notin S} Q_{ij}}{\Pi(S)}$, which we can denote $\frac{Q(S,\bar{S})}{\Pi(S)}$. Suppose we try picking $S = \{q_1, \ldots, q_k\}$. Then we can leave the set only by the edge (q_k, q_{k+1}). All adjacent pairs of nodes in the model have transition probabilities $p_{ij} = p_{ji}$, which tells us that all states have the same stationary probability, $\frac{1}{n}$. Therefore, we can establish the capacity of the edge (q_k, q_{k+1}) to be $\pi_k p_{k,k+1} = \frac{1}{n} \times \frac{1}{4}$, since we cut non-self-transition probabilities in half. Furthermore, $\Pi(S)$ will be $\frac{k}{n}$. Thus,

$$\frac{Q(S,\bar{S})}{\Pi(S)} = \frac{1/n \times 1/4}{k/n} = \frac{1}{4k}$$

for $S = \{q_1, \ldots, q_k\}$.

This is a nice bound, but we cannot use it yet because we do not know if we chose the right S. With the canonical path method, we can choose our canonical paths, but with the conductance method we cannot choose our subset S. We need to find the S that gives us the minimum conductance. It may be, for example, that S is some arbitrary collection of "islands" of state space like that in figure 10.2(c). It may be hard to determine rigorously which of the possible configurations will give us minimum conductance.

Fortunately, though, we are only seeking to put an upper bound on mixing time, which requires only that we find a lower bound on the conductance. Suppose that we choose some arbitrary set of k "islands" containing m total states. Then our conductance will be the sum over the edges exiting from the endpoints of the islands. There must be at least $2k - 2$ such endpoints (in the worst case, our islands include the first and last node), and there cannot be zero if there is any state not in S. Each island endpoint contributes a value of $\pi_i p_{i,i+1} = \frac{1}{4n}$ to the conductance. Thus,

$$Q(S,\bar{S}) \geq \frac{2k-2}{4n}.$$

We need to consider $k = 1$ as a special case, or we will get a bound of zero. There must be at least one edge out of a single state, so for the $k = 1$ case

$$Q(S,\bar{S}) \geq \frac{1}{4n}.$$

Furthermore, $\Pi(S) = \frac{m}{n}$. Therefore, we can set the following lower bounds on the possible conductance Φ:

$$\begin{cases} \frac{Q(S,\bar{S})}{\Pi(S)} \geq \frac{2k-2}{4n} \times \frac{n}{m} = \frac{2k-2}{4m}, & k > 1 \\ \frac{Q(S,\bar{S})}{\Pi(S)} \geq \frac{1}{4n} \times \frac{n}{m} = \frac{1}{4m}, & k = 1 \end{cases}$$

We can minimize this by choosing $k = 1$ and setting m as large as possible. $\Pi(S)$ can be at most $\frac{1}{2}$, so the worst case is $m = \frac{n}{2}$. Choosing $k = 1$, $m = \frac{n}{2}$, we get that the conductance is bounded by

$$\Phi \geq \frac{1}{2n}.$$

We can then bound the mixing time as follows:

$$\tau_q(\varepsilon) \leq 2\left(\frac{1}{2n}\right)^{-2}\left(\ln\left(\frac{1}{n}\right)^{-1} + \ln(\varepsilon)^{-1}\right) = 8n^2\left(\ln\left(\frac{1}{n}\right)^{-1} + \ln(\varepsilon)^{-1}\right)$$

$$= O(n^2\ln n + n^2\ln\varepsilon^{-1}).$$

Recall that we slowed our Markov chain down by a factor of 2 in order to satisfy the condition of self-loop probabilities at least $\frac{1}{2}$, so our original chain should actually mix twice as fast as this. In any event, though, our chain is rapidly mixing, and we can show that its mixing time is at worst $O(n^2\log n)$.

In general, we will not know the stationary distribution for this kind of problem, so we will need to use bounds for that as well. As long as we can lower-bound conductance (or upper-bound maximum edge loading), we can put some sort of bound on the mixing time.

10.4   Final Comments

We have been using these theorems on the assumption that we are provided the Markov model and wish to analyze it. In practice, we usually design the model, and our task is to design one that is rapidly mixing. For example, we may know ratios of equilibrium values between states, as with thermodynamic equilibrium distributions, and will want to design a Markov model that rapidly approaches the right equilibrium. If we use a Metropolis model, then we get to choose the edge set, even though the transition probabilities are fixed for us, given the edge set. We can also apply importance sampling to adjust the transition probabilities for a given edge set. We can consider the bounding methods covered in this chapter as guidelines for how to make a good (i.e., rapidly mixing) model, rather than just a way to prove things about a model given to us. For example, if we design a model, we want to design it so we do not place too much load on any edge, avoiding bottleneck edges that can slow down mixing. We also want to design it so there are no "trapped" subsets of states that are hard to exit once the model enters them.

References and Further Study

The standard reference for this topic is a chapter in Hochbaum [56] written by two of the inventors of this field, Alisdair Sinclair and Mark Jerrum; it was the primary source in developing the material presented here. That chapter was the source of the mixing time notation and of the two theorems cited here establishing mixing time bounds from edge loading and conductance. Sinclair and Jerrum focus on mixing time bounds in the context of finding solutions to computationally intractable optimization and counting problems, but the techniques are nonetheless more broadly applicable. For example, they include an extended example using the techniques to study equilibria of monomer–dimer systems, which may be of direct interest to readers of this text.

Those interested in the primary literature can refer to Sinclair and Jerrum [115] for the conductance method and to Sinclair [116] for the canonical path method. There is also an extensive literature on uses of the methods for various specific applications, which may be useful to readers looking for additional illustrative examples of their use in practice. Hochbaum provides a good starting point for a search for these methods. The field has advanced considerably since that text was written, though, and a manual literature search may be necessary for those interested in learning the state of the art in the use of these and other mixing-time bounding techniques.

11 Continuous-Time Markov Models

So far, we have seen Markov models that describe purely discrete processes. That is, they move between discrete sets of states at discrete points in time. In this chapter, we will see a way of generalizing Markov models to consider at least continuous evolution of time, even though the state set is still discrete. A continuous-time Markov model (CTMM) moves through a state set like a standard Markov model, but allows the time per step to vary according to a continuous distribution. The time change between states can be thought of as either the time required to leave the current state or the time required to enter the next state. We will refer to this as a *waiting time*. The result is a class of model in which we can consider not just which sequence of steps we take, but also when we take them. As we will see in this chapter and chapters 12 and 17, these models are very useful in describing several important systems in biology, ranging from molecular evolution to reaction chemistry.

11.1 Definitions

In order to describe the evolution of Markov models over continuous time, we need to generalize the notion of a transition matrix to incorporate the times at which transitions occur. Instead of defining a constant transition probability p_{ij} as the probability of going from q_i to q_j, we instead have a function $p_{ij}(t)$ representing the probability of being in q_j at time t, given that we were in q_i at time 0. That is:

$$p_{ij}(t) = Pr\{q(t) = q_j \,|\, q(0) = q_i\}.$$

For a CTMM, the functions p_{ij} cannot be chosen arbitrarily, however. We want to define our models in such a way that they are "memoryless," meaning that what happens next in the model depends only on the current state and not on any prior state or on the current time. This is an extension of the standard Markov model property that what the model does next depends only on its current state. In other words,

$$p_{ij}(t) = Pr\{q(t) = q_j \,|\, q(0) = q_i\} = Pr\{q(s + t) = q_j \,|\, q(s) = q_i\} \quad \forall s > 0.$$

It turns out that this property can be satisfied only if the waiting times are described by exponential random variables. Thus the model will move through its states with some set of transition probabilities, just like a standard Markov model, but will also have attached to each transition an exponential random variable describing how long it needs to make the transition. The parameter of the exponential random variable and the probability of the transition will have to be related to one another, though, in ways that will become apparent shortly.

One way to think of how such a model works is to imagine each edge (i, j) having a characteristic rate λ_{ij}. If the model starts at state q_1, then we can characterize its behavior a short time later by considering how likely it is to move to any neighbor state j over a very short span of time. We define this probability, $p_{ij}(\Delta t)$ for $i \neq j$, to be $\lambda_{ij}\Delta t$ for sufficiently small Δt. This implies that $p_{ii}(\Delta t) = 1 - (\sum_{j \neq i} \lambda_{ij})(\Delta t)$. The waiting times for any nonzero Δt in this model will be geometrically distributed with probability $1 - p_{ii}(\Delta t)$. In the limit, however, as Δt goes to zero (shrinking the probability of movement and the time elapsed per step), these discrete geometric waiting time distributions will approach continuous exponential distributions.

This realization leads to an alternative but equivalent way to understand CTMMs. We can think of each possible transition out of a node as having its own waiting time distribution, an exponential random variable. This exponential random variable will have parameter λ_{ij}, the same λ_{ij} as in the previous representation. A move is accomplished by sampling from all of the waiting time distributions and choosing the smallest one. Suppose we are at state q_1 of our model and we examine the transitions out of that state. Then we can imagine that each neighbor state has an independent timer attached to it with its own characteristic rate. State q_2's timer waits an amount of time described by an $Exp(\lambda_{12})$ random variable, state q_3's timer waits an amount of time described by an $Exp(\lambda_{13})$ random variable, and state q_4's timer waits an amount of time described by an $Exp(\lambda_{14})$ random variable. When the first timer goes off, we move to the corresponding state and start a new set of timers based on transition rates out of the new state. That gives us a simple way to simulate a CTMM, shown in figure 11.1.

Example Suppose we have the CTMM of figure 11.2. Then we may start in state q_1. We sample two exponential random variables:

$$t_{12} \leftarrow Exp(\lambda_{12})$$

$$t_{14} \leftarrow Exp(\lambda_{14})$$

If $t_{12} < t_{14}$, we move to q_2 and update the time to t_{12}. Otherwise we move to q_4 and update the time to t_{14}.

Suppose we end up in state q_2 at time t_2. Then we sample waiting times for the two neighbors

```
1. t ← 0
2. qᵢ ← q₀
3. repeat
   A. t_min ← ∞
   B. for each transition (i, j) from qᵢ
      i. t_ij ← Exp(λ_ij)
      ii. if t_ij < t_min
          a. t_min ← t_ij
          b. q_next ← q_j
   C. qᵢ ← q_next
   D. t ← t + t_min
```

Figure 11.1
Pseudocode for simulating a CTMM.

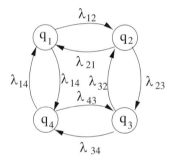

Figure 11.2
A CTMM with associated edge rates.

$$t_{21} \leftarrow Exp(\lambda_{21})$$

$$t_{23} \leftarrow Exp(\lambda_{23}).$$

If $t_{21} < t_{23}$, we move to q_1 and update the timer to $t_{12} + t_{21}$. Otherwise we move to q_3 and update the time $t_{12} + t_{23}$.

11.2   Properties of CTMMs

This second definition of CTMMs is also convenient for studying some of their properties. Suppose we are in state q_i and we have a set of k transitions out of q_i with rates $\lambda_1, \ldots, \lambda_k$. We can then ask some questions about how our CTMM will behave.

How long we can expect to remain in q_i?   We will leave q_i at a time described by the minimum of k exponential random variables. This minimum, $\min\{Exp(\lambda_1), Exp(\lambda_2), \ldots, Exp(\lambda_k)\}$, is itself a random variable and has a simple closed-form expression:

$$Pr\{\min\{Exp(\lambda_1), Exp(\lambda_2), \ldots, Exp(\lambda_k)\} > t\}$$

$$= Pr\{Exp(\lambda_1) > t \wedge Exp(\lambda_2) > t \wedge \cdots \wedge Exp(\lambda_k) > t\}.$$

Because the distributions are independent, we can say this is equal to the following:

$$Pr\{Exp(\lambda_1) > t\} \times Pr\{Exp(\lambda_2) > t\} \times \cdots \times Pr\{Exp(\lambda_k) > t\}$$

$$= e^{-\lambda_1 t} \times e^{-\lambda_2 t} \times \cdots \times e^{-\lambda_k t}$$

$$= e^{-(\lambda_1 + \lambda_2 + \cdots + \lambda_k)t}.$$

This final expression is equal to

$$Pr\{Exp(\lambda_1 + \lambda_2 + \cdots + \lambda_k) > t\},$$

which is exactly the distribution for an $Exp(\lambda_1 + \lambda_2 + \cdots + \lambda_k)$ random variable. That is,

$$\min\{Exp(\lambda_1), Exp(\lambda_2), \ldots, Exp(\lambda_k)\} = Exp\left(\sum_{j=1}^{k} \lambda_j\right).$$

So we know that the time we spend in state q_i is exponentially distributed with parameter $\sum_{j=1}^{k} \lambda_j$. The mean of an exponential with parameter λ is λ^{-1}, so we expect to spend $(\sum_{j=1}^{k} \lambda_j)^{-1}$ units of time in q_i before moving to some q_j.

What is the probability we will go to any given q_j next?   What we are asking is the probability that a particular $Exp(\lambda_j)$ has the minimum value over all of the exponential distributions we sample at a given step. We can first simplify a bit by noting that this is equivalent to asking the probability that $Exp(\lambda_j) < \min_{j' \neq j}\{Exp(\lambda_{j'})\} = Exp(\sum_{j' \neq j} \lambda_{j'})$. Let us define $\lambda^* = \sum_{j' \neq j} \lambda_{j'}$. Then we want to know $Pr\{Exp(\lambda_j) < Exp(\lambda^*)\}$, which we can evaluate as follows:

$$\int_0^\infty \left(\int_x^\infty \lambda^* e^{-\lambda^* y}\, dy\right) \lambda_j e^{-\lambda_j x}\, dx = \int_0^\infty (e^{-\lambda^* x}) \lambda_j e^{-\lambda_j x}\, dx$$

$$= \int_0^\infty \lambda_j e^{-(\lambda_j + \lambda^*)x}\, dx = \frac{\lambda_j}{\lambda_j + \lambda^*} = \frac{\lambda_j}{\lambda_1 + \cdots + \lambda_k}.$$

Thus, the probability of going to any given state j out of the k possible states is proportional to the rate of the transition to state j relative to the other rates.

Example   Let us use these facts to look at a real-world system we may model by a CTMM: particle interactions in a simple trimer system. Suppose we have a solution with three proteins in it: A, B, and C. The proteins are capable of forming a trimer, ABC, and can do this by any of three possible reaction pathways:

1. A binds B to form AB, then AB binds C to form ABC.
2. A binds C to form AC, then AC binds B to form ABC.
3. B binds C to form BC, then BC binds A to form ABC.

Let us suppose that we have a series of rate constants for each of these possible transitions: λ_{A+B}, λ_{A+C}, λ_{B+C}, λ_{AB+C}, λ_{AC+B}, and λ_{BC+A}. Then the time for that interaction to occur will be distributed as the minimum of the times for each of the three possible dimerization reactions to occur. In general, the time to the next reaction in such a system will be distributed as the minimum of the times to all possible next reactions. Thus, a CTMM provides a good description of such a system. This particular system is represented by the CTMM of figure 11.3.

Given this model, we can ask about some properties of the system. For example, what are the probabilities of following each of the pathways? We can evaluate these probabilities by noting that if we do not allow reversible bonds, then the pathway chosen is entirely determined by the first step in the pathway. The probability of choosing the first pathway is then the probability of taking transition $A + B \rightarrow AB$ first. This is given by

$$\frac{\lambda_{A+B}}{\lambda_{A+B} + \lambda_{A+C} + \lambda_{B+C}}.$$

Similarly, the probability of taking the second pathway is

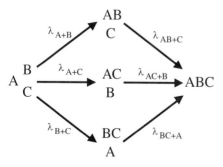

Figure 11.3
A CTMM model for a heterotrimer system.

$$\frac{\lambda_{A+C}}{\lambda_{A+B} + \lambda_{A+C} + \lambda_{B+C}},$$

and the probability of taking the third pathway is

$$\frac{\lambda_{B+C}}{\lambda_{A+B} + \lambda_{A+C} + \lambda_{B+C}}.$$

We can also ask what the overall expected rate of trimer formation for the model will be. The first step has time distributed as $Exp(\lambda_{A+B} + \lambda_{A+C} + \lambda_{B+C})$. If we take the first option, which happens with probability $\frac{\lambda_{A+B}}{\lambda_{A+B}+\lambda_{A+C}+\lambda_{B+C}}$, then the remaining time required will be described by an $Exp(\lambda_{AB+C})$ random variable. Likewise, if we take the second option, which happens with probability $\frac{\lambda_{A+C}}{\lambda_{A+B}+\lambda_{A+C}+\lambda_{B+C}}$, then the remaining time required will be described by an $Exp(\lambda_{AC+B})$ random variable; and if we take the third option, which happens with probability $\frac{\lambda_{B+C}}{\lambda_{A+B}+\lambda_{A+C}+\lambda_{B+C}}$, then the remaining time required will be described by an $Exp(\lambda_{BC+A})$ random variable. Putting this together, the total expected time for the trimer formation is described by

$$E[Exp(\lambda_{A+B} + \lambda_{A+C} + \lambda_{B+C})] + \frac{\lambda_{A+B}}{\lambda_{A+B} + \lambda_{A+C} + \lambda_{B+C}} E[Exp(\lambda_{AB+C})]$$

$$+ \frac{\lambda_{A+C}}{\lambda_{A+B} + \lambda_{A+C} + \lambda_{B+C}} E[Exp(\lambda_{AC+B})] + \frac{\lambda_{B+C}}{\lambda_{A+B} + \lambda_{A+C} + \lambda_{B+C}} E[Exp(\lambda_{BC+A})]$$

$$= \frac{1}{\lambda_{A+B} + \lambda_{A+C} + \lambda_{B+C}} + \frac{\lambda_{A+B}}{(\lambda_{A+B} + \lambda_{A+C} + \lambda_{B+C})\lambda_{AB+C}}$$

$$+ \frac{\lambda_{A+C}}{(\lambda_{A+B} + \lambda_{A+C} + \lambda_{B+C})\lambda_{AC+B}} + \frac{\lambda_{B+C}}{(\lambda_{A+B} + \lambda_{A+C} + \lambda_{B+C})\lambda_{BC+A}}$$

$$= \frac{\lambda_{AB+C}\lambda_{AC+B}\lambda_{BC+A} + \lambda_{A+B}\lambda_{AC+B}\lambda_{BC+A} + \lambda_{A+C}\lambda_{AB+C}\lambda_{BC+A} + \lambda_{B+C}\lambda_{AB+C}\lambda_{AC+B}}{\lambda_{A+B} + \lambda_{A+C} + \lambda_{B+C}}.$$

That tells us that the overall rate of the reaction is

$$\frac{\lambda_{A+B} + \lambda_{A+C} + \lambda_{B+C}}{\lambda_{AB+C}\lambda_{AC+B}\lambda_{BC+A} + \lambda_{A+B}\lambda_{AC+B}\lambda_{BC+A} + \lambda_{A+C}\lambda_{AB+C}\lambda_{BC+A} + \lambda_{B+C}\lambda_{AB+C}\lambda_{AC+B}}.$$

11.3 The Kolmogorov Equations

The sort of analysis we did above will help us learn some basic properties of relatively simple CTMMs, but there is a more general way to evaluate the time evolution of a CTMM, using the *Kolmogorov equations*. To present them, we first need to define some terms. Assume we have a CTMM with n states, with rate λ_{ij} between any pair of states q_i and q_j. Further define for each state q_i a self-transition rate λ_{ii}, where

$$\lambda_{ii} = -\sum_{j \neq i} \lambda_{ij}.$$

Finally, define

$$p_{ij}(t) = Pr\{q(t) = q_j \mid q(0) = q_i\}$$

as at the beginning of the chapter. Then the time evolution of the Markov model is described by the following sets of differential equations:

$$\frac{dp_{ij}(t)}{dt} = \sum_{k=1}^{n} p_{ik}(t)\lambda_{kj} \quad \text{(forward Kolmogorov equations)}$$

or, equivalently,

$$\frac{dp_{ij}(t)}{dt} = \sum_{k=1}^{n} \lambda_{ik} p_{kj}(t) \quad \text{(backward Kolmogorov equations)}.$$

Essentially, the forward Kolmogorov equations say that the rate at which the model moves into state q_j is determined by the sum over all intermediate states q_k of the probability of being in q_k times the instantaneous rate of movement from q_k to q_j. The backward Kolmogorov equations say that the rate of movement from q_i to q_j is the sum over instantaneous rates of movement from q_i to any intermediate state q_k times the probability of moving from q_k to q_j in time t. The Kolmogorov equations are essentially a continuous-time extension of the Chapman–Kolmogorov equations we saw in the context of discrete-time Markov models.

The Kolmogorov equations can be represented more concisely in a matrix format:

$$\frac{dP(t)}{dt} = \Lambda P(t),$$

where $P(t)$ is a matrix whose entries are the $p_{ij}(t)$ functions and Λ is the transition rate matrix, where entry (i, j) is λ_{ij}. In the scalar case,

$$\frac{dp(t)}{dt} = \lambda p(t)$$

is solved by

$$p(t) = p(0)e^{\lambda t}.$$

In the matrix case, the system is similarly solved by

$$P(t) = P(0)e^{\Lambda t},$$

except that we have to define what it means to use a matrix as an exponent. We define $e^{\Lambda t}$ for matrix Λ using the Taylor series for e^x centered on $x = 0$:

$$e^{\Lambda t} = \sum_{i=0}^{\infty} \frac{(\Lambda t)^i}{i!}.$$

Therefore, the solution to the Kolmogorov equations is

$$P(t) = P(0)\sum_{i=0}^{\infty} \frac{(\Lambda t)^i}{i!}.$$

Unfortunately, that summation is generally going to be difficult to evaluate. We can approximate the series using the following identity:

$$e^{\Lambda t} = \lim_{n \to \infty}\left(I + \Lambda\frac{t}{n}\right)^n.$$

Plugging in a sufficiently large n will give a reasonable estimate of the distribution at any given point in time. We may also want to numerically integrate the Kolmogorov equations, a topic we will cover in a subsequent chapter. It may also be possible to work with the model more efficiently by using more advanced linear algebra concepts than we assume in this text. Nonetheless, we can sometimes get a useful closed-form expression directly from the Kolmogorov equations, as we will see from the following example.

Example   Proline cis-trans isomerization. The amino acid proline is unusual in that its side chain loops back from the alpha carbon to connect to its amino group. (It is therefore technically an *imino acid*, not an *amino acid*.) It is possible for proline to take on two isomeric forms, depending on whether the loop is pointing in the same direction as or in the opposite direction from the carboxyl oxygen. These are called the cis and trans isomers, and are illustrated in figure 11.4.

The trans isomer is favored by a factor of about 1000:1 thermodynamically, but there is a high energy barrier to the conversion and it therefore happens slowly. This isomerization is the rate-limiting step in some protein-folding reactions; the protein cannot fold properly until the prolines all shift into the right isomers. We can represent the behavior of a single proline as a CTMM by declaring that state q_0 is the cis isomer and q_1 is the trans isomer. We can then define $\lambda_f = \lambda_{01}$ to be the rate

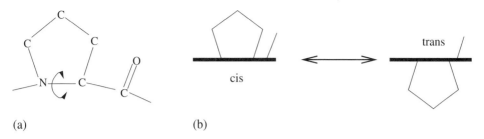

Figure 11.4
Cis and trans isomerization of proline. (a) Molecular model showing the proline ring. (b) Model of the cis and trans states.

of cis-trans isomerization and $\lambda_r = \lambda_{10}$ to be the rate of trans-cis isomerization. We can then ask, If we start out in the trans state, what is the probability we are still in the trans state at some arbitrary time t? Or, if we start out in the cis state, what is the probability we have moved and are in the trans state at time t? We can answer these questions with the Kolmogorov equations.

First, we need to know the transition rate parameters of the model. The non-self-transitions are already known to us:

$$\lambda_{01} = \lambda_f$$

$$\lambda_{10} = \lambda_r.$$

The self-transitions are defined to be $\lambda_{ii} = -\sum_{j \neq i} \lambda_{ij}$, giving us the following:

$$\lambda_{00} = -\lambda_f$$

$$\lambda_{11} = -\lambda_r.$$

These values give us the following forward Kolmogorov equations:

$$\frac{dp_{00}(t)}{dt} = p_{00}(t)\lambda_{00} + p_{01}(t)\lambda_{10} = -\lambda_f p_{00}(t) + \lambda_r p_{01}(t)$$

$$\frac{dp_{01}(t)}{dt} = p_{00}(t)\lambda_{01} + p_{01}(t)\lambda_{11} = \lambda_f p_{00}(t) - \lambda_r p_{01}(t)$$

$$\frac{dp_{10}(t)}{dt} = p_{10}(t)\lambda_{00} + p_{11}(t)\lambda_{10} = -\lambda_f p_{10}(t) + \lambda_r p_{11}(t)$$

$$\frac{dp_{11}(t)}{dt} = p_{10}(t)\lambda_{01} + p_{11}(t)\lambda_{11} = \lambda_f p_{10}(t) - \lambda_r p_{11}(t).$$

If we look at two of these in isolation, we can simplify a bit:

$$\frac{dp_{00}(t)}{dt} = -\lambda_f p_{00}(t) + \lambda_r p_{01}(t) = -\lambda_f p_{00}(t) + \lambda_r(1 - p_{00}(t)) = -(\lambda_f + \lambda_r)p_{00}(t) + \lambda_r$$

$$\frac{dp_{11}(t)}{dt} = \lambda_f p_{10}(t) - \lambda_r p_{11}(t) = \lambda_f(1 - p_{11}(t)) - \lambda_r p_{11}(t) = -(\lambda_f + \lambda_r)p_{11}(t) + \lambda_f.$$

In the above forms, we can solve for these two functions. In general, $\frac{dy}{dt} = ay + b$ is solved by $y(t) = Ce^{at} - \frac{b}{a}$. Plugging into that form gives us

$$p_{00}(t) = C_0 e^{-(\lambda_f + \lambda_r)t} + \frac{\lambda_r}{\lambda_f + \lambda_r}$$

$$p_{11}(t) = C_1 e^{-(\lambda_f + \lambda_r)t} + \frac{\lambda_f}{\lambda_f + \lambda_r}.$$

If we then use the initial condition that $p_{00}(0) = p_{11}(0) = 1$ to set C_0 and C_1, we get the following:

$$p_{00}(t) = \left(1 - \frac{\lambda_r}{\lambda_f + \lambda_r}\right)e^{-(\lambda_f + \lambda_r)t} + \frac{\lambda_r}{\lambda_f + \lambda_r}$$

$$p_{11}(t) = \left(1 - \frac{\lambda_f}{\lambda_f + \lambda_r}\right)e^{-(\lambda_f + \lambda_r)t} + \frac{\lambda_f}{\lambda_f + \lambda_r}.$$

And since $p_{01} = 1 - p_{00}$ and $p_{10} = 1 - p_{11}$:

$$p_{01}(t) = \left(1 - \frac{\lambda_r}{\lambda_f + \lambda_r}\right) - \left(1 - \frac{\lambda_r}{\lambda_f + \lambda_r}\right)e^{-(\lambda_f + \lambda_r)t} = \frac{\lambda_f}{\lambda_f + \lambda_r}\left(1 - e^{-(\lambda_f + \lambda_r)t}\right)$$

$$p_{10}(t) = \left(1 - \frac{\lambda_f}{\lambda_f + \lambda_r}\right) - \left(1 - \frac{\lambda_f}{\lambda_f + \lambda_r}\right)e^{-(\lambda_f + \lambda_r)t} = \frac{\lambda_r}{\lambda_f + \lambda_r}\left(1 - e^{-(\lambda_f + \lambda_r)t}\right).$$

We can therefore determine the exact distribution of the system at any point in time from its initial distribution. If we take the limit as t goes to infinity, these equations also tell us that at equilibrium, the states are occupied proportionally to their transition rates from one to another, no matter where we start, which is as we would expect. Knowing the thermodynamic equilibrium thus tells us the ratios of the rates to one another, although it cannot tell us the actual values of the rates.

We can extend this kind of model to a model of the states of N prolines in a larger protein. We then need 2^N states to represent the isomerizations of all N prolines. The Kolmogorov equations can once again be used to analyze the time evolution of the model, although the math becomes somewhat more involved.

References and Further Study

There are many places one can learn more about continuous-time Markov models. Two texts by Ross that we have previously encountered [103], [102], as well as a third we have not yet seen [117], provide excellent coverage of the topic. Some of the classic references for the topic include Doob [118], Cox and Miller [119], Karlin [120], and Resnick [121]. We have only scratched the surface of the theory of these models here, and these references will provide greater depth. Readers may benefit from seeing coverage of this material specifically for scientific applications. Three options for more applied treatments of this material are Wilkinson [110], van Kampen [122], and Benedek and Villars [123]. CTMM models have many applications beyond biological modeling, including modeling in other natural sciences, finance, and computer systems, and readers interested in greater depth in these areas can turn to the literatures for modeling in those fields as well.

12 Case Study: Molecular Evolution

This is the second of our case study chapters, in which we look at how some of the methods we have been studying have been applied to real-world systems in biology. We will specifically look at some of the ways Markov models are used in studies of molecular evolution. Markov models show up in many contexts in understanding molecular evolution, from the level of single DNA bases up to whole populations of organisms.

12.1 DNA Base Evolution

At a small scale, Markov models provide a way to describe how a single DNA base behaves across multiple generations. Suppose we isolate one single DNA base in a single-cell organism, then ask what base is found in the corresponding position of each descendant of that organism along a single evolutionary lineage. That is, we look at the corresponding base in one child of the original organism, one child of that child, one child of that grandchild, and so on. For simplicity, we will assume that the base is not under any selective pressure, meaning that there is no survival advantage to having one allele (base value) rather than another. Usually, if we look one or a few generations later, the ancestor and the descendant will have the same allele. Many generations later, though, the base in the descendant will be independent of that in the ancestor. We want to understand how we move between these two extremes.

12.1.1 The Jukes–Cantor (One-Parameter) Model

One way to study this problem is to treat it as a purely discrete process. Each generation, we have a new organism with its own base. It would seem reasonable to assume that the base in one generation depends only on the base in the immediately previous generation. The sequence of bases over a single line from ancestor to descendants is then a Markov chain. The Markov model has four states corresponding to the four possible bases, which we can call q_A, q_C, q_G, and q_T.

   The simplest such model of base evolution assumes that on each generation we are equally likely to change the current base to any of the three remaining possibilities. This is known as the *Jukes–Cantor model*. The Jukes–Cantor model is described by a single parameter, λ. In terms of this parameter, the model has the following transition matrix:

$$M = \begin{bmatrix} 1 - 3\lambda & \lambda & \lambda & \lambda \\ \lambda & 1 - 3\lambda & \lambda & \lambda \\ \lambda & \lambda & 1 - 3\lambda & \lambda \\ \lambda & \lambda & \lambda & 1 - 3\lambda \end{bmatrix}$$

In other words, we have probability λ of moving to any particular different base and $1 - 3\lambda$ of staying with the current base. This matrix gives us the Markov model graph of figure 12.1(a).

   Once we have a Markov model describing the base evolution process, we can apply our tools for analyzing Markov models to understand how this model behaves over long periods of time. For example, we can look at the eigenvalues and eigenvectors of the matrix:

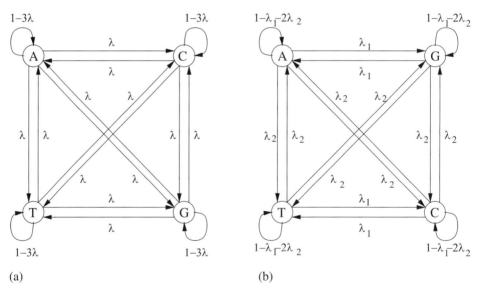

(a)   (b)

Figure 12.1
Markov graphs for the (a) Jukes–Cantor and (b) Kimura models of base evolution.

$$
x_1 = \begin{bmatrix} 1/4 \\ 1/4 \\ 1/4 \\ 1/4 \end{bmatrix}, \quad x_2 = \begin{bmatrix} 1/2 \\ -1/2 \\ 0 \\ 0 \end{bmatrix}, \quad x_3 = \begin{bmatrix} 1/4 \\ 1/4 \\ -1/2 \\ 0 \end{bmatrix}, \quad x_4 = \begin{bmatrix} 1/6 \\ 1/6 \\ 1/6 \\ -1/2 \end{bmatrix},
$$

$$
\lambda_1 = 1, \quad \lambda_2 = 1 - 4\lambda, \quad \lambda_3 = 1 - 4\lambda, \quad \lambda_4 = 1 - 4\lambda.
$$

We can then determine the probability of observing any given base after k generations by explaining the initial distribution in terms of eigenvectors:

$$
M^k \begin{bmatrix} 1 \\ 0 \\ 0 \\ 0 \end{bmatrix} = 1^k \begin{bmatrix} 1/4 \\ 1/4 \\ 1/4 \\ 1/4 \end{bmatrix} + (1 - 4\lambda)^k \begin{bmatrix} 1/2 \\ -1/2 \\ 0 \\ 0 \end{bmatrix}
$$

$$
+ \frac{2}{3}(1 - 4\lambda)^k \begin{bmatrix} 1/4 \\ 1/4 \\ -1/2 \\ 0 \end{bmatrix} + \frac{1}{2}(1 - 4\lambda)^k \begin{bmatrix} 1/6 \\ 1/6 \\ 1/6 \\ -1/2 \end{bmatrix}
$$

We can then infer that as k goes to infinity, all the components except the first will die away and the model will approach an equal probability of each base. We can also determine the rate at which the transients die away from the preceding solution. All eigenvalues but the first are $1 - 4\lambda$, so the transients should decay geometrically by a factor of $1 - 4\lambda$ per generation.

For a process like this, where the probability of base mutation per generation is very small (perhaps 10^{-9}), a discrete model is not usually particularly appropriate because so many generations are needed to move away from the starting state. We can therefore also consider a continuous version of the Jukes–Cantor model. For example, we may say that one "unit" of time is 10^6 generations, and then we can mostly ignore the fact that the process becomes discrete if we look at extremely small numbers of time units.

In this variant, we say that there is an instantaneous rate λ of moving from any base to any other, giving us a continuous-time Markov model. This assumption then fixes our self-transition rates at -3λ. Given this model, we can use the Kolmogorov equations to determine the behavior over long time scales. For example, we can find the probability that a base that was initially A remains A at time t as follows:

$$
\frac{dp_{AA}}{dt} = -3\lambda p_{AA} + \lambda p_{AC} + \lambda p_{AG} + \lambda p_{AT}
$$

$$
= -3\lambda p_{AA} + \lambda(p_{AC} + p_{AG} + p_{AT})
$$

$$= -3\lambda p_{AA} + \lambda(1 - p_{AA})$$

$$= -4\lambda p_{AA} + \lambda.$$

This differential equation is solved by

$$p_{AA}(t) = \frac{1}{4} + Ce^{-4\lambda t}.$$

Applying the initial condition $p_{AA}(0) = 1$ gives us

$$p_{AA} = \frac{1}{4} + \frac{3}{4}e^{-4\lambda t}.$$

Thus, we can conclude that the model approaches an equilibrium with a $\frac{1}{4}$ stationary probability of being in q_A and that it approaches this equilibrium with a rate of 4λ. By symmetry, p_{CC}, p_{GG}, and p_{TT} will behave identically. Similarly, the probability that A evolves into any other particular base, say T, at time t can be determined as follows:

$$\frac{dp_{AT}}{dt} = \lambda p_{AA} + \lambda p_{AC} + \lambda p_{AG} - 3\lambda p_{AT}$$

$$= \lambda(1 - p_{AT}) - 3\lambda p_{AT}$$

$$= -4\lambda p_{AT} + \lambda.$$

This differential equation has the solution

$$p_{AT}(t) = \frac{1}{4} + Ce^{-4\lambda t}.$$

Applying the initial condition $p_{AT}(0) = 0$ then yields

$$p_{AT}(t) = \frac{1}{4} - \frac{1}{4}e^{-4\lambda t}.$$

By symmetry, any p_{ij} where $i \neq j$ will behave identically, approaching an equilibrium probability of $\frac{1}{4}$ with rate 4λ. Figure 12.2 plots these equations for $\lambda = 1$. All probabilities converge on $\frac{1}{4}$, with p_{ii} decaying from 1 to $\frac{1}{4}$ and p_{ij} increasing from 0 to $\frac{1}{4}$, both at rate 4λ.

12.1.2   Kimura (Two-Parameter) Model

As we can see, Jukes–Cantor is a very tractable model mathematically, but it is not very realistic biologically. There are many ways in which real molecular evolution

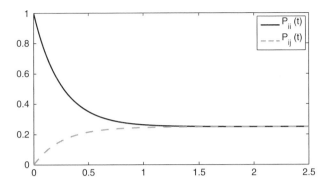

Figure 12.2
Time progress of $p_{ii}(t)$ and $p_{ij}(t)$ as a function of time for any base i and base $j \neq i$ in the continuous Jukes–Cantor model.

differs from the simple assumptions of Jukes–Cantor. For example, bases are often under selection. They may have very different mutation rates from one another, depending on local sequence context. They may have a bias toward high GC or high AT content, depending on organism or sequence type. Addressing these problems, though, will require a lot of other information that may not be available to us. One flaw in the Jukes–Cantor model that can, however, be addressed without too much case-specific knowledge is that a base is not equally likely to mutate into each of the three other possibilities. In particular, mutations of bases are classified into two types: *transitions*, which involve either A $\leftrightarrow$ G or C $\leftrightarrow$ T conversions, and *transversions*, which involve any other conversions. Transitions occur at a significantly higher rate than transversions, although the ratio varies between different types of DNA. For example, animal nuclear DNA has about a 2:1 ratio, whereas mitochondrial DNA has about a 20:1 ratio.

We can capture this bias for transitions over transversions with the Kimura model, which uses two separate rates, λ_1 for transitions and λ_2 for transversions. The discrete Kimura model has the following transition matrix:

$$
\begin{array}{c}
\\
A \\
G \\
C \\
T
\end{array}
\begin{array}{cccc}
A & G & C & T \\
\left[\begin{array}{cccc}
1 - \lambda_1 - 2\lambda_2 & \lambda_1 & \lambda_2 & \lambda_2 \\
\lambda_1 & 1 - \lambda_1 - 2\lambda_2 & \lambda_2 & \lambda_2 \\
\lambda_2 & \lambda_2 & 1 - \lambda_1 - 2\lambda_2 & \lambda_1 \\
\lambda_2 & \lambda_2 & \lambda_1 & 1 - \lambda_1 - 2\lambda_2
\end{array}\right]
\end{array}
$$

This matrix gives rise to the graph of figure 12.1(b).

As with the Jukes–Cantor model, we can use powers of the matrix to find the mutation probabilities after multiple generations. For long time scales, we can use also use a continuous version of the Kimura model much like the one we saw for the Jukes–Cantor model. The continuous Kimura model has the following rates:

$$\lambda_{AG} = \lambda_{GA} = \lambda_{CT} = \lambda_{TC} = \lambda_1$$

$$\lambda_{AA} = \lambda_{GG} = \lambda_{CC} = \lambda_{TT} = -\lambda_1 - 2\lambda_2$$

$$\lambda_{ij} = \lambda_2 \quad \text{otherwise.}$$

We can again analyze the long-term behavior of the model using the Kolmogorov equations. The Kolmogorov equations for conversions starting from A are as follows:

$$\frac{dp_{AA}}{dt} = (-\lambda_1 - 2\lambda_2)p_{AA} + \lambda_1 p_{AG} + \lambda_2(p_{AC} + p_{AT})$$

$$\frac{dp_{AG}}{dt} = \lambda_1 p_{AA} + \lambda_2(p_{AC} + p_{AT}) + (-\lambda_1 - 2\lambda_2)p_{AG}$$

$$\frac{dp_{AC}}{dt} = \lambda_2(p_{AA} + p_{AG}) + \lambda_1 p_{AT} + (-\lambda_1 - 2\lambda_2)p_{AC}$$

$$\frac{dp_{AT}}{dt} = \lambda_2(p_{AA} + p_{AG}) + \lambda_1 p_{AC} + (-\lambda_1 - 2\lambda_2)p_{AT}.$$

Equations for other starting bases would proceed similarly.

Solving for these equations analytically requires more advanced math than we cover in this text. Essentially, we have to find the eigenvalues of the transition matrix and then guess solutions of the form $\sum c_i e^{\lambda_i t}$. We will not go through how this is done here, but the results are the following:

$$p_{AA} = \frac{1}{4} + \frac{1}{4}e^{-4\lambda_2 t} + \frac{1}{2}e^{-2(\lambda_1 + \lambda_2)t}$$

$$p_{AG} = \frac{1}{4} + \frac{1}{4}e^{-4\lambda_2 t} - \frac{1}{2}e^{-2(\lambda_1 + \lambda_2)t}$$

$$p_{AT} = p_{AC} = \frac{1}{4} - \frac{1}{4}e^{-4\lambda_2 t}.$$

Again, solutions for the other starting bases will follow by symmetry.

The probability distribution thus evolves toward equilibrium at two rates: a fast rate, $2(\lambda_1 + \lambda_2)$, at which we establish an equilibrium between the original base and its transition base, and a slower rate, $4\lambda_2$, at which we establish an equilibrium with the other two bases. Note that when $\lambda_1 = \lambda_2$, both rates are $4\lambda_1$, just as we would expect from our analysis of the Jukes–Cantor model.

12.2 Simulating a Strand of DNA

If we trace a single evolutionary lineage, it is likely we are concerned about more than just a single mutating base. We want a way of simulating mutations in a segment of many DNA bases. One way to simulate multiple bases is to simulate each independently, according to its own Kimura or Jukes–Cantor Markov model. For example, if we have the starting strand $AAAAA$, we may represent the homologous strand in descendant organisms by assuming it is produced by five independent Markov models, each with initial state q_A. We can then sample independently from each model at any time point at which we want to choose a random descendant.

We can, however, exploit some properties of the probability models we are using to derive another model that can be more efficient, especially for simulating large pieces of DNA over relatively short time scales. One of the consequences of the continuous Kimura and Jukes–Cantor models is that the time until a mutation appears in any particular base will be described by an exponential random variable. This means that the time until a mutation first appears anywhere in a set of n bases is also exponential, with n times the rate of the single-base model. A process in which some event occurs repeatedly, with the time between events described by independent identically distributed (i.i.d.) exponential random variables, is known as a *Poisson process*. The number of events occurring in any span of time t in a Poisson process with waiting time parameter λ is described by a Poisson random variable with parameter λt. Therefore, if bases accumulate mutations at rate λ per generation per base, then the number of mutations in k generations for n bases will be Poisson-distributed with parameter $\lambda k n$. We cannot quite consider that to be a good model of the number of mutations in a short strand, though, because a base can mutate more than once. We can get a base that mutates from A to C to T, which will give us only one change even though there are two mutation events. Or we can get A to T to A, which will give us no changes at all.

Suppose, though, that we assume the limit of a large number of bases and a low probability of mutation. We assume that we have some new parameter λ^*, which is the limit of λn as $\lambda \to 0$ and $n \to \infty$. In this case, the probability of a mutation occurring twice in the same base will approach zero. We can therefore treat the number of mutations accumulated in a particular span of time t as purely Poisson-distributed with parameter $\lambda^* t$. Furthermore, we can treat our DNA strand as a continuous segment running from, say, position 0 to position 1. Then, as long as we assume the same mutation rate across the strand, the mutations will be equally likely to occur anywhere on the strand. So if we sample our *Poisson*($\lambda^* t$) random variable and find that there are k mutations, we can simply pick k $U[0, 1]$ uniform random numbers to represent their positions. This model, in which we assume we have so many bases that multiple mutations in the same base never happen, is called the *infinite sites model*.

12.3 Sampling from Whole Populations

So far, we have been discussing how to simulate DNA changes along a single evolutionary line. But what if we want to study changes throughout a population? We can still use the methods we just learned to simulate molecular evolution along any particular line, but we need to embed those DNA-level models into a model of the behavior of the population. It turns out that this population-level behavior can also be described by Markov models. We will see here how to derive a particular very useful Markov model, called the *coalescent*, widely used in population genetics to simulate possible population histories.

Before we can describe the model, though, we need to make some assumptions. In particular, we will make a collection of assumptions called the *Wright–Fisher neutral model*. The model assumes the following:

1. *Discrete generations*   a population goes through distinct generations where every organism in generation i has parents only from generation $i - 1$.
2. *Random mating*   each organism in a generation selects its parent(s) uniformly at random from the previous generation and independently from all other organisms in its own generation.
3. *No selection*   no organism is more likely than any other to survive and reproduce.
4. *Random mutations*   mutations accumulate with equal probability in all bases at all times.

None of these assumptions is exactly true, but they can be a good model for studying the evolution of selectively neutral bases. For the moment, we will also assume that we are looking at a haploid organism and that we have a fixed population size throughout time. (We will see how to discard those assumptions later in the chapter.)

Our goal will be to create a model in which we can sample a collection of k individuals from a population of size N after the population has evolved for t generations, where generally $N \gg k$. This model is meant to simulate what we will observe if we sequence the DNA of a small number of individuals from a large population. We will eventually see the coalescent method for this problem. But in order to explain how the coalescent works, it will be helpful to work through two simpler models that we will not actually want to use in practice.

First, imagine that we simulate a population by simulating every individual in that population for t generations, then picking k individuals at random from the final generation as our sample. The model will work as follows:

1. Start with a group of N *founders* with random DNA sequences to fill the first generation.
2. Pick N random members of the current generation, allowing repeats, to be the parents of the next generation.
3. Mutate each chosen parent according to whatever mutation model we want to create the next generation.
4. Return to step 2 until we have reached t generations.
5. Choose a sample of k individuals at random from the final generation and return their sequences.

This method is illustrated by figure 12.3(a).

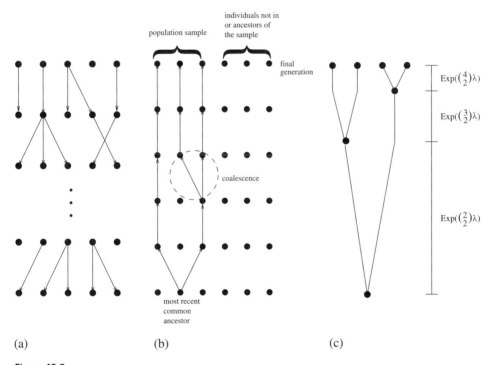

(a)   (b)   (c)

Figure 12.3
Derivation of the coalescent model. (a) Initial description of a discrete simulation forward in time from a population of N founders to the final population from which samples are drawn. (b) Simulation backward in time from the sample of k to the ancestors of those k. (c) Replacement of discrete sampling steps with continuous-time coalescence steps.

This first method is a fine approach in the sense that it is a correct simulation of our model. But it is also very inefficient. The runtime will vary linearly with the population size and number of generations. We will, for example, probably be unable to simulate a bacterial population with perhaps 10^{12} members over thousands of generations—realistic numbers for studying molecular evolution in a real laboratory setting.

We can improve on the approach by making two realizations. First, we really do not care about anyone in the final generation who is not one of the k we sample. We also do not care about anyone in any previous generation who is not an ancestor of one of those k, anyone in the generation before that who is not an ancestor of one of those ancestors, and so on. We can use these insights to create a more efficient simulation by essentially turning the problem upside down. Instead of starting with the founders and working forward in time, we can start with the k individuals we care about in the final generation and then go backward in time to figure out from whom they are descended in the previous generations.

In this revised model, we only need to represent k people in the final generation. Furthermore, we need at most k in any previous generation because the k in the last generation cannot have more than k ancestors in any other generation. In fact, each time that two members of generation i happen to share an ancestor in generation $i - 1$, the number of individuals we need to keep track of goes down by 1 for all prior generations. This process by which two lineages converge into one when they share a common ancestor is called *coalescence*. If we have a population of size N, then the probability that any two given lineages will coalesce in the previous generation is $\frac{1}{N}$. When all the lineages coalesce into one, called the most recent common ancestor (MRCA), we can generate a random DNA strand for that one MRCA, who will be a common ancestor of everyone in our sample, and then go back down the tree, simulating the mutations acquired in each subsequent generation. Figure 12.3(b) illustrates this revised model, tracing a lineage from the final generation (top) down to the point where every lineage has coalesced into one.

This approach will generally be much more practical than our first attempt, since we only have to simulate organisms and mutations in direct evolutionary paths to the members of our final sample from their common ancestor. The runtime will therefore depend on k and t, but not on N, except indirectly through N's influence on the number of mutations observed.

We can actually do even better, though, at least approximately, by noting a few things:

$$Pr\{\text{coalesce 2 given lineages in one generation}\} = \frac{1}{N};$$

therefore

---

1. Start with a sample of size $k = k_0$ in the final generation.
2. While $k > 1$
   A. Sample the time to the next coalescence, $t_i = Exp\left(\binom{k}{2}\right)$ in scaled time.
   B. Choose two lineages to coalesce uniformly at random and decrement k.
3. Choose a random sequence for the MRCA and then go forward in time, inserting mutations into each lineage based on the amount of scaled time elapsed on each path.

Figure 12.4
Simplified pseudocode for the basic coalescent method for simulating a population sample of size k.

$$Pr\{\text{do not coalesce 2 given lineages in one generation}\} = 1 - \frac{1}{N}$$

and

$$Pr\{\text{do not coalesce 2 given lineages in } N\tau \text{ generations}\} = \left(1 - \frac{1}{N}\right)^{N\tau}.$$

$\left(1 - \frac{1}{N}\right)^{N\tau}$ is approximately $e^{-\tau}$ for large N. This means that we can treat coalescences as an approximately exponential process with parameter N. In fact, it is often easier to adopt a notion called "scaled time" and sample coalescence times from a distribution with parameter 1, then later scale the number of generations elapsed by a factor of N. These observations lead to a third attempt at simulating a population, which is illustrated in figure 12.3(c) and presented as pseudocode in figure 12.4. In this revised model, we simulate the sample set, as in our second attempt, but jump directly to successively coalescence events rather than simulating all the noncoalescing generations in between.

This third version is the full coalescent model. The runtime of this method is independent of N (except in the number of mutations observed) and does not require a parameter t. Simulating the coalescences requires time dependent only on k. Because it jumps between discrete states with exponentially distributed times, we can represent the coalescent model as a kind of continuous-time Markov model in which we start in some state k and transition to a state $k - 1$, then $k - 2$, and so on, terminating when we reach state 1.

12.4   Extensions of the Coalescent

The basic coalescent model we just described provides a means to simulate molecular evolution within a single haploid population of fixed size over time. The coalescent model is very versatile, though, and can accommodate many extensions. The remainder of this chapter considers some ways we can generalize the method to handle other sequence types or to relax some of our starting assumptions.

12.4.1   Variable Population Sizes

First, suppose we want to relax the assumption of fixed population size. Instead of assuming a constant population N, suppose that at any point in time, t, we have a population size $N(t)$. Population size enters into the coalescent model only in how much we have to scale coalescent time to get the number of generations on any given edge. For example, in a population that is doubling with each generation, we will cut the scaled time in half with each generation, as illustrated in figure 12.5(a).

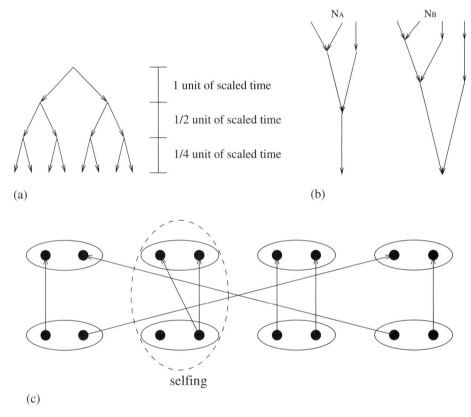

Figure 12.5
Various extensions of the coalescent model. (a) Variable population size, as in this example of a population doubling with each generation. (b) Population substructure, illustrated with two populations with sizes N_A and N_B. (c) Diploid organisms, illustrated by a hermaphroditic organism capable of undergoing *selfing*.

In terms of $N(t)$, the scaled time that elapses in t generations is

$$g(t) = \sum_{i=1}^{t} \frac{1}{N(i)}.$$

If we assume a continuous model of time, we will instead have

$$g(t) = \int_0^t \frac{1}{N(s)} ds.$$

We can then translate a coalescent time τ into the number of generations elapsed, t, by applying the inverse function:

$$t = g^{-1}(\tau).$$

This procedure allows us to run the coalescent model as before, ignoring $N(t)$, then after the fact convert coalescent time into elapsed number of generations along each edge of the resulting tree. We can then sample the mutations along each edge based on generations and get a distribution that is correct for the true variable population size.

Example   For the doubling population of figure 12.5(a):

$$g(t) = \int_0^t \frac{1}{2^s} ds = \frac{e^{s \ln s} - 1}{N(0) \ln 2}$$

$$g^{-1}(t) = \frac{\ln(1 + N(0) \ln 2t)}{\ln 2}.$$

To sample from this model of doubling population, we first run the coalescent as if there were a fixed population size. We then find the start time τ_0 and the end time τ_1 for each edge in the tree. Next, we scale that edge into generations with the formula $t = g^{-1}(\tau_1) - g^{-1}(\tau_0)$. Then we sample mutations for the edge, based on the assumption that it represents t generations of elapsed time.

12.4.2   Population Substructure

Suppose we have have two separate populations with sizes N_A and N_B. If at some point in time we have k_A lineages in population A and k_B lineages in population B, then the time to coalesce in either one is $Exp\left(\frac{1}{N_A}\binom{k_A}{2} + \frac{1}{N_B}\binom{k_B}{2}\right)$. We can generalize this to m populations by saying the time to coalesce is $Exp\left(\sum_{i=1}^{m} \frac{1}{N_i}\binom{k_i}{2}\right)$. This situation is illustrated in figure 12.5(b).

We can also throw in "migration probabilities" by which someone can move from one population to another. For example, if we allow an exponential time with rate β_{ij} of someone moving from population i to population j, then the time to any event (either coalescence or migration) will be distributed as

$$Exp\left(\sum_{i=1}^{m}\frac{1}{N_i}\binom{k_i}{2} + \sum_{i=1}^{m}\sum_{j \neq i}k_i\beta_{ij}\right).$$

12.4.3   Diploid Organisms

The coalescent model can usually be generalized in a straightforward manner to diploid organisms. If we are simulating haploid DNA with diploid organisms (e.g., Y chromosome or mitochondrial DNA), then we can simulate it exactly as with a haploid population. For truly diploid autosomal DNA, we can approximately treat the system as if it were a population of $2N$ individuals and group them after the fact into pairs corresponding to organisms. Some slight corrections may be needed to account for a phenomenon known as *selfing*. Selfing occurs when an organism reproduces with itself and creates a child with two copies of a single ancestral strand. Selfing is illustrated in figure 12.5(c). Though some organisms are incapable of selfing, others, such as many flowering plants, may undergo selfing more frequently than reproduction with others of their species. We thus may need to adjust coalescence rates to account for the bias introduced by having a differential selfing rate. Readers can refer to the sources in References and Further Reading for details on how these adjustments may be made for different cases of diploid reproduction.

12.4.4   Recombination

When we are looking at diploid organisms, we generally have to worry about recombination, in which segments of chromosome swap between homologous chromosomes in each organism. Recombination is illustrated in figure 12.6(a), which shows segments of DNA swapping between the two chromosome copies. Simulating recombination is a bit challenging, since it means that a given piece of DNA can have more than one ancestor. We can add recombination to the coalescent model by allowing the number of lineages to both expand and contract as we go back in time. The model will still allow coalescences, in which two lineages join together because they have a common ancestor. It will also allow recombinations, in which one lineage splits into two because two ancestors contributed to that descendant lineage's DNA. In such cases, we assign some random position in the sequences at which we switch from the first ancestor to the second ancestor. This is illustrated in figure 12.6(b). We then need to treat recombination and coalescence as two possible transitions from the current state of the coalescent model, each with its own intrinsic rate. We will have a waiting time of $Exp(\frac{1}{N}\binom{k}{2})$ until the next coalescence, exactly as with the stan-

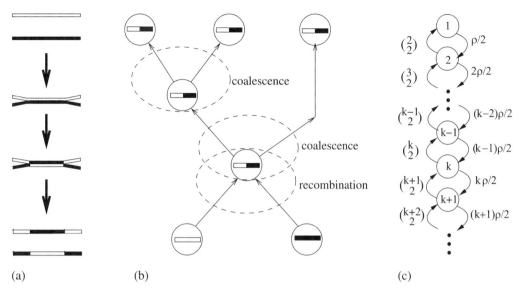

Figure 12.6
Incorporating recombination into the coalescent model. (a) The process of recombination between homologous chromosomes. (b) The coalescent process in the presence of recombination. As we go backward in time, pairs of lineages can merge due to common ancestry (coalescence) or a single lineage can divide to reflect multiple ancestry (recombination). (c) The bidirectional CTMM created by the coalescent with recombination.

dard coalescent, and a waiting time of $Exp(k\rho/2)$ until the next recombination for some rate parameter ρ. ρ is derived from a per-sequence recombination rate r by the formula $\rho = 4Nr$. Combining these two rates gives us a total waiting time to any next event of $Exp\left(\frac{1}{N}\binom{k}{2} + k\rho/2\right)$. Figure 12.6(c) shows the CTMM describing this process.

Since the number of lineages can both increase and decrease, we may wonder how we can be sure we will ever get a common ancestor. In fact, we can guarantee that the model will eventually find an MRCA by using the CTMM representation of the coalescent with recombination. This CTMM is a special kind of Markov model known as a *branching process*, in which we have a set of elements that can divide or disappear independently of one another. Informally, we know that the model must eventually get to just one lineage because the probability of moving toward larger k grows proportionally to k while the probability of moving toward smaller k grows proportionally to k^2. This means that the model is extremely unlikely to get to very large k and is certain to eventually reach $k = 1$. It may increase again afterward, but for the purposes of simulating the process, we only need it to get to one lineage once. Once we have an MRCA, we can create its sequence and insert mutations along the lineages forward in time from there.

References and Further Study

The Jukes–Cantor and Kimura models are standard models for studies of molecular evolution and are covered adequately in a wide variety of sources on these topics. Graur and Li [124] provides a clear coverage of these issues, as well as many others likely to be of interest to readers of this chapter. The best presentation of the coalescent model of which I am aware is found in a review chapter by Nordborg [125] in the *Handbook of Statistical Genetics*, which was an important source in preparing this chapter's discussion of the coalescent and its extensions. For more depth on the general topics covered here, the reader may refer to a more general text on population genetics, such as Hartl and Clark [126].

The Jukes–Cantor and Kimura models are, of course, originally due to Jukes and Cantor [127] and Kimura [128]. The coalescent model is due to Kingman [129]. The various extensions of the basic coalescent that were covered here are derived from Nordborg [125]. Many other coalescent extensions are available in the literature, and a current search may therefore prove helpful for those requiring more specialized coalescent variants.

13 Discrete Event Simulation

As we have seen, one way of representing continuous-time Markov models is to repeatedly consider every transition that may happen next, determine the time at which each will happen, and pick the one with minimum time. This representation of CTMMs is a special case of a more general class of models called *discrete event models*. In a discrete event model, we have a set of discrete states, just as in a Markov model, and move between states in continuous time, as in a CTMM. However, instead of insisting that all transitions have exponential times, we will allow for any possible waiting time distributions.

Example   Suppose we have a cell with a set of channel proteins. Each channel type takes in or expels some given type of ion. We will assume we have three ions, A, B, and C, and counts of each, n_A, n_B, and n_C. As a first pass, we will assume waiting times are in fact exponential. We will assume three channel types moving ions into the cell, with rates λ_{A+}, λ_{B+}, and λ_{C+}, and three channel types moving the ions out of the cell, with rates $n_A\lambda_{A-}$, $n_B\lambda_{B-}$, and $n_C\lambda_{C-}$. The model is illustrated in figure 13.1(a). Because waiting times are exponential, we can treat this system as a CTMM. The underlying graph will be a three-dimensional cubic grid where each point represents a count of each of the three ions. Using this representation, we can simulate the system using the pseudocode of figure 13.1(b). We can also simulate the system by using the Kolmogorov equations to establish the distribution of states over time and sample from that distribution.

This approach will work in limited cases, but it has problems. The most important one is that we must have exponential distributions for all waiting times to get a CTMM and to be able to use the tools we have available for working with CTMMs. It can also be inefficient, even if we have a proper CTMM, if we cannot solve the Kolmogorov equations.

Suppose we now assume instead that the input channels have waiting times of the form $U[\alpha_A, \beta_A]$, $U[\alpha_B, \beta_B]$, and $U[\alpha_C, \beta_C]$. This may be a reasonable approximation if the channels need a certain amount of time to "recover" after transporting an ion

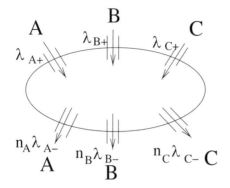

(a)

(b)

1. compute waiting times for possible transitions from the current state:

$$t_{A+} \leftarrow Exp(\lambda_{A+}), t_{B+} \leftarrow Exp(\lambda_{B+}), t_{C+} \leftarrow Exp(\lambda_{C+})$$

$$t_{A-} \leftarrow Exp(n_A\lambda_{A-}), t_{B-} \leftarrow Exp(n_B\lambda_{B-}), t_{C-} \leftarrow Exp(n_C\lambda_{C-})$$

2. find $dt = \min\{t_{A+}, t_{B+}, t_{C+}, t_{A-}, t_{B-}, t_{C-}\}$
3. if $dt = t_{A+}$ then $n_A \leftarrow n_A + 1$
4. if $dt = t_{B+}$ then $n_B \leftarrow n_B + 1$
5. if $dt = t_{C+}$ then $n_C \leftarrow n_C + 1$
6. if $dt = t_{A-}$ then $n_A \leftarrow n_A - 1$
7. if $dt = t_{B-}$ then $n_B \leftarrow n_B - 1$
8. if $dt = t_{C-}$ then $n_C \leftarrow n_C - 1$
9. if $t \leftarrow t + dt$
10. go to step 1

(c)

1. compute waiting times for possible transitions from the current state:

$$t_{A+} \leftarrow U[\alpha_A, \beta_A], t_{B+} \leftarrow U[\alpha_B, \beta_B], t_{C+} \leftarrow U[\alpha_C, \beta_C]$$

$$t_{A-} \leftarrow Exp(n_A\lambda_{A-}), t_{B-} \leftarrow Exp(n_B\lambda_{B-}), t_{C-} \leftarrow Exp(n_C\lambda_{C-})$$

2. find $t = \min\{t_{A+}, t_{B+}, t_{C+}, t_{A-}, t_{B-}, t_{C-}\}$
3. if $t = t_{A+}$ then $n_A \leftarrow n_A + 1$, $t_{A+} \leftarrow t + U[\alpha_A, \beta_A]$, $t_{A-} \leftarrow t + Exp(n_A\lambda_{A-})$
4. if $t = t_{B+}$ then $n_B \leftarrow n_B + 1$, $t_{B+} \leftarrow t + U[\alpha_B, \beta_B]$, $t_{B-} \leftarrow t + Exp(n_B\lambda_{B-})$
5. if $t = t_{C+}$ then $n_C \leftarrow n_C + 1$, $t_{C+} \leftarrow t + U[\alpha_C, \beta_C]$, $t_{C-} \leftarrow t + Exp(n_C\lambda_{C-})$
6. if $t = t_{A-}$ then $n_A \leftarrow n_A - 1$, $t_{A-} \leftarrow t + Exp(n_A\lambda_{A-})$
7. if $t = t_{B-}$ then $n_B \leftarrow n_B - 1$, $t_{B-} \leftarrow t + Exp(n_B\lambda_{B-})$
8. if $t = t_{C-}$ then $n_C \leftarrow n_C - 1$, $t_{C-} \leftarrow t + Exp(n_C\lambda_{C-})$
9. go to step 2

before they can send the next one, but never wait too long after that. We will assume that the output channels still have exponential waiting times. If we do this, the Kolmogorov equations are no longer valid. Furthermore, our previous simulation algorithm no longer works. In particular, we cannot simply sample all the times again on each step because we need to know how much time has elapsed since each channel was last used. We need to modify our algorithm as in figure 13.1(c).

By recomputing only some of the events on each step, we can keep track of elapsed time for the remaining ones since they were first sampled. This is not necessary when the events are exponentially distributed, since the exponential distribution is memoryless. When we are dealing with other distributions, such as uniform, though, it is needed. Then the t_+ values will be sampled correctly since we calculate the waiting time when the channel is last used and keep track of that sampled value until the channel is next used.

13.1   Generalized Discrete Event Modeling

We can derive a more general simulation method by examining what we needed to do to adapt our model to uniform waiting times. The algorithm we developed did not depend on the fact that our waiting times were uniform for some events, and in fact it would have worked fine for any strictly positive random variables. In an abstract sense, we created a method for simulating this system by stepping between *events* that could have any waiting times between them. We can generalize further by saying that an event may be anything that changes the state of a simulation. The behavior of an event is characterized by the following properties:

· The time when it occurs
· How it changes the simulation state
· Other events it *invalidates* (deletes)
· Other events it creates.

For instance, in the current example, if we have an event of the t_{A+} kind, that event changes the simulation state by incrementing n_{A+}. It invalidates the preexisting t_{A-} event. And it creates a new t_{A+} and a new t_{A-} event.

Suppose we define *e.time* to be the time at which event e should occur, *e.invalid*(E) to be a function that identifies events in E invalidated by e, *e.add*( ) to be a function

Figure 13.1
Model of ion entry and exit through a set of channel proteins in a cell. (a) Graphical illustration of the model. Ions enter with fixed rates per ion type and leave with rates proportional to the current concentration in the cell. (b) Pseudocode for a first version of the model, assuming all waiting times are exponentially distributed. (c) Pseudocode for a variant of the model, assuming ions enter the cell with uniformly distributed waiting times.

1. generate an initial event list E and system state S
2. while not done
    A. find the event $e \in E$ with minimum $e.time$
    B. update the current time to $e.time$
    C. $S \leftarrow e.update(S)$ (e.g. $n_A \leftarrow n_A + 1$)
    D. $E \leftarrow E/e.invalid(E)$ (e.g. erase the old t_{A-})
    E. $E \leftarrow E \cup e.add()$ (e.g. resample t_{A+}, t_{A-})

Figure 13.2
Pseudocode for generic discrete event simulation showing the correspondence to the channel example. The
pseudocode assumes we have procedures to find the time at which a given event occurs (`e.time`), to determine how a given event e updates the event list (`e.update()`), to identify the other events e it invalidates
(`e.invalidate()`), and to enumerate the other events e creates (`e.add()`).

that outputs the new events created by event e, and $e.update(S)$ to be a function that
updates system state S to reflect the action of event e. Then we can summarize this
process of discrete event simulation generically by the pseudocode of figure 13.2.
This approach is less restrictive than CTMMs, although much harder to analyze
in general. It is, however, usually easy to implement, in that it lends itself well to
object-oriented design. Specifically, we can think of an event as an object that has
an activation time, that has a method that acts on the simulation state to produce
an updated state, that has a method for identifying some events as invalid, and that
has a method for creating some new events that may occur later. Then we can simulate the system by repeatedly finding the minimum-time event, activating its methods, and using them to update the system state and the set of pending events.

13.2 Improving Efficiency

This discrete event approach gives us a very general simulation framework, but if we
want to use it in practice, we will need to consider how we can make it efficient. To
examine this issue, suppose we look at our previous example but imagine that we
have N types of ions instead of three. What will the efficiency of each step of the
method be? Sampling among all possible events requires $O(N)$ time, once per simulation. Picking the minimum of the t_{i+} and t_{i-} events requires $O(N)$ time per simulation step. Updating the state and picking new events requires $O(1)$ time per
simulation step. The initial sampling step will have negligible cost if we run the simulation long enough, so we are mainly concerned with the per-step costs in step 2 of
the pseudocode. This step constitutes the *event loop*, the set of operations repeatedly
performed to activate successive events. Picking the minimum event is therefore the
bottleneck, with a cost of $O(N)$ for each pass through the event loop.

Table 13.1
Runtimes for priority queue operations for five queue data structures

	Extract-min	Insert	Delete	Notes
Unsorted list	$O(N)$	$O(1)$	$O(N)$	
Sorted list	$O(1)$	$O(N)$	$O(N)$	
Binary heap	$O(\log N)$	$O(\log N)$	$O(\log N)$	
Fibonacci heap	$O(\log N)$	$O(1)$	$O(\log N)$	Amortized
Calendar queue	$O(1)$	$O(1)$	$O(N)$	Expected, with caveats

To speed up this method, we need a way to choose the minimum of N items quickly. We can accomplish this by maintaining a data structure that lets us choose its minimum element quickly, without requiring too much additional time to maintain the data structure as we update the simulation. Specifically, we need to be able to perform the following three operations efficiently:

1. Extract-min: remove the minimum element from the data structure.
2. Insert: add a new element to the data structure.
3. Delete: remove an element from the data structure.

These are the operations that define a *priority queue*. Different priority queue implementations have different times for these operations, as summarized in table 13.1.

Using this table, we can find ways to speed up our N-ion channel example. For instance, if we use a Fibonacci heap to maintain our event set, then our event loop will require $O(\log N)$ time per step to pick the minimum-time event, plus $O(\log N)$ time to update the state and pick new events. Although we increased the time of the updating step, we have reduced the total amortized runtime from $O(N)$ to $O(\log N)$ per pass through the event loop. This assumption will be valid provided the number of steps is $\Omega\left(\frac{N}{\log N}\right)$, which will cause the amortized cost of the first step to be $O(\log N)$ per pass through the event loop. We could apply a similar analysis to any other application of discrete event methods to select a queue method that gives us optimal efficiency for that application with minimum implementation difficulty.

One practical issue worth mentioning is that it is often more efficient not to delete invalidated events from the queue but simply to have a way to check whether an event is invalid when it reaches the top of the queue. This strategy is particularly useful if we are using a queue type for which extract-min is a faster operation than delete. A queuing method that defers the test for validity until an event is ready for processing is called a *lazy* queuing strategy. With our channel protein example, we could extend our data structures by adding a *posting time* to each event that says when it was placed in the queue. We could then maintain an array, *valid*, of times at which events referring to particular ions were invalidated. That is, when we add

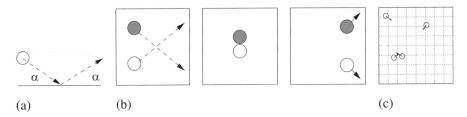

Figure 13.3
The hard-sphere collision dynamics model. (a) An inelastic collision between a particle and a boundary, producing an angle of reflection equal to the incident angle. (b) A collision between two particles. (c) Grid method for accelerating computation by dividing the space into artificial grid boxes.

an ion of type i, we would set $valid[i]$ to the current time. We would then check whether an event is still "valid" by comparing the time it was placed on the queue with the time at which its ion type was last invalidated. Invalid events could thus be discarded with constant cost, as opposed to the typically larger cost of actively removing them from the queue.

13.3   Real-World Example: Hard-Sphere Model of Molecular Collision Dynamics

One real application of discrete event simulations is simulating collision interactions in a molecular system. One common simplified model for this system is the hard-sphere model, in which we ignore the exact interaction forces among particles and treat them as if they were simply solid spheres moving through space. A sphere is assumed to have inertia, so if it is in motion, it will continue moving with the same velocity until it collides with something. At that point, it will deflect with a perfectly elastic collision. This is sometimes referred to as a *billiard ball* model because the particles are assumed to behave like billiard balls moving on a table. The model is illustrated by figure 13.3. Spheres can collide either with boundaries of the system or with other spheres.

Although this is a continuous system, in which we could track particle positions over any point in continuous time, it can in fact be implemented more efficiently as a discrete event system. Specifically, we simulate the system only through discrete changes in state produced by collision events, jumping over all of the time between these events. If we need to know what is going on at some point in time between collisions, we can linearly interpolate particle positions between the states at the surrounding collisions.

If we have two particles, this is easy enough to simulate, but what if we have N particles? To develop a discrete event model, we need to define the allowed event types of the simulation. There are two possible event types we need to consider:

1. Collision of sphere i with a wall:
• The time is calculated by solving for

$$\min_{t}\{x_i(t) = x_{min} + r, x_i(t) = x_{max} - r, y_i(t) = y_{min} + r, y_i(t) = y_{max} - r\},$$

where x_{min}, x_{max}, y_{min}, and y_{max} are the boundaries of the space; r is the sphere radius; and $(x_i(t), y_i(t))$ is the position of sphere i at time t. Since the spheres have constant velocity between steps, $x_i(t)$ and $y_i(t)$ are simple linear equations of one variable, making this an easy calculation.
• Upon the event, we find the new trajectory of sphere i, invalidate any existing events involving sphere i, determine new times for collisions by sphere i with the walls or with any other spheres, and place the events in the queue.
2. Collision of spheres i and j:
• The collision time is calculated by solving for

$$(x_i(t) - x_j(t))^2 + (y_i(t) - y_j(t))^2 = (2r)^2$$

(i.e., when the centers of the two spheres are exactly two sphere radii apart).
• Upon the event, we update both sphere trajectories, invalidate any existing events referring to either sphere, and create new events for all possible collisions involving either sphere.

We can then simulate the hard-sphere model using a discrete event loop. We first initialize by finding all $O(N^2)$ possible ways two sphere would collide, given their initial positions and velocities and all $O(N)$ ways a sphere could collide with a wall, creating an event for each. We then repeatedly pick the minimum-time event and implement the event behavior described above. If we want to know the simulation state at a specific point in continuous time, we can apply the discrete event loop until we pass that time, then interpolate back to the desired time point.

At any given time, we are likely to have events in our queue that will never happen because they will be invalidated before they reach the top of the queue. For example, it might be that we initially create an event for spheres 1 and 2 colliding at time 10 and spheres 3 and 4 colliding at time 5. When spheres 3 and 4 collide, we may compute new events for them based on their new trajectories, and decide that now spheres 1 and 3 will collide at time 8. When spheres 1 and 3 collide, the event for collision between 1 and 2 that has been sitting on the queue will become invalid. We will need to compute a new possible collision time for those two spheres. We can handle this situation by invalidating events in the event queue that refer to those spheres or by using a lazy strategy to recognize that their collision was invalid at the time it reached the top of the queue.

The efficiency of this method will depend on what kind of data structures we use to implement it. We can implement this model without a queue by simply computing all possible next events on each step and picking the minimum. That will require $O(N^2)$ work per step, trying every possible next event to find the one with minimum time. Using a priority queue implemented as a binary heap would allow us to reduce that to $O(N \log N)$ work per step. A calendar queue can potentially reduce this to $O(N)$ work per step, although that is not guaranteed (see supplementary material below).

It is possible to improve on this method by creating a new kind of "artificial event." The bottleneck in our existing implementation is the need on each event to find possible collisions between the affected sphere(s) and all $O(N)$ others. Suppose we artificially create new boundaries within our simulation space. That is, we break up the space into a set of boxes, as illustrated in figure 13.3(c). Then we create a new kind of event, a box change event, representing the time at which the center of a sphere moves from one box to another. The event has the following properties:

· The time to the event is computed in the same way as for boundary collisions, but using the current box boundaries as x_{max}, x_{min}, y_{max}, and y_{min}.
· Upon a box change event for sphere i, we invalidate all existing events for sphere i, create new box change events for the ways i might leave its current box, create new wall collision events for i if its box borders a wall, and create collision events for i with particles in the same or adjacent boxes.

This modification in principle will require more events per unit time, since we are simulating the same process as before and still need an event for every collision in addition to these extra box change events that do not change the physical system. However, the method will generally do much less work per event since the bottleneck with the old method—sampling $O(N)$ new sphere–sphere collision events every time we have a valid event—is reduced to sampling only among spheres in the same or neighboring boxes. We can do this so long as the box width is at least twice the sphere radius, because then a sphere cannot collide with another sphere before the next event if there is at least a box width separating them. There must be a box change event for at least one of them before any collision.

The exact runtime change will of course depend on how many boxes we have. If we have very small boxes, then our simulation will be overwhelmed by box change events and will make very little progress. If we have very large boxes, then our method will reduce to the original method and we will get no advantage from having boxes. If we assume that the spheres are, on average, uniformly distributed in the space, then having $O\left(\frac{N}{\log N}\right)$ boxes is likely to work well. That will reduce the average number of spheres per box to $O(\log N)$, reducing the time per event to logarithmic while not generally adding too many additional events due to box changes. To choose the exact optimal box size will require a more sophisticated analysis involving the relative sphere and space sizes and the velocities of the particles, which we will not go into here.

13.4 Supplementary Material: Calendar Queues

We referred above to several data structures for implementing priority queues. Readers with introductory discrete algorithms training are likely to be familiar with all of them, with the possible exceptions of Fibonacci heaps and calendar queues. Those two can be very effective in practice, though, so it is useful to know at least what their performance bounds are, and preferably how to implement them. One can read about Fibonacci heaps in various algorithms texts, such as Cormen et al. [14]. Information on calendar queues is harder to find, though. We will therefore briefly cover them here.

The basic idea behind a calendar queue is to divide our queue into a set of "buckets" chosen so that approximately a constant number of events sits in each bucket. Each bucket is treated like a day of the week on a calendar. If we have a Monday bucket, then any event occurring this Monday goes into the bucket, as does any event occurring next Monday, the Monday after that, and so on. The Tuesday bucket contains all events occurring this Tuesday, next Tuesday, the Tuesday after that, and so on. Hence the name "calendar queue." In general, we assume that we have n buckets, each with width w. All events from time span $[0, w]$ go in bucket 1, $[w, 2w]$ go in bucket 2, $[2w, 3w]$ go in bucket 3, . . . , and $[(n-1)w, nw]$ go in bucket n. We then wrap around and place events from time span $[nw, (n+1)w]$ in bucket 1, those from $[(n+1)w, (n+2)w]$ in bucket 2, and so on. This process is illustrated in figure 13.4. Within each bucket, we can store the events as a sorted linked list.

We can perform the basic priority queue operations on a calendar queue as follows:

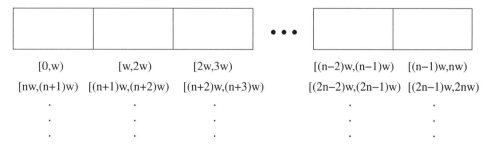

Figure 13.4
Calendar queue data structure. A set of buckets represents intervals of time stored in the queue. Times beyond the last bucket wrap around to the first.

- Insert: given an event e with time t, find the bucket $b = \left\lceil \frac{t \text{ "mod" } nw}{w} \right\rceil$, where we are somewhat abusing notation to have "mod" mean the number left over after taking the largest possible integer number of factors of nw from t. Place e into bucket b, then insert it into the sorted list.
- Extract-min: given current time t, do the following:

while (first element of current bucket is greater than $t + w$)

$\quad t \leftarrow t + w$

go to the next bucket, wrapping around from bucket n to bucket 1

return the first element of the current bucket.

- Delete: search all buckets for the event, to delete and remove it.

In the worst case, all three operations take linear time in the queue size. In practice, though, the calendar queue tends to take constant time for insert and extract-min operations. This can actually be proved in the average case for many common distributions of event times (e.g., uniform, normal, exponential), provided n and w are chosen intelligently for the distribution. The two parameters can be chosen from prior knowledge of the event time distributions or dynamically adjusted, based on observed event distributions as the method runs. The method is therefore a little complicated to use, but is often the fastest method in practice.

References and Further Study

Discrete event simulation, though widely used in practice, is not often taught as a methodology in itself in the simulation literature since it is simply a way of conceptualizing any of a broad class of simulation methods. Nonetheless, any standard introductory algorithms text will provide basic information on most of the data structures and methods seen in this chapter, as well as other queue methods we have not covered. Cormen et al. [14] remains an excellent choice, as is Kozen [17], although many others will serve well. The extended example on hard-sphere collisions was based on work by Rapaport [130] on discrete event methods for this problem. Calendar queues are a relatively recent addition to the data structure literature, and one may refer to the primary reference by Brown [131] for more information on their theory and use.

14 Numerical Integration 1: Ordinary Differential Equations

At this point, we will shift our discussion from discrete to continuous simulation models. When working with continuous models, we will assume that we have a system whose state is described by a vector of continuous functions of time:

$$\vec{v}(t) = (v_1(t), v_2(t), \ldots, v_k(t)).$$

One example of such a continuous simulation model is molecular motion. If we have a single atom moving freely in a vacuum, then we may describe our system as a vector of three variables representing the spatial position of the atom over time:

$$\vec{v}(t) = (x(t), y(t), z(t)) = (a_x t + b_x, a_y t + b_y, a_z t + b_z).$$

Here (a_x, a_y, a_z) represents the velocity of the atom, and (b_x, b_y, b_z) its initial position. More generally, we may imagine that we have n atoms that are under the influence of some force field:

$$\vec{f}(x_1(t), y_1(t), z_1(t), x_2(t), y_2(t), z_2(t), \ldots)$$

$$= \left(m_1 \frac{d^2 x_1}{dt^2}, m_1 \frac{d^2 y_1}{dt^2}, m_1 \frac{d^2 z_1}{dt^2}, m_2 \frac{d^2 x_2}{dt^2}, m_2 \frac{d^2 y_2}{dt^2}, m_2 \frac{d^2 z_2}{dt^2}, \ldots \right).$$

The force is a vector in which each element of the vector corresponds to one degree of freedom of one atom, and is itself a function of the positions of all atoms. Force is related to the second derivative of position (i.e., acceleration) by Newton's second law of motion, $\vec{f} = m\vec{a}$.

Another example of a continuous simulation model we often use in biology applications is the representation of a chemical reaction system by the law of mass action. Suppose we have the reaction system

$$A + B \underset{k_2}{\overset{k_1}{\rightleftharpoons}} I \underset{k_4}{\overset{k_3}{\rightleftharpoons}} C + B.$$

This can be a model of an enzymatic reaction, in which enzyme B binds to substrate A, forming intermediate I, then catalyzes a conversion from A to C before releasing C. This system can be described by a vector of concentrations of the various reactants:

$$\vec{v}(t) = ([A], [B], [C], [I]).$$

The behavior of this system is then described in the limit of large numbers of molecules by the following differential equations:

$$\frac{d[A]}{dt} = k_2[I] - k_1[A][B]$$

$$\frac{d[B]}{dt} = (k_2 + k_3)[I] - k_1[A][B] - k_4[C][B]$$

$$\frac{d[C]}{dt} = k_3[I] - k_4[C][B]$$

$$\frac{d[I]}{dt} = k_1[A][B] + k_4[C][B] - (k_2 + k_3)[I].$$

Yet another example of a continuous simulation system in biology is population dynamics on large scales. For example, suppose we have a predator A that reproduces at rate $\lambda_1 A$ and a prey B that reproduces at rate $\lambda_2 B$. The predator consumes prey at a rate $\lambda_3 AB$. Predators also die of starvation at a rate $\lambda_4 e^{-B} A$. If we have just a few predators and preys, then we may model this as a CTMM. If we have a lot, though, then we can model this system by the following differential equations:

$$\frac{dA}{dt} = \lambda_1 A - \lambda_4 e^{-B} A$$

$$\frac{dB}{dt} = \lambda_2 B - \lambda_3 AB.$$

The common theme in all of these models is differential equations. For nontrivial systems, we often will have a description of how the system instantaneously changes at any given point in time, and we need to translate that into a simulation of the time evolution of the system. Such simulations are described by systems of differential equations. Therefore, simulating continuous systems is typically the same problem as integrating differential equations.

This is one of those topics, like continuous optimization, for which we will only be able to scratch the surface. There is an enormous body of literature on the theory of numerical integration and on special-purpose methods particularly well suited for certain various kinds of systems, which we unfortunately must skip due to space limitations. The goal of this chapter is to introduce some of the basic principles of numerical integration, present some broad general-purpose tools, and show how to apply them in practice and assess their performance. Readers who end up working extensively with such systems will do well to undertake more advanced study on differential equations and numerical integration. A few sources providing more advanced coverage of various topics are in References and Further Study.

14.1   Finite Difference Schemes

It is usually impossible to analytically integrate a system of differential equations. There are some special cases, such as those where a system is linear (i.e., $\frac{d\vec{v}}{dt} = A\vec{v}$), where analytical solutions are possible. Generally, though, we will need to use *numerical integration schemes*, which approximate the integration of a specific system for a specific amount of time. The most common way to do this is by using *finite difference schemes*, in which we approximately integrate by summing over approximations for short time steps. In general, a finite difference scheme will involve successive iterations of the form

$$\vec{v}_{n+1} = \vec{f}(\vec{v}_n, \vec{v}_{n-1}, \ldots).$$

That is, the estimated value of the integral at each successive step is some function of the values at the previous steps. We then compute successive approximations to $\vec{v}$ over some range $t = [0, \tau]$ by computing approximations at different time steps:

$$\vec{v}_0 = \vec{v}(0)$$

$$\vec{v}_1 \approx \vec{v}(\Delta t)$$

$$\vec{v}_2 \approx \vec{v}(2\Delta t)$$

$$\vdots$$

$$\vec{v}_{\tau/\Delta t} \approx \vec{v}(\tau).$$

For most of this discussion we will examine different methods by assuming we have only a single dependent variable, x, and a single independent variable, t. We can generalize to multiple dependent variables by having one iterator formula for each variable for each time step. For a system of k dependent variables where $v_{i,j}$ is the approximation to variable i at time step j, the approximation for the full system will take the following form:

$$v_{1,n+1} = f_1(v_{1,n}, v_{2,n}, \ldots, v_{k,n}, v_{1,n-1}, v_{2,n-1}, \ldots, v_{k,n-1}, \ldots)$$

$$v_{2,n+1} = f_2(v_{1,n}, v_{2,n}, \ldots, v_{k,n}, v_{1,n-1}, v_{2,n-1}, \ldots, v_{k,n-1}, \ldots)$$

$$\vdots$$

$$v_{k,n+1} = f_k(v_{1,n}, v_{2,n}, \ldots, v_{k,n}, v_{1,n-1}, v_{2,n-1}, \ldots, v_{k,n-1}, \ldots).$$

14.2 Forward Euler

The most basic finite difference method is the *forward Euler* method. With the forward Euler method, we approximate the differential equation

$$\frac{dx}{dt} = f(x)$$

at successive time steps, using the approximation

$$x_{n+1} = x_n + \Delta t f(x_n).$$

Example   Suppose we want to approximate the chemical reaction system

$$A + B \underset{k_2}{\overset{k_1}{\rightleftharpoons}} C,$$

described by the differential equations

$$\frac{d[A]}{dt} = k_2[C] - k_1[A][B]$$

$$\frac{d[B]}{dt} = k_2[C] - k_1[A][B]$$

$$\frac{d[C]}{dt} = k_1[A][B] - k_2[C].$$

To perform this approximation by forward Euler, we will use the iterators

$$[A]_{n+1} = [A]_n + \Delta t(k_2[C]_n - k_1[A]_n[B]_n)$$

$$[B]_{n+1} = [B]_n + \Delta t(k_2[C]_n - k_1[A]_n[B]_n)$$

$$[C]_{n+1} = [C]_n + \Delta t(k_1[A]_n[B]_n - k_2[C]_n).$$

We will apply this iterator from our initial state for $\tau/\Delta t$ steps to estimate the system state at time τ.

For the forward Euler scheme to work, it must be true that $x(t) + \Delta t \frac{dx}{dt}$ is an approximation to $x(t + \Delta t)$. Why might this be so? We can understand why the scheme approximates the integral using Taylor series. If we expand the function we are trying to approximate, $x(t + \Delta t)$, around t we get the following:

$$x(t + \Delta t) = x(t) + x'(t)\Delta t + \frac{x''(\xi)}{2!}\Delta t^2.$$

In other words, $x(t + \Delta t)$ is approximated by $x(t) + x'(t)\Delta t$ with an error that is $O(\Delta t^2)$. Since we add an error of $O(\Delta t^2)$ on each step of our numerical integration and we require $O(1/\Delta t)$ steps to integrate to some fixed τ, the total error of the integration is $O(\Delta t)$. A method for which the accumulated error from integrating for any fixed time τ is $O(\Delta t^1)$ is called a *first-order method*. The order of accuracy of a method is one of the most important issues we need to consider in deciding whether a given method is appropriate for a given problem.

Another property we need to understand in choosing a numerical integration scheme for a particular problem is the *stability* of the scheme. Stability describes whether over large numbers of steps the values will increase, fade away to a constant, or do something in between. Stability of a scheme is generally classified into one of three possibilities:

- Stable: values decay to a constant.
- Unstable: values continue to blow up indefinitely.
- Semistable: oscillatory behavior that does not fade away.

The stability of a scheme will depend on both the specific system being examined and the time step size used. If we want to run a simulation for a long time, then we need to make sure that the numerical integration scheme's stability behavior matches that of the system being simulated.

We can evaluate the conditions under which a given scheme is stable using a technique called *von Neumann analysis*:

1. Guess a solution of the form $e^{\omega t}$.
2. Find some $g(\omega\Delta t)$ for which $x_{n+1} = g(\omega\Delta t)x_n$.
3. Find the region of the complex plane for which $|g(\omega\Delta t)| < 1$. This is where the scheme is stable. It is unstable where $|g(\omega\Delta t)| > 1$ and semistable where $|g(\omega\Delta t)| = 1$.

For example, to apply von Neumann analysis to the forward Euler method, we first state the basic iterator for the scheme:

$$x_{n+1} = x_n + \Delta t \frac{dx_n}{dt}.$$

We then guess that $x_n = e^{\omega t}$, and thus $\frac{dx_n}{dt} = \omega e^{\omega t}$. Therefore,

$$x_{n+1} = e^{\omega t} + \omega \Delta t e^{\omega t}.$$

We then rearrange to express x_{n+1} in terms of x_n:

$$x_{n+1} = (1 + \omega \Delta t)e^{\omega t}$$

$$x_{n+1} = (1 + \omega \Delta t)x_n.$$

This finally tells us the function $g(\omega \Delta t)$:

$$g(\omega \Delta t) = (1 + \omega \Delta t).$$

We then want to know where $|g(\omega \Delta t)| = |(1 + \omega \Delta t)|$ is less than 1 in the complex plane. For forward Euler, it happens that $|g(\omega \Delta t)| < 1$ in a circle of radius 1 centered on -1, as illustrated in figure 14.1(a).

This tells us where the scheme is stable if we happen to be looking at a system whose solution is an exponential $e^{\omega t}$. It is stable if $\omega \Delta t$ falls in the circle of figure 14.1(a). But what if the solution is not an exponential? In fact, everything is an exponential. Whatever the actual solution to our system is, we can always break it into some Fourier series, expressing it as a sum of exponentials. It may not be easy to determine the ω values, but we know that there are some such ωs. If all of these ω

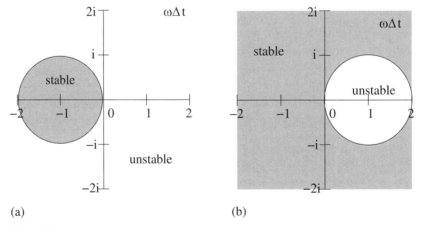

(a)   (b)

Figure 14.1
Stability diagrams for some of the finite difference schemes examined here, derived by von Neumann analysis. Each diagram shows the region of the ωt plane for which the method is stable. Semistable solutions fall in the boundaries of the region. (a) Stability of the forward Euler method. (b) Stability of the backward Euler method.

values have a strictly negative real component, then there is some step size Δt for which the forward Euler scheme will be stable. If any ω is strictly imaginary or has a positive real component, then there is no step size that makes the scheme stable.

Note that stability is not necessarily a good thing. If ω has a positive real component, then the actual system really will blow up, so one can argue that a finite difference scheme that fails to increase is not an accurate description of the system. An unstable scheme is "correct" in that case because it behaves like the real system. If ω is strictly imaginary, then the real system will oscillate, so a semistable scheme is "correct." It is only when ω has a negative real component that our scheme should actually be stable.

In the case of our chemical example above, it is not obvious what values of Δt will lead to stability. The real physical system should approach an equilibrium, though, so we know all of the Fourier components have negative real parts. We can therefore be confident that if we make the time step small enough, the numerical scheme will be stable. In practice, we may initially select a step size based on the accuracy analysis so as to give an acceptable error, then adjust it smaller if that step size turns out to yield instability.

14.3   Backward Euler

Forward Euler is often a serviceable scheme for solving simple problems. But there are alternatives that may be better for some systems. One similarly simple alternative scheme is *backward Euler*. Backward Euler is specified by the iterator:

$$x_{n+1} = x_n + \Delta t \frac{dx_{n+1}}{dt} = x_n + \Delta t f(x_{n+1}).$$

We can immediately see a problem with this scheme, in that the formula for determining x_{n+1} depends on x_{n+1}. A scheme for which the iterator depends on the value it is calculating is called an *implicit method*, in contrast to an *explicit method* like forward Euler, in which x_{n+1} is computed strictly from earlier iterates. To use an implicit method, we need to solve for x_{n+1}. For backward Euler, that means we need to solve

$$x_{n+1} - \Delta t f(x_{n+1}) = x_n$$

for x_{n+1}. In other words, we need to invert $x - \Delta t f(x)$.

Example   Suppose we want to integrate the equation

$$\frac{dx}{dt} = -3x = f(x).$$

Then we need to solve

$$x_{n+1} = x_n - 3\Delta t x_{n+1}.$$

This equation is solved by

$$(3\Delta t + 1)x_{n+1} = x_n$$

$$x_{n+1} = \frac{x_n}{3\Delta t + 1}.$$

The preceding formula is then the iterator we will use to compute the approximation for each successive time step, given the approximation at the previous step.

Note that we do not necessarily need to be able to analytically invert an equation in order to use an implicit method. If we do not know how to invert $x - \Delta t f(x)$, we can still use backward Euler by numerically inverting the equation. For example, if we want x satisfying $x - \Delta t f(x) = y$, then we can look for a zero of the function $h(x) = x - \Delta t f(x) - y$. We can find a zero of $h(x)$ using any of the zero-finding schemes we learned in chapter 5.

Now that we know how to use backward Euler, we can ask how we will expect it to perform. The first thing we want to know is if it is accurate. The scheme is based on the assertion that

$$x(t + \Delta t) \approx x(t) + \Delta t x'(t + \Delta t),$$

which is equivalent to saying

$$x(t) \approx x(t - \Delta t) + \Delta t x'(t),$$

or, with some rearrangement,

$$x(t - \Delta t) \approx x(t) - \Delta t x'(t).$$

From Taylor series, we know that

$$x(t - \Delta t) = x(t) - \Delta t x'(t) + \frac{x''(\xi)}{2!}\Delta t^2.$$

This tells us that the approximation is valid and has an error per step that is $O(\Delta t^2)$. Therefore, the total error accumulated over the $\tau/\Delta t$ steps needed to integrate for a fixed time τ is $O(\Delta t)$. Backward Euler is thus a first-order method like forward Euler.

If backward Euler does not give us an accuracy improvement over forward Euler, we may infer that it must give some advantage in stability behavior. We can assess that by von Neumann analysis. We guess $x(t) = e^{\omega t}$, so $x'(t) = \omega e^{\omega t} = \omega x(t)$. Then

$$x_{n+1} = x_n + \Delta t \omega x_{n+1}$$

$$(1 - \Delta t \omega) x_{n+1} = x_n$$

$$x_{n+1} = \frac{1}{1 - \Delta t \omega} x_n.$$

Therefore, our scaling factor per step is

$$g(\Delta t \omega) = \frac{1}{1 - \Delta t \omega},$$

and we need to know where

$$|g(\Delta t \omega)| = \left| \frac{1}{1 - \Delta t \omega} \right| < 1.$$

It turns out that this condition is satisfied everywhere except in a circle of radius 1 centered on $\Delta t \omega = 1$, as illustrated in figure 14.1(b).

What is particularly useful about this scheme is that it is stable in the entire left half of the complex plane, $\mathscr{R}e\{\Delta t \omega\} < 0$. Any scheme that is stable whenever the real part of $\Delta t \omega$ is negative is called *unconditionally stable* because it is stable for every genuinely stable system. Unconditional stability is a general property of implicit schemes and is the reason we will go to the extra trouble required to use an implicit scheme. Note that the scheme is also stable in some regions where the real system would be unstable. If we were simulating a genuinely unstable system with backward Euler, we would need to be careful to choose a small enough step size to give the correct instability. We would probably not use an implicit method to simulate an unstable system, though.

14.4 Higher-Order Single-Step Methods

It will often be necessary in practice to use methods that have higher than first-order accuracy if we want our simulations to run in a reasonable amount of time. There are two main methods for achieving better than first-order accuracy with finite difference methods. The first involves computing intermediate partial steps between each of our major time steps. A scheme that does this is known as a *single-step method*. One example is the *midpoint method*. For each time step, the midpoint method first computes an intermediate estimate $x_{n+1/2}$ and then uses $x_{n+1/2}$ to compute the approximation for the full step, x_{n+1}. It uses the following iterator:

$$x_{n+1/2} = x_n + \frac{\Delta t}{2} f(x_n)$$

$$x_{n+1} = x_n + \Delta t f(x_{n+1/2}).$$

That is, we take half an Euler step, then use the derivative at this midpoint instead of at x_n to more accurately estimate the difference between x_n and x_{n+1}.

We can analyze the accuracy of the method by again using Taylor series. We first merge our two iteration steps into a single formula:

$$x_n + \Delta t f(x_{n+1/2}) = x_n + \Delta t f\left(x_n + \frac{\Delta t}{2} f(x_n)\right).$$

We then perform a Taylor expansion of $f\left(x_n + \frac{\Delta t}{2} f(x_n)\right)$ to transform our approximation to the following:

$$= x_n + \Delta t f(x_n) + \frac{\Delta t^2}{2} f(x_n) \frac{\partial f(x_n)}{\partial x} + O(\Delta t^3).$$

We can then apply the *chain rule* of derivation, which says that $\frac{d^2 x}{dt^2} = \frac{df}{dt} = \frac{\partial f}{\partial x} \frac{dx}{dt} = \frac{\partial f}{\partial x} f(x)$, to get the following:

$$= x_n + \Delta t f(x_n) + \frac{\Delta t^2}{2} \frac{d^2 x}{dt^2} + O(\Delta t^3)$$

$$= x(t + \Delta t) + O(\Delta t^3).$$

The error per step is $O(\Delta t^3)$, so the midpoint method is second-order accurate.

We can also analyze the stability of the midpoint method by von Neumann analysis. We first use the fact that $\frac{dx}{dt} = \omega x$ to get

$$x_{n+1/2} = x_n + \frac{\Delta t}{2} \omega x_n$$

and

$$x_{n+1} = x_n + \Delta t \omega x_{n+1/2}.$$

Hence

$$x_{n+1} = x_n + \Delta t \omega \left(x_n + \frac{\Delta t}{2} \omega x_n\right)$$

$$= \left(1 + \Delta t \omega + \frac{(\Delta t \omega)^2}{2}\right) x_n.$$

Therefore, the method is stable where $\left|1 + \Delta t\omega + \frac{(\Delta t\omega)^2}{2}\right| < 1$. This region does not correspond to a simple shape as those for the Euler methods do. Essentially, though, it is a slightly expanded and distorted version of the forward Euler stability region, covering a bit more of the left half of the complex plane.

The midpoint method is the simplest example of a class of single-step methods called *Runge–Kutta methods*, which compute a series of intermediate values and use a combination of them to find an approximation at each step. There is a Runge–Kutta method for any desired integer order of accuracy, although it is rarely used beyond about sixth order. In general, the stability regions of Runge–Kutta methods become slightly larger with each successive increase in order of accuracy. Runge–Kutta methods with varying levels of accuracy are commonly used by prepackaged numerical integration schemes, such as we might find in the popular Matlab system. For example, the fourth-order Runge–Kutta method is defined by the following iterator:

$$k_1 = f(x_n)$$

$$k_2 = f\left(x_n + \frac{\Delta t}{2}k_1\right)$$

$$k_3 = f\left(x_n + \frac{\Delta t}{2}k_2\right)$$

$$k_4 = f(x_n + \Delta t k_3)$$

$$x_{n+1} = x_n + \frac{\Delta t}{6}(k_1 + 2k_2 + 2k_3 + k_4).$$

We will not bother analyzing this method, but it is fourth-order accurate and has a somewhat larger stability region than does the midpoint method. The fourth-order Runge–Kutta method is the closest thing there is to a "default" numerical integration scheme, and is a good practical choice for generic problems in which one has no reason to favor any particular special-purpose scheme. There are similar schemes for higher order, but the fourth-order one is the most frequently used.

14.5   Multistep Methods

The major alternatives to single-step methods are multistep methods. Instead of computing several intermediate values for each time step, multistep methods do just one calculation but use several past time points to make it. The simplest example is the *leapfrog method*, defined by the iterator

$$x_{n+1} = x_{n-1} + 2\Delta t f(x_n).$$

This scheme estimates x_{n+1} using both x_n and x_{n-1}.

We can determine the accuracy of the method by using two Taylor expansions:

$$x(t + \Delta t) = x(t) + \Delta t x'(t) + \frac{\Delta t^2}{2} x''(t) + \frac{\Delta t^3}{3!} x'''(\xi_1)$$

$$x(t - \Delta t) = x(t) - \Delta t x'(t) + \frac{\Delta t^2}{2} x''(t) - \frac{\Delta t^3}{3!} x'''(\xi_2).$$

Subtracting the second expansion from the first gives us the following:

$$x(t + \Delta t) - x(t - \Delta t) = 2\Delta t x'(t) + \frac{\Delta t^3}{3!} [x'''(\xi_1) + x'''(\xi_2)]$$

$$x(t + \Delta t) = x(t - \Delta t) + 2\Delta t x'(t) + O(\Delta t^3).$$

In other words, $x(t - \Delta t) + 2\Delta t x'(t)$ is an approximation to $x(t + \Delta t)$ that has error $O(\Delta t^3)$, making leapfrog a second-order scheme.

There is a small catch in the error analysis here, with which we need to be careful. We need two initial points to get the leapfrog started (x_0 and x_1). Our problems are generally posed so that we have only one initial point, x_0. We therefore need to use some other estimate to get the initial x_1 before we can start the leapfrog iterations. We can use forward or backward Euler to go from x_0 to x_1, introducing an $O(\Delta t^2)$ error for just that one step, then use leapfrog for $x_2, \ldots, x_n$ for a total accumulated error of $O(\Delta t^2)$. If we were using a higher-order multistep method, we would need to use a higher-order initial scheme to get it started. For example, if we used a third-order multistep scheme but used forward Euler to get the initial points needed to start it, then the second-order error of a fixed number of Euler iterations would overwhelm the third-order error of the multistep iterations. We would need to initialize the iterator with the midpoint method or some other second-order single-step scheme to get the benefit of a third-order multistep scheme.

The stability of leapfrog is a little harder to analyze than the accuracy, using the tools we have covered. We have to treat it as a *linear recurrence relation*, a general term for an iterator in which we compute x_{n+1} from a linear combination of prior iterates. Using the guess $x(t) = e^{\omega t}$ gives us

$$x_{n+2} = x_n + 2\Delta t f(x_{n+1}) = x_n + 2\Delta t \omega x_{n+1}.$$

Solutions to a linear recurrence relation usually have the form $x_n = \sum_j \alpha_j \gamma_j^n$ for some sets of constants α_j and γ_j. Therefore, we can try guessing $x_n = \gamma^n$. Our iterator then becomes

$$\gamma^{n+2} = \gamma^n + 2\Delta t \omega \gamma^{n+1}$$

$$\gamma^2 = 1 + 2\Delta t \omega \gamma$$

(or $\gamma = 0$)

$$\gamma^2 - 2\Delta t \omega \gamma - 1 = 0$$

$$\gamma = \Delta t \omega \pm \sqrt{(\Delta t \omega)^2 + 1}.$$

The preceding formula tells us the possible values for the growth factor $g(\omega \Delta t)$. While it is a little hard to see what this means, we can learn an important feature of this $g(\omega \Delta t)$ by assuming $\omega \Delta t = \sqrt{\beta} i$ for some constant β. Making the substitution gives us

$$\gamma = \sqrt{\beta} i \pm \sqrt{1 - \beta}$$

$$|\gamma| = \sqrt{1 - \beta + \beta} = 1, \quad \text{where } \beta < 1.$$

What does this mean? If ω is strictly imaginary (meaning the system is oscillatory) and $\Delta t < |\frac{1}{\omega}|$, then the leapfrog method will be semistable. It will produce an oscillatory motion that does not grow or decay over time. This makes leapfrog a good choice when simulating a system that we know should produce oscillatory motion.

Just as midpoint was an example of the broader class of Runge–Kutta methods, so leapfrog is an example of a broader class of multistep methods called the *Adams–Bashforth* methods. Adams–Bashforth methods can similarly be found for any desired order of accuracy. Unlike Runge–Kutta methods, though, Adams–Bashforth methods have smaller stability regions as we go to higher orders of accuracy. Multistep methods are less widely used in practice than single-step methods even though they are superficially more efficient, in large part because they do not deal as well with some practical issues relating to step size selection that we will consider in the next section.

14.6   Step Size Selection

All of the above discussion assumes we have some fixed Δt. If we know the behavior of our problem well, we can often choose a good value of Δt that will maintain some desired error bound. For example, suppose we are using forward Euler and want an error bounded by some ε. We will need $\tau/\Delta t$ steps with error $\frac{x''(\xi)}{2!}\Delta t^2$ per step. If we can bound $x''(\xi)$ by some constant C, we can insist that

$$\frac{\tau}{\Delta t} \times \frac{C}{2!} \times \Delta t^2 < \varepsilon$$

$$\Delta t < \frac{2\varepsilon}{\tau C}.$$

We can therefore exactly set a time step to guarantee our desired error bound. We can perform similar analyses for other finite difference schemes.

For many real-world problems, though, we may not have a bound, or at least a good one, for the derivatives of x. Or it may be that the derivatives of x vary so much from one domain of the problem to another that a bound that is good for a few time points is much too conservative for most others. We may then waste an enormous amount of work using a much smaller step size than we need for most of our problem. We can get around this issue by using *adaptive methods*. Adaptive methods automatically adjust the step size based on empirical estimates of the error as they run, shrinking the step size in hard problem regions and expanding it in easy ones.

One simple variant of adaptive step size selection works by computing two different approximations for each time step. For example, we can compute the following two approximations:

$$x_{n+1} = x_n + \Delta t f(x_n)$$

$$\hat{x}_{n+1/2} = x_n + \frac{\Delta t}{2} f(x_n), \quad \hat{x}_n = \hat{x}_{x+1/2} + \frac{\Delta t}{2} f(x_{n+1/2}).$$

These are the standard forward Euler approximation and a second forward Euler approximation with half the step size. We can then use the fact that the error in $\hat{x}_{n+1}$ should be about half of that in x_{n+1} to approximately bound the error in the current step by $|x_{n+1} - \hat{x}_{n+1}|$. If $|x_{n+1} - \hat{x}_{n+1}| < \frac{\varepsilon \Delta t}{\tau}$, then we use $\hat{x}_{n+1}$ as our approximation for the current step. Otherwise, we cut the step size in half and try recomputing $\hat{x}_{n+1}$ and x_{n+1} with the new step size, repeating until the estimated error is sufficiently small. The result will be a method that picks some appropriate step size for each step, based on a baseline maximum step size Δt.

We can improve on the approach a bit by using the following two approximations:

$$x_{n+1} = x_n + \Delta t f(x_n)$$

(forward Euler)

$$\hat{x}_{n+1} = x_n + \frac{\Delta t}{2} (f(x_n) + f(x_{n+1}))$$

(a second-order approximation called *Huen's method*.)

Then, if $|x_{n+1} - \hat{x}_{n+1}| < \frac{\varepsilon \tau}{\Delta t}$, we use $\hat{x}_{n+1}$ as our approximation; otherwise, we cut the step size in half and try again. Note that the work we did in computing x_{n+1} is used in computing $\hat{x}_{n+1}$. This adaptive method is an example of an *embedded method*, so called because we embed the computation of a lower-order accuracy approximation inside the computation of a higher-order accuracy approximation. We then compare the solutions at different orders of accuracy to estimate the error. Many embedded Runge–Kutta schemes with varying levels of accuracy have been defined in the literature. References and Further Study provides pointers to references in which one can find some of these methods. Later on, in chapter 22, we will see a method called Richardson extrapolation that can be used to create adaptive schemes mixing any desired orders of accuracy.

There are various other practical tricks we can use to speed up numerical integration. For example, if our function is reasonably well-behaved, it will usually be more efficient to use a step size only slightly larger than that from the prior step as a best guess to the next step in an adaptive integrator. We may start with double the prior step size as our first guess instead of going back to Δt as the first guess at each step, which allows us to make use of the fact that the problem characteristics are likely to be similar in local regions while still giving us a way to increase step sizes when we move out of hard regions.

Developing adaptive methods is much harder to do correctly for multistep schemes, since changing the time step essentially renders most of the prior points useless. It is always possible to combine approximations at variable distances in time to get a desired order of accuracy using Taylor series. But we are very unlikely to get a method with good stability properties by combining points at some arbitrary set of distances from one another. There is a class of multistep schemes called *predictor-corrector schemes* that do manage to implement step-size selection within a multistep scheme without compromising stability. For example, we can combine an Adams–Bashforth iterator with a kind of implicit multistep scheme called an Adams–Moulton scheme to create an adaptive multistep method called an Adams–Bashforth–Moulton scheme. In the interests of space, though, we will not study these more advanced methods here and will recommend sticking with Runge–Kutta methods when an adaptive scheme is needed.

References and Further Study

Probably the handiest practical reference for using the techniques we have covered here is the Numerical Recipes series [82], which presents most of the methods examined above. It also contains a broader selection of Runge–Kutta, Adams–Bashforth, and Adams–Moulton schemes, including other adaptive, embedded, and predictor-corrector variations. Greater depth may be found in any of many texts available on numerical analysis or differential equations. Stoer and Bulirsch [132] is highly recommended for numerical analysis topics in general. It has extensive coverage of numerical integration methods, including many specialized methods and aspects of the underlying theory we do not have room to cover here. Kythe and Schäferkotter [133] provides extensive coverage of numerical integration specifically, including many kinds of techniques we did not cover in this chapter. Readers may also refer to Hamming [134] for some additional schemes and some of the theory behind the methods we have seen and their analysis.

For biological applications, especially in biophysics, it is often helpful to use more specialized integration schemes specifically designed so that their errors obey physical conservation laws. Those with need for greater depth on this topic can refer to LeVeque [135].

Though we looked only at finite difference schemes, there are some entirely different ways of doing numerical integration that we did not consider here. After finite difference schemes, the most widely used methods are the *spectral methods*, which use Fourier transforms to perform integration in the frequency domain. The reader can refer to Gottlieb and Orszag [136] for more information on these methods. Similar methods have been developed based on wavelet transforms, for which the reader can refer to Kythe and Schäferkotter [133].

15 Numerical Integration 2: Partial Differential Equations

In the last chapter, we saw a few schemes for numerical integration of systems of differential equations. As we saw, it is relatively easy to deal with systems of multiple dependent variables. In this chapter, we will examine the somewhat harder problem of systems with multiple independent variables. Typically, time will be one of the independent variables. The others are likely to be spatial dimensions, although they can be many other things.

An important example of a problem with multiple independent variables is diffusion of a chemical in a solution. Imagine we have a solution divided into two parts by a removable barrier. In one half, we have a high concentration of some solute and in the other a low concentration, as illustrated in figure 15.1. What will happen to this solution if we remove the barrier? The answer will be a function of x, y, z, and t, which we can denote by $C(x, y, z, t)$, representing the concentration of solute at point (x, y, z) at time t.

We then want to know how this system behaves. Starting from some initial state $C(x, y, z, 0)$, the behavior of a diffusive system is described by the following partial differential equation (PDE):

$$\frac{\partial C}{\partial t} = D\left(\frac{\partial^2 C}{\partial x^2} + \frac{\partial^2 C}{\partial y^2} + \frac{\partial^2 C}{\partial z^2}\right).$$

This equation says that the partial derivative of concentration with respect to time varies with the sum of the second partial derivatives with respect to each spatial dimension. The equation can be abbreviated

$$\frac{\partial C}{\partial t} = D\nabla^2 C.$$

The operator ∇^2 is called a *Laplacian*. The particular equation above is known as the *heat equation* because it was originally formulated to model how heat applied to the surface of a material will diffuse through the material. It is, however, a good description of many kinds of diffusion-related phenomena, including chemical diffusion

removable barrier

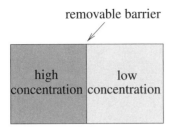

Figure 15.1
Setup of a chemical diffusion problem. Solutions at two concentrations are separated by a barrier. We are interested in how concentration varies as a function of time and space after we remove the barrier.

processes. In the remainder of this chapter, we will look at how to numerically integrate systems, like this one, involving partial derivatives with respect to multiple independent variables.

15.1 Problems of One Spatial Dimension

To begin with, we will simplify a bit and assume we have just one time dimension t and one spatial dimension x. Then we will examine the problem

$$\frac{\partial C}{\partial t} = D\frac{\partial^2 C}{\partial x^2}.$$

Let us further assume that we are looking at a system on the time range $t = [0, 1]$ and the spatial range $x = [0, 1]$.

Just as we found it useful before to break continuous time into discrete time steps, so we will find it useful to break continuous space into discrete spatial points. We will therefore assume that we are evaluating C at specific time points $0, \Delta t, 2\Delta t, 3\Delta t, \ldots$ and at specific spatial points $0, \Delta x, 2\Delta x, 3\Delta x, \ldots$. Then we can convert our problem into a system of a finite number of coupled differential equations:

$$\frac{\partial C(0, t)}{\partial t} = D\frac{\partial^2 C(0, t)}{\partial x^2}$$

$$\frac{\partial C(\Delta x, t)}{\partial t} = D\frac{\partial^2 C(\Delta x, t)}{\partial x^2}$$

$$\frac{\partial C(2\Delta x, t)}{\partial t} = D\frac{\partial^2 C(2\Delta x, t)}{\partial x^2}$$

$\vdots$

That is, we treat the concentration at each discretized point in space as if it were a different variable. Then we can integrate the system in time just as we did before for systems of one independent and multiple dependent variables. For example, suppose we define $x_i = i\Delta x$ and $C_j(x_i)$ to be our approximation to $C(i\Delta x, j\Delta t)$. We can then numerically integrate this system with forward Euler by the following iterator:

$$C_{n+1}(x_i) = C_n(x_i) + \Delta t \frac{\partial C_n(x_i)}{\partial t} = C_n(x_i) + \Delta t D \frac{\partial^2 C_n(x_i)}{\partial x^2}.$$

This formula raises a problem, though. Now that we have discretized x, how do we evaluate $\frac{\partial^2 C_n(x_i)}{\partial x^2}$? The answer is that we can estimate spatial derivatives from spatially discretized values using a numerical derivative formula. We saw some formulas for numerically estimating derivatives in chapter 5. We will now similarly use combinations of a few adjacent grid points to estimate the second derivative at a given grid point. One way to perform this estimate is through a *centered difference* approximation:

$$\frac{\partial^2 f}{\partial x^2} \approx \frac{f(x + \Delta x) + f(x - \Delta x) - 2f(x)}{\Delta x^2}.$$

We can analyze the accuracy of this approximation in much the same way as we analyze the accuracy of a finite difference scheme: Taylor series. Applying the Taylor series for $f(x + \Delta x)$ and $f(x - \Delta x)$ gives us the following:

$$f(x + \Delta x) = f(x) + \Delta x f'(x) + \frac{\Delta x^2}{2!} f''(x) + \frac{\Delta x^3}{3!} f'''(x) + O(\Delta x^4)$$

$$+f(x - \Delta x) = f(x) - \Delta x f'(x) + \frac{\Delta x^2}{2!} f''(x) - \frac{\Delta x^3}{3!} f'''(x) + O(\Delta x^4)$$

$$\frac{-2f(x) = -2f(x)}{f(x + \Delta x) + f(x - \Delta x) - 2f(x) = 0 + 0 + \Delta x^2 f''(x) + 0 + O(\Delta x^4)}$$

Therefore,

$$\frac{f(x + \Delta x) + f(x - \Delta x) - 2f(x)}{\Delta x^2} = f''(x) + O(\Delta x^2).$$

That is, $\frac{f(x+\Delta x)+f(x-\Delta x)-2f(x)}{\Delta x^2}$ is an approximation to $\frac{\partial^2 f}{\partial x^2}$ with error $O(\Delta x^2)$. We can use this approximation to the second derivative to construct an iterator with which we can numerically integrate our diffusion system:

$$C_{n+1}(x_i) = C_n(x_i) + \Delta t D \frac{C_n(x_{i+1}) + C_n(x_{i-1}) - 2C_n(x_i)}{\Delta x^2}.$$

At each time step, we apply the preceding iterator at all discrete spatial points x_i. Since we calculate $C_{n+1}(x_i)$ using only $C_n(x_j)$ values, we can compute the new value at a given grid point x_i for time step $n+1$ using only values from time step n.

15.2   Initial Conditions and Boundary Conditions

Our scheme is not yet complete, though. The next thing we need to know is the *initial conditions* of the problem. When there is one independent variable, the initial condition is just a single number. For example, if we are solving $\frac{dx}{dt} = f(x)$, we need to know $x(0)$ to get started. When we have spatial dimensions to consider, the initial condition specifies the starting point at time zero for all spatial positions. For our one-dimensional example, we need to know $C(x, 0)$ for all x. Since x is discretized into $x_i \in \{0, \Delta x, 2\Delta x, \ldots\}$, we need to know

$$C(x_0, 0) = C(0, 0)$$

$$C(x_1, 0) = C(\Delta x, 0)$$

$$C(x_2, 0) = C(2\Delta x, 0)$$

$$\vdots$$

These values will usually be inferred from our model. For example, if we are simulating diffusion into a channel that is initially empty of our solute, we will say $C(x_i, 0) = 0$ for all i.

A trickier issue is the specification of *boundary conditions*. Boundary conditions describe the behavior of the model at the edges of the space. For example, our centered difference scheme for the second derivative of $C_n(x_i)$ with respect to x requires that we know $C_n(x_{i-1})$ and $C_n(x_{i+1})$. We need a way to find these values if x_i is at the edge of our grid.

The simplest form of boundary condition is one where we directly specify the value of the function at the boundary. That is, we may declare that $C(0, t) = C_0$ and $C(1, t) = C_1$ for fixed C_0 and C_1 and for all time. This may be the case if we assume our tube is a thin capillary connecting two compartments with different concentrations, as illustrated in figure 15.2(a). Then we may assume that the capillary allows so little solution to flow that the two compartments essentially remain at constant concentrations. A boundary condition specifying the value at the boundary is called a *Dirichlet boundary condition*. Given this boundary condition, we will run our meth-

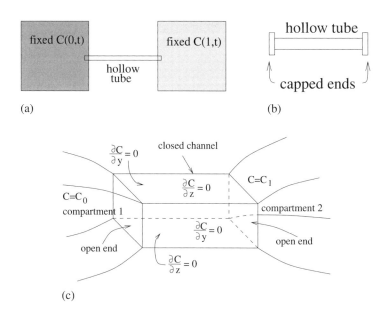

Figure 15.2
Boundary conditions for variants of diffusive system integration. (a) A Dirichlet boundary condition created by assuming that we are studying a thin tube whose ends are immersed in large volumes of fixed concentration. (b) A Neumann boundary condition created by assuming the ends of the tube are capped. (c) A more complicated three-dimensional model of diffusion through a channel where we might assume a combination of constraints on the value and the derivative of the concentration on different boundaries of the channel.

ods exactly as before, except that we plug in C_0 for $C_{n+1}(0)$ and C_1 for $C_{n+1}(1/\Delta t)$ in the numerical derivative formula when updating points adjacent to the boundaries.

It is also possible that the boundary condition may itself be a function of time. For example, we may imagine that we are steadily diluting the solution in one compartment while holding the other fixed. Then we may declare that our boundary conditions are

$$C(0, t) = C_0, \quad C(1, t) = \frac{C_1}{t},$$

and therefore

$$C_{n+1}(0) = C_0, \quad C_{n+1}(1/\Delta x) = \frac{C_1}{(n + 1)\Delta t}.$$

Often we will have a boundary condition on the derivative of concentration instead of the value of concentration. This is known as a *Neumann boundary condition*.

This kind of condition may occur if we were to cap the ends of our tube, as in figure 15.2(b). In this case, the closed ends of the tube mean that there must be no net diffusion across the boundary. We will impose that constraint on the model with the boundary conditions

$$\frac{\partial C(0)}{\partial x} = \frac{\partial C(1)}{\partial x} = 0.$$

This kind of boundary condition raises a small problem for us, since we can no longer apply our finite difference scheme at the boundaries. For example, if we want to update the concentration at the leftmost point in the tube, $C_n(\Delta x)$, our scheme requires us to know $C_n(0)$, which is now undefined. We therefore need to replace our centered difference scheme with a different approximation for $\frac{\partial^2 C(x_i)}{\partial x^2}$:

$$\frac{\partial^2 C(x_i)}{\partial x^2} \approx \frac{C_n(2\Delta x) - C_n(\Delta x) - \Delta x C_n'(0)}{(3/2)\Delta x^2}.$$

We can understand why this approximation works by again applying Taylor series:

$$f(2\Delta x) = f(0) + 2\Delta x f'(0) + 4\Delta x^2 \frac{f''(0)}{2!} + O(\Delta x^3)$$

$$-f(\Delta x) = f(0) + \Delta x f'(0) + \Delta x^2 \frac{f''(0)}{2!} + O(\Delta x^3)$$

$$\frac{-\Delta x f'(0) = -\Delta x f'(0)}{f(2\Delta x) - f(\Delta x) - \Delta x f'(0) = 0 + 0 + (3/2)\Delta x^2 f''(0) + O(\Delta x^3)}$$

Therefore, $\frac{C_n(2\Delta x) - C_n(\Delta x) - \Delta x C_n'(0)}{(3/2)\Delta x^2}$ is an approximation to $f''(\Delta x)$ that has error $O(\Delta x)$. We can, of course, find other schemes that cancel out more error terms and give us higher accuracy. For example, we may look at a fourth point to get another equation we can use to cancel out the $O(\Delta x^3)$ term.

It is also possible to mix boundary conditions. For example, we may have a tube open at one end and capped at the other, giving boundary conditions such as

$$C(0, t) = C_0, \quad \frac{\partial C(1, t)}{\partial x} = 0.$$

In higher dimensions, a boundary may be a surface rather than a point. In such cases, we need to specify the boundary condition for every point on each boundary surface. This condition may itself be a function of space. For example, if one boundary is an $x - y$ plane, then there may be a concentration gradient within the plane expressed as some function of x and y.

15.3 An Aside on Step Sizes

When choosing step sizes for time and space for a multidimensional scheme, it is important to consider the accuracy of the approximations in both dimensions. Suppose we rearrange our finite difference scheme from above as follows:

$$\frac{C_{n+1}(x_i) - C_n(x_i)}{\Delta t} = \frac{C_n(x_{i+1}) + C_n(x_{i-1}) - 2C_n(x_i)}{\Delta x^2}.$$

Then we can interpret each side of the equation as an approximation to a derivative:

$$\frac{\partial C}{\partial t} + O(\Delta t) = \frac{\partial^2 C}{\partial x^2} + O(\Delta x^2).$$

We would say that this is a first-order scheme in time and a second-order scheme in space.

We can infer from these error terms that an extremely small Δx is unlikely to be useful with this scheme, since the error from the x discretization will be dominated by the error from the t discretization. In general, we want $\Delta t \sim \Delta x^2$, so that the two sources of error will have similar magnitudes. Any decrease of either step size beyond that will lead to wasted work. Furthermore, if we decide to use a higher-order scheme in the x dimension, say $O(\Delta x^3)$, then we need to increase the spatial step size or shrink the time step size to bring them back into balance if we want to benefit from the new scheme.

15.4 Multiple Spatial Dimensions

If we now move back to three dimensions, then our diffusive system will take the form

$$\frac{\partial C}{\partial t} = D\left(\frac{\partial^2 C}{\partial x^2} + \frac{\partial^2 C}{\partial y^2} + \frac{\partial^2 C}{\partial z^2}\right).$$

To approximately integrate this system, we discretize in time and in all three spatial dimensions, which can give us the following finite difference scheme:

$$\frac{C_{n+1}(x_i, y_j, z_k) - C_n(x_i, y_j, z_k)}{\Delta t}$$

$$= D \frac{C_n(x_{i+1}, y_j, z_k) + C_n(x_{i-1}, y_j, z_k) - 2C_n(x_i, y_j, z_k)}{\Delta x^2}$$

$$+ D \frac{C_n(x_i, y_{j+1}, z_k) + C_n(x_i, y_{j-1}, z_k) - 2C_n(x_i, y_j, z_k)}{\Delta y^2}$$

$$+ D \frac{C_n(x_i, y_j, z_{k+1}) + C_n(x_i, y_j, z_{k-1}) - 2C_n(x_i, y_j, z_k)}{\Delta z^2}.$$

Simulating now involves a somewhat more complicated set of iterations because we must increment over successive time steps, and at each time step must update all spatial points. Our numerical integration scheme will then be coded as a loop over x, y, and z dimensions for each time point.

For this three-dimensional version we need initial conditions $C_0(x_i, y_j, z_k)$ for every discrete point (x_i, y_j, z_k). We also need boundary conditions for all faces of the space we are considering. We may again have mixed kinds of boundary conditions. For example, we may have some boundaries of our space sealed off, in which case we will fix the derivative to 0 as our boundary condition. We may have others where we assume we are connecting our space to some compartment of uniform concentration. Figure 15.2(c) illustrates a possible way the boundary conditions may be set for this system.

15.5   Reaction–Diffusion Equations

One particularly useful application of multidimensional integration for biology and biochemistry is for simulating chemistry in an inhomogeneous solution. For example, suppose we have the reaction $A + B \rightleftharpoons C$. Suppose further that we start with a beaker full of A and add a drop of concentrated B to the top of the beaker. That drop will begin reacting with the A when it hits the solution and will simultaneously diffuse outward from where it landed. The time progress of a system like this is described by a special kind of partial differential equation model called a *reaction–diffusion equation*. The general form of a reaction–diffusion equation is the following:

$$\frac{\partial \vec{C}}{\partial t} = \vec{D} \nabla^2 \vec{C} + \vec{f}(\vec{C}).$$

Here, $\vec{C}$ is a vector of possibly several reactant concentrations. $\vec{D}$ is a diagonal matrix describing the diffusion rate of each reactant. $\vec{D} \nabla^2 \vec{C}$ is collectively called the *diffusion term*. $\vec{f}(\vec{C})$ is called the *reaction term* and is simply our standard system of mass action differential equations for the reaction neglecting space.

Example Suppose we have a simple dimer system $A + A \rightleftharpoons_{k_-}^{k_+} B$. Further suppose that A has diffusion rate D_A and B has diffusion rate D_B. Then we will have the following components for our reaction–diffusion system:

$$\vec{C} = \begin{bmatrix} [A] \\ [B] \end{bmatrix}$$

$$\vec{D} = \begin{bmatrix} D_A & 0 \\ 0 & D_B \end{bmatrix}$$

$$\vec{f}(\vec{C}) = \begin{bmatrix} f_A(\vec{C}) \\ f_B(\vec{C}) \end{bmatrix} = \begin{bmatrix} -2k_+[A]^2 + 2k_-[B] \\ k_+[A]^2 - k_-[B] \end{bmatrix}.$$

Putting these pieces together, we get the following reaction–diffusion equations:

$$\frac{\partial [A]}{\partial t} = D_A \left(\frac{\partial^2 [A]}{\partial x^2} + \frac{\partial^2 [A]}{\partial y^2} + \frac{\partial^2 [A]}{\partial z^2} \right) + (-2k_+[A]^2 + 2k_-[B])$$

$$\frac{\partial [B]}{\partial t} = D_B \left(\frac{\partial^2 [B]}{\partial x^2} + \frac{\partial^2 [B]}{\partial y^2} + \frac{\partial^2 [B]}{\partial z^2} \right) + (k_+[A]^2 - k_-[B]).$$

To convert these equations into a numerical scheme, we have to discretize t, x, y, and z. Using forward Euler in time and a centered difference in space will give us the following scheme:

$$\frac{[A]_{n+1}(x_i, y_j, z_k) - [A]_n(x_i, y_j, z_k)}{\Delta t} = D_A \left(\begin{array}{l} \frac{[A]_n(x_{i+1}, y_j, z_k) + [A]_n(x_{i-1}, y_j, z_k) - 2[A]_n(x_i, y_j, z_k)}{\Delta x^2} \\ + \frac{[A]_n(x_i, y_{j+1}, z_k) + [A]_n(x_i, y_{j-1}, z_k) - 2[A]_n(x_i, y_j, z_k)}{\Delta y^2} \\ + \frac{[A]_n(x_i, y_j, z_{k+1}) + [A]_n(x_i, y_j, z_{k-1}) - 2[A]_n(x_i, y_j, z_k)}{\Delta z^2} \end{array} \right)$$

$$+ (-2k_+([A]_n(x_i, y_j, z_k))^2 + 2k_-[B]_n(x_i, y_j, z_k))$$

$$\frac{[B]_{n+1}(x_i, y_j, z_k) - [B]_n(x_i, y_j, z_k)}{\Delta t} = D_B \left(\begin{array}{l} \frac{[B]_n(x_{i+1}, y_j, z_k) + [B]_n(x_{i-1}, y_j, z_k) - 2[B]_n(x_i, y_j, z_k)}{\Delta x^2} \\ + \frac{[B]_n(x_i, y_{j+1}, z_k) + [B]_n(x_i, y_{j-1}, z_k) - 2[B]_n(x_i, y_j, z_k)}{\Delta y^2} \\ + \frac{[B]_n(x_i, y_j, z_{k+1}) + [B]_n(x_i, y_j, z_{k-1}) - 2[B]_n(x_i, y_j, z_k)}{\Delta z^2} \end{array} \right)$$

$$+ (k_+([A]_n(x_i, y_j, z_k))^2 - k_-[B]_n(x_i, y_j, z_k)).$$

We can implement this scheme using several nested loops: an outer one iterating over time steps and three nested inner loops iterating over the three spatial dimensions for each time step.

   We can similarly simulate this system with higher-order schemes. For example, if we want a higher-order multistep scheme in time, we can substitute a higher-order approximation to $\frac{\partial C}{\partial t}$ on the left-hand side of the iterators above. For example, to do a leapfrog scheme in time with the same spatial scheme, we can rewrite the leapfrog iterator

$$x_{n+1} = x_{n-1} + 2\Delta t f(x_n)$$

as

$$\frac{x_{n+1} - x_{n-1}}{2\Delta t} = f(x_n),$$

giving us the following reaction–diffusion scheme:

$$\frac{[A]_{n+1}(x_i, y_j, z_k) - [A]_{n-1}(x_i, y_j, z_k)}{2\Delta t} = D_A \left(\begin{array}{c} \frac{[A]_n(x_{i+1}, y_j, z_k) + [A]_n(x_{i-1}, y_j, z_k) - 2[A]_n(x_i, y_j, z_k)}{\Delta x^2} \\ + \frac{[A]_n(x_i, y_{j+1}, z_k) + [A]_n(x_i, y_{j-1}, z_k) - 2[A]_n(x_i, y_j, z_k)}{\Delta y^2} \\ + \frac{[A]_n(x_i, y_j, z_{k+1}) + [A]_n(x_i, y_j, z_{k-1}) - 2[A]_n(x_i, y_j, z_k)}{\Delta z^2} \end{array} \right)$$

$$+ \left(-2k_+([A]_n(x_i, y_j, z_k))^2 + 2k_-[B]_n(x_i, y_j, z_k) \right)$$

$$\frac{[B]_{n+1}(x_i, y_j, z_k) - [B]_{n-1}(x_i, y_j, z_k)}{2\Delta t} = D_B \left(\begin{array}{c} \frac{[B]_n(x_{i+1}, y_j, z_k) + [B]_n(x_{i-1}, y_j, z_k) - 2[B]_n(x_i, y_j, z_k)}{\Delta x^2} \\ + \frac{[B]_n(x_i, y_{j+1}, z_k) + [B]_n(x_i, y_{j-1}, z_k) - 2[B]_n(x_i, y_j, z_k)}{\Delta y^2} \\ + \frac{[B]_n(x_i, y_j, z_{k+1}) + [B]_n(x_i, y_j, z_{k-1}) - 2[B]_n(x_i, y_j, z_k)}{\Delta z^2} \end{array} \right)$$

$$+ \left(k_+([A]_n(x_i, y_j, z_k))^2 - k_-[B]_n(x_i, y_j, z_k) \right)$$

Note that it is written in the above form for convenience in separating the temporal and spatial schemes. We would actually use the scheme to find A_{n+1} and B_{n+1} from the prior time points, as follows:

$$[A]_{n+1}(x_i, y_j, z_k) = [A]_{n-1}(x_i, y_j, z_k)$$

$$+ 2\Delta t D_A \left(\begin{array}{c} \frac{[A]_n(x_{i+1}, y_j, z_k) + [A]_n(x_{i-1}, y_j, z_k) - 2[A]_n(x_i, y_j, z_k)}{\Delta x^2} \\ + \frac{[A]_n(x_i, y_{j+1}, z_k) + [A]_n(x_i, y_{j-1}, z_k) - 2[A]_n(x_i, y_j, z_k)}{\Delta y^2} \\ + \frac{[A]_n(x_i, y_j, z_{k+1}) + [A]_n(x_i, y_j, z_{k-1}) - 2[A]_n(x_i, y_j, z_k)}{\Delta z^2} \end{array} \right)$$

$$+ 2\Delta t(-2k_+([A]_n(x_i, y_j, z_k))^2 + 2k_-[B]_n(x_i, y_j, z_k))$$

$$[B]_{n+1}(x_i, y_j, z_k) = [B]_{n-1}(x_i, y_j, z_k)$$

$$+ 2\Delta t D_B \left(\begin{array}{c} \frac{[B]_n(x_{i+1}, y_j, z_k) + [B]_n(x_{i-1}, y_j, z_k) - 2[B]_n(x_i, y_j, z_k)}{\Delta x^2} \\ + \frac{[B]_n(x_i, y_{j+1}, z_k) + [B]_n(x_i, y_{j-1}, z_k) - 2[B]_n(x_i, y_j, z_k)}{\Delta y^2} \\ + \frac{[B]_n(x_i, y_j, z_{k+1}) + [B]_n(x_i, y_j, z_{k-1}) - 2[B]_n(x_i, y_j, z_k)}{\Delta z^2} \end{array} \right)$$

$$+ 2\Delta t(k_+([A]_n(x_i, y_j, z_k))^2 - k_-[B]_n(x_i, y_j, z_k)).$$

If we want a higher-order spatial scheme, we can substitute a different approximation scheme for $\nabla^2 \vec{C}$ on the right-hand side. If we want to use a higher-order Runge–Kutta scheme in time, we can treat the right-hand side as a black box for computing $\frac{\partial C}{\partial t}$ and plug that into any Runge–Kutta scheme. There is an infinite variety of combinations of schemes one can construct, and many specialized versions are known for these systems to have particularly good stability behavior, to obey conservation laws, or to exhibit various other useful features for different sorts of systems. The References and Further Study section at the end of the chapter provides some guidance for learning more about these well-studied schemes for PDEs.

15.6 Convection

Partial differential equations also provide a means for modeling directed movement of a solute through a solution, a process known as *convection*. For example, we may wish to model the movement of a substance that is being actively transported from one pole of a cell to the other. The change in local concentration due to convective motion is described by the first spatial derivative of the system. In the one-dimensional case, we can describe a solute C being transported in the positive x direction by the equation

$$\frac{\partial C}{\partial t} = -\frac{\partial C}{\partial x}.$$

More generally, we may have components of the x, y, and z dimensions to create convection in the direction of some arbitrary spatial vector (v_x, v_y, v_z), as follows:

$$\frac{\partial C}{\partial t} = -v_x \frac{\partial C}{\partial x} - v_y \frac{\partial C}{\partial x} - v_z \frac{\partial C}{\partial x}.$$

We can numerically integrate a convective system, much like a diffusive one, by combining a numerical spatial derivative with a numerical scheme for time integration. For example, forward Euler with a first-order forward difference approximation in space will give the following scheme for the preceding convection problem:

$$C_{n+1}(x_i, y_i, z_i) = C_n(x_i, y_i, z_i) - \Delta t \left(v_x \frac{C_n(x_{i+1}, y_i, z_i) - C_n(x_i, y_i, z_i)}{\Delta x} \right.$$

$$+ v_y \frac{C_n(x_i, y_{i+1}, z_i) - C_n(x_i, y_i, z_i)}{\Delta y}$$

$$\left. + v_z \frac{C_n(x_i, y_i, z_{i+1}) - C_n(x_i, y_i, z_i)}{\Delta z} \right).$$

This scheme will have first-order accuracy in both Δt and Δx.

Just as we can combine reaction and diffusion terms to get a reaction–diffusion equation, so we can combine convection and diffusion terms to get a *convection–diffusion* equation. Or we can combine convection, reaction, and diffusion terms for a model of chemistry in which both passive diffusion and active transport can occur:

$$\frac{\partial C}{\partial t} = -c\frac{\partial C}{\partial x} + d\frac{\partial^2 C}{\partial x^2} + f(C).$$

These terms are additive, so we can evaluate them independently when computing the right-hand side of a numerical iterator and then add up their individual contributions to get the estimates for the next time point.

There are many other special cases of partial differential equations important to biological systems that one may consider. For example, the *wave equation*

$$\frac{\partial^2 C}{\partial t^2} = a^2 \nabla^2 C$$

is important to various applications in biophysics, as well as many other contexts in physics, optics, and fluid dynamics. The principles of working with this or any other

PDE are the same as those we studied above for diffusive systems, although well-studied equations like this one will generally have their own special-purpose schemes in addition to the default ones we can derive.

References and Further Study

Numerical integration of partial differential equations, like numerical integration in general, is a big field of which we have seen only a brief sample. Although one can get far with the techniques discussed above for combining standard temporal ODE methods with simple spatial derivative approximations, advanced work in the field relies on a large body of specialized schemes developed for particular kinds of systems. For example, stability questions can get quite complicated with partial differential equations, and a variety of special-purpose integration schemes are available for handling different PDEs in practice. Some of this material is covered in the same references as we saw for ODEs, such as Stoer and Bulirsch [132], Kythe and Schäfer-kotter [133], LeVeque [135], and the Numerical Recipes series [82]. More depth about PDEs specifically can be found in Strickwerda [137]. Just as with ODEs, there are alternatives to finite difference methods for PDEs. For example, spectral methods can apply to PDEs just as they do to ODEs, and one can often productively combine a spectral method in some independent variables with a finite difference method in others. Readers can refer to Gottlieb and Orszag [136] for more information.

The primary literature on the specific problems covered here is of some curious historical interest. The heat equation was developed by Fourier in the same paper in which he first developed the theory of Fourier series, "Théorie analytique de la chaleur." In fact, Fourier theory was originally developed specifically to describe the solutions to the heat equation. The reaction–diffusion equation was first developed by Alan Turing, one of the most important figures in the history of computer science but someone not generally remembered for his contributions to biology, as part of a mostly forgotten theory of biological morphogenesis [138]. Information on other important PDEs we have not covered in depth may be found in the references above.

16 Numerical Integration 3: Stochastic Differential Equations

So far, we have seen various kinds of discrete models one can use to represent stochastic or "noisy" systems, and we have seen continuous methods one can use for smoothly evolving, predictable systems. In some cases, though, we may want to simulate a system that is both continuous and noisy. The canonical example of this is a particle acting under the influence of Brownian motion. If we watch one particle of some sort in a solution undergoing Brownian motion, that particle will clearly have a continuous path; it occupies some position at any point in time and must transition between those positions at intermediate times. Its path is also stochastic, however. In this chapter, we will see how we can model a system that is both continuous and stochastic. We will begin by considering the case of Brownian motion and then show how to generalize to broader classes of stochastic continuous functions.

16.1 Modeling Brownian Motion

Physically, Brownian motion describes how a particle moves in a solution subject to random thermal fluctuations within the solution. In order to model the process, however, we need a more precise mathematical model of Brownian motion. There is a formal definition of Brownian motion, referred to as a *standard Weiner process* $W(t)$, which is a random variable on time t satisfying the following properties:

1. $W(0) = 0$.
2. For any s and t satisfying $0 \leq s < t$, $W(t) - W(s)$ is a normal random variable with mean 0 and variance $t - s$.
3. For any s, t, u, and v satisfying $0 \leq s < t < u < v$, $W(t) - W(s)$ and $W(v) - W(u)$ are independent random variables.

This process will describe a continuous motion, but one that is stochastic. Like a discrete Markov model, there is no state except the current position, so we do not need to know where the particle has been to figure out where it is likely to go next. This process also has a kind of fractal property in that we can zoom in to any resolution and the process will still exhibit the same stochastic behavior.

This definition suggests a numerical method for simulating Brownian motion at any desired degree of resolution. We first discretize time, deciding that we will look at increments of some size Δt. We then simulate movements between successive steps of size Δt, using the three properties above. Property 1 tells us our initial conditions (i.e., that we start at position zero). Property 2 tells us that we can update our particle positions at successive time points by generating normally distributed random variables. Property 3 tells us that we can sample the variables for successive time steps independently from one another. Putting all that together, we get the following simple pseudocode for simulating a one-dimensional Brownian walk for some time T:

$x \leftarrow 0$

for $i = 1$ to $T/\Delta t$

$\quad x \leftarrow x + \sqrt{\Delta t} N(0, 1).$

Figure 16.1(a) shows a path generated from this procedure for $T = 1$ and $\Delta t = 0.001$.

We can generalize this model pretty easily to some more complicated Brownian motion models. For example, if we want a particle moving in three dimensions, we can use the same procedure for each dimension. If we want to track multiple particles, then we can keep an array of x variables, as long as the particles do not collide. Collisions, though, or many other kinds of forces that may act on these particles, do not clearly follow from the above pseudocode. We therefore will need to consider a more general approach to stochastic continuous simulation.

16.2 Stochastic Integrals and Differential Equations

We can come up with a more general formulation by recognizing that our procedure for simulating Brownian motion is very similar to how we numerically integrate a differential equation. In particular, we are using a sort of finite difference scheme. When we looked at finite difference schemes, we saw that we can integrate the differential equation

$$\frac{dx}{dt} = f(x)$$

using the forward Euler iterator

$$x_{n+1} = x_n + \Delta t f(x).$$

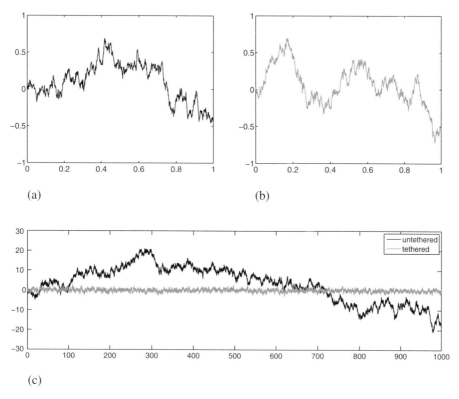

Figure 16.1
Numerical simulations of Brownian motion models. For each figure, the x axis represents time and the y axis, offset of the Brownian motion process from the origin. (a) Pure one-dimensional Brownian motion simulated with $\Delta t = 0.001$. (b) Simulation of a tethered particle acting under Brownian motion with a restoring force bringing it back toward the origin. (c) Side-by-side comparison of the two systems revealing their divergence over long time scales.

We can restate the forward Euler iterator as

$$x_{n+1} = x_n + dx,$$

where dx is the deterministic instantaneous change in x over one time step, $\Delta t f(x)$. With our Brownian motion simulations, we use a similar iterator:

$$x_{n+1} = x_n + dW,$$

where dW is a random variable $\sqrt{\Delta t} N(0, 1)$ representing the instantaneous change in position due to stochastic fluctuations from Brownian motion. In the former case, what we are really doing is using the fundamental theory of calculus to say that

$$\frac{dx}{dt} = f(x)$$

$$dx = f(x(t))\,dt$$

$$x = \int f(x(t))\,dt.$$

We can analogously say in the latter case that

$$dx = dW$$

$$x = \int dW.$$

Our pseudocode above for simulating Brownian motion is really just applying the forward Euler method to this latter "integral," except that we are integrating over a random variable dW in place of the fixed quantity dt.

This observation introduces the concept of *stochastic integrals*, in which we perform integration with respect to a random process instead of a uniform time. We can more generally think of integrating an arbitrary function $x(t)$ with respect to a Brownian motion process $W(t)$:

$$X(T) = \int_0^T x(t)\,dW(t).$$

There are different ways to define exactly what this notation means, though. We will examine one particular definition of stochastic integration called an *Itô integral*, in which the above stochastic integral is defined as

$$\lim_{\Delta t \to 0} \sum_{j=0}^{T/\Delta t - 1} x(j\Delta t)(W((j+1)\Delta t) - W(j\Delta t))$$

$$= \sum_{j=0}^{T/\Delta t - 1} x(j\Delta t)(\sqrt{\Delta t}N(0,1)).$$

There are alternative ways to define a stochastic integral, such as the *Stratonovich integral*, in which we use the limit of a slightly different sum, but it does not really matter which definition we choose as long as we are consistent about it. We will therefore somewhat arbitrarily confine ourselves to the Itô integral.

The reason it is worth going to the trouble of defining our Brownian motion process with the notation of integrals and differential equations is that this conversion provides a straightforward way to combine stochastic and deterministic components in the same differential equation. If we want to model a variable X that is affected by a deterministic force $f(X)$ and a stochastic force $g(X)$, we can describe X as

$$dX(t) = f(X(t)) \, dt + g(X(t)) \, dW(t).$$

Or, equivalently,

$$X(t) = X_0 + \int_0^t f(X(s)) \, ds + \int_0^t g(X(s)) \, dW(s).$$

For example, suppose we want to model the motion of a molecule acting under Brownian motion, but tethered to some much larger, immobile structure. This situation may arise if we have a protein anchored into a cell membrane, but one terminus of that protein is disordered and flops around in solution. If we want to track the movement of the endpoint of the disordered terminus, we may propose that it acts as if it were subject to Brownian motion, but with an additional ideal spring restoring force drawing it back to the point $x = 0$. Then we can describe its motion with the following stochastic differential equation (SDE):

$$dx(t) = -kx(t) \, dt + D \, dW(t),$$

where k is a spring constant for the tether and D is a diffusion constant. We can equivalently describe the system by the stochastic integral equation

$$x(t) = x_0 + \int_0^t -kx(s) \, ds + \int_0^t D \, dW(t).$$

16.3 Integrating SDEs

The above formulation of stochastic integrals and differential equations gives us a way of describing continuous stochastic processes, but it does not exactly tell us how to simulate them. We can derive a stochastic integration method by combining the stochastic finite difference method we derived for simulating Brownian motion above with one of our previous deterministic finite difference methods. For example, suppose we want to integrate our equation for Brownian motion of a tethered molecule. We previously simulated pure Brownian motion with a method analogous to forward Euler, so we can combine that Euler-like method for the stochastic part of the equation with standard forward Euler for the deterministic part. That is, given our equation

$$dx(t) = -kx(t) \, dt + D \, dW(t),$$

we would generate the following pseudocode:

$x_0 \leftarrow x(0)$

for $i = 1$ to $T/\Delta t$

$\quad x_i = x_{i-1} - kx_{i-1}\Delta t + D\sqrt{\Delta t}N(0, 1).$

This procedure of using forward Euler for the deterministic part and the Euler-like stochastic sum for the stochastic part is called the *Euler–Maruyama method*.

Figure 16.1(b) shows a simulation of the tethered molecule's position over time, using $k = D = 1$. This picture may superficially look similar to figure 16.1(a), which lacks the restoring force of the tether. When we superimpose simulations of the two over a sufficiently long time scale in figure 16.1(c), though, we can see that the tether substantially changes the long-term behavior.

We can use the same basic Euler–Maruyama method for any SDE

$$dx(t) = f(x(t))\,dt + g(x(t))\,dW(t)$$

by plugging in f_i and g_i, the current estimates of $f(x)$ and $g(x)$, in the appropriate places in the formula:

$$x_{i+1} \leftarrow x_i + f(x_i)\Delta t + g(x_i)\sqrt{\Delta t}N(0, 1).$$

Example   Suppose we are interested in a noisy model of a chemical reaction system. We will assume we have a simple one-step reaction, $A + B \rightarrow^k C$, but propose that there is a certain amount of noise in the system. It may be that molecules get degraded and new ones get created, or that our reactants participate in other, rarer reactions that occasionally consume or release copies of them. To model this system with SDEs, we can start with our standard ODE system of equations, rewritten in our SDE notation:

$$dA = -kAB\,dt$$

$$dB = -kAB\,dt$$

$$dC = kAB\,dt.$$

We then have to add in noise terms. There are many ways one can add noise to this system by SDEs, and we need a model of the noise just as we develop a model of the deterministic part of the process. Suppose we decide the noise creates and removes copies of the three reactants independently of one another. This may be a reasonable model if the source of the noise is the reactants' participating in other reactions. Under that assumption, we may propose that the amount of noise in each reactant should be proportional to its current concentration. Then our full SDEs might look like the following:

$$dA = -kAB\,dt + \kappa A\,dW_1(t)$$

$$dB = -kAB\,dt + \kappa B\,dW_2(t)$$

$$dC = kAB\,dt + \kappa C\,dW_3(t),$$

where κ is a constant of proportionality for the noise term. We can then simulate this system by the Euler–Maruyama method with the following pseudocode:

$A_0 \leftarrow A(0)$

$B_0 \leftarrow B(0)$

$C_0 \leftarrow C(0)$

for $i = 1$ to $T/\Delta t$

$\quad A_i = A_{i-1} - kA_{i-1}B_{i-1}\Delta t + \kappa A_{i-1}\sqrt{\Delta t}N(0,1)$

$\quad B_i = B_{i-1} - kA_{i-1}B_{i-1}\Delta t + \kappa B_{i-1}\sqrt{\Delta t}N(0,1)$

$\quad C_i = C_{i-1} + kA_{i-1}B_{i-1}\Delta t + \kappa C_{i-1}\sqrt{\Delta t}N(0,1).$

Figure 16.2 shows a single trajectory of this system for ten units of time, with the rate constant set to 1 and the noise constant set to 0.01, and with $A(0) = B(0) = 1$, $C(0) = 0$.

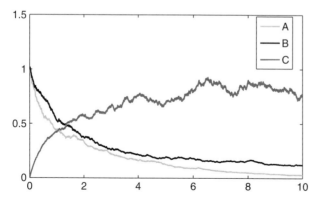

Figure 16.2
Stochastic simulation of the reaction system $A + B \rightarrow C$ with rate constant 1, noise constant 0.01, and initial conditions $A(0) = 1$, $B(0) = 1$, $C(0) = 0$.

Note that different assumptions about the source of the noise may lead to different SDEs. For example, if we assume the noise is produced by chance fluctuations in the number of reaction events rather than by independent changes in reactant counts, then we need a noise model that conserves molecules in the system. We may, for example, instead use the following SDEs:

$$dA = -kAB\,dt - \kappa A\,dW_1(t)$$

$$dB = -kAB\,dt - \kappa B\,dW_1(t)$$

$$dC = kAB\,dt + \kappa C\,dW_1(t).$$

Here, we have only a single Brownian process, representing excess reaction events above the expected deterministic reaction rate. To simulate this version of the system, we will sample only a single $N(0,1)$ random variable per loop iteration and then apply it to each of the three reactants, rather than sampling one random variable per reactant.

16.4   Accuracy of Stochastic Integration Methods

Analyzing accuracy of SDEs is more complicated than for deterministic differential equations. We still need some understanding of how it is done, though, if we want to be able to use these methods in practice. With deterministic differential equations, the error of the scheme is the difference between the answer given by the scheme and the answer given by the true process. With SDEs, we cannot define error so easily because the outputs of the scheme and the true process are both random variables. Even a perfect simulation of the stochastic process will produce different results each time we use it. We therefore need to measures errors based on the distribution of possible outputs rather than any individual output. Just as with deterministic equations, we will seek to define an "order of accuracy" of a scheme, representing how its error varies with step size Δt. Two definitions of order of accuracy are widely used for this purpose.

The more important definition, and the one we usually will want to consider as the order of accuracy of our method, is known as the *strong order of convergence*. Suppose we define x_n to be the value of the variable we are integrating as of step n, and $x(t_n)$ to be the value of the real process at time point n. x_n and $x(t_n)$ are then both random variables. If the method is a perfect simulation, they will have the same distribution. We can measure how far apart they are by considering the expectation of their difference. In particular, we say a method has strong order of convergence γ if

$$Ex[|x_n - x(t_n)|] \le C\Delta t^\gamma$$

for some constant C specific to the problem, sufficiently large n, and sufficiently small Δt.

We can also define a *weak order of convergence* to be γ if

$$|Ex[x_n] - Ex[x(t_n)]| \leq C\Delta t^\gamma$$

for some constant C specific to the problem, sufficiently large n, and sufficiently small Δt. This is a less stringent definition because it neglects effects from the variances of the two processes that are captured in the strong order definition.

Even given a precise definition, though, order of convergence for stochastic integration methods is much harder to analyze than the order of accuracy of deterministic schemes. In practice, one may need to assess the order of convergence empirically by choosing some test function that is analytically solvable and seeing how the error of the scheme varies with Δt. It happens that the Euler–Maruyama method has strong order $\frac{1}{2}$ and, with our more specialized definition, weak order 1, but we will not be able to go into any more depth than stating those results.

There is a stochastic integration method with strong order 1 derivable by adding a correction term to the Euler–Maruyama approximation:

$$x_{n+1} = x_n + \Delta t f(x_n) + g(x_n)\sqrt{\Delta t}\,v + \frac{1}{2}g(x_n)g'(x_n)(\Delta t v^2 - \Delta t),$$

where v is an $N(0,1)$ normal random variable. This is called Milstein's method. There is even an analogue to the Taylor series for stochastic functions, called the Itô-Taylor series, that can be used to perform the analyses described here and to derive methods of arbitrarily high order. The Itô-Taylor series does not have a simple statement like the standard Taylor series, though, and would have little practical value without a lot more mathematical depth than we can provide here. We will therefore omit it as well.

16.5   Stability of Stochastic Integration Methods

Stability is also a trickier concept for stochastic versus deterministic finite difference schemes. Because each dependent variable at any point in time is a random variable, we need to consider the probable growth of the function over some distribution. It is not adequate to consider just the expected value of the variable, though. For example, pure Brownian motion is an unstable process, since a particle acting under Brownian motion is expected to have moved a net distance $O(\sqrt{T})$ in any finite time T. Yet the expected position of a particle moving under pure Brownian motion is the origin at all times. We therefore need to define stability in terms of some measure of expected deviation from zero.

The simplest definition of stability for us to use is based on the expectation of the variable squared. That is, if we have a stochastic variable $x(t)$, we want to say that $x(t)$ is stable if $\lim_{t\to\infty} Ex[(x(t))^2]$ is a constant. We can perform this test with a method analogous to the von Neumann analysis we used for deterministic finite difference schemes. Recall that previously we examined stability behavior by assuming solutions of the form

$$x(t) = e^{\omega t}.$$

This is equivalent to saying that our differential equation was

$$\frac{dx}{dt} = \omega x.$$

We will generalize von Neumann analysis to the stochastic case by saying that our test functions will have the form

$$dx = \omega x \, dt + \mu \, dW(t).$$

In the deterministic case, the true system is stable when $\mathcal{R}e\{\omega\} < 0$, unstable when $\mathcal{R}e\{\omega\} > 0$, and semistable when $\mathcal{R}e\{\omega\} = 0$. For the stochastic situation, there is an equivalent test for stability:

$$\lim_{t\to\infty} Ex[(x(t))^2] = 0 \Leftrightarrow \mathcal{R}e\{\omega\} + \frac{1}{2}|\mu|^2 < 0.$$

That is, the system should be stable when $\mathcal{R}e\{\omega\} + \frac{1}{2}|\mu|^2 < 0$ and unstable when $\mathcal{R}e\{\omega\} + \frac{1}{2}|\mu|^2 > 0$.

This test tells us how the real system should behave, but what about the finite difference scheme applied to it? In general, for this definition of stability the calculations will not be too hard. We need to evaluate $\lim_{n\to\infty} Ex[x_n^2]$. The Euler–Maruyama scheme for our test function uses the following iterator:

$$x_{n+1} = x_n + \omega\Delta t x_n + \mu\sqrt{\Delta t}N(0,1)$$

$$= (1 + \omega\Delta t)x_n + \mu\sqrt{\Delta t}N(0,1).$$

Suppose we have a set of independent $N(0,1)$ normal random variables, which we will call $N_1, N_2, N_3, \ldots$. Then we can derive the following values for the test function at successive iterations of the scheme:

$$x_1 = (1 + \omega\Delta t)x_0 + \mu\sqrt{\Delta t}N_1$$

$$x_2 = (1 + \omega\Delta t)^2 x_0 + (1 + \omega\Delta t)\mu\sqrt{\Delta t}N_1 + \mu\sqrt{\Delta t}N_2$$

$$x_3 = (1 + \omega\Delta t)^3 x_0 + (1 + \omega\Delta t)^2 \mu\sqrt{\Delta t}N_1 + (1 + \omega\Delta t)\mu\sqrt{\Delta t}N_2 + \mu\sqrt{\Delta t}N_3$$

$$\vdots$$

$$x_k = (1 + \omega\Delta t)^k x_0 + \sum_{j=1}^{k}(1 + \omega\Delta t)^{j-1}\mu\sqrt{\Delta t}N_j.$$

We really want to know about $Ex[x_k^2]$, but this is actually very easy to evaluate. When we take the pairwise square of terms from the above function, any term with the form CN_j for constant C or CN_iN_j for $i \neq j$ will have expectation zero and can be ignored. $Ex[N_j^2] = 1$, so we can drop N_j^2 factors from the expectation. Thus, we need only the sum of squares of the individual terms of x_k:

$$Ex[x_k^2] = (1 + \omega\Delta t)^{2k}x_0^2 + \sum_{j=1}^{k}(1 + \omega\Delta t)^{2j-2}\mu^2\Delta t.$$

If we look at the growth factor of this function, it is no longer a simple multiplicative factor $x_{k+1} = g(\omega\Delta t)x_k$, as it was in the deterministic case. Rather:

$$Ex[x_{k+1}^2] = (1 + \omega\Delta t)^2 x_k + \mu^2\Delta t.$$

The expectation is therefore a more complicated linear recurrence of the form $y_{k+1} = ay_k + b$. Recall that solutions to a linear recurrence relation generally have the form $y_k = \gamma^k$ for some γ. So to find the stability region, we need to find where $|\gamma| = 1$ for

$$\gamma^{k+1} = a\gamma^k + b.$$

This is true only where $|a| + |b| = 1$. Thus, our scheme is stable where

$$|1 + \omega\Delta t|^2 + \mu^2\Delta t < 1.$$

This condition clearly cannot be satisfied if $\mathcal{R}e\{\omega\} > 0$, indicating that our method will be unstable in the right half of the complex plane no matter what the noise coefficient μ is. If the deterministic part is oscillatory (ω purely imaginary), then it will also be impossible to get stability regardless of the noise term. Where $\mathcal{R}e\{\omega\} < 0$, it is a little harder to characterize. But if the real problem is stable, then we should be able to get a stable numerical simulation by choosing sufficiently small Δt, just as with the deterministic forward Euler method. The same sort of analysis can be applied to the Milstein scheme or any other stochastic integration scheme, although the analysis will in general get much messier for the higher-order schemes.

References and Further Study

Though stochastic differential equations are not a recent invention and have seen wide use in other fields, they are still not well known in the biological modeling community. The most lucid description of the topic I have found is a review article on stochastic differential equations by Higham [139] that was very helpful in preparing this chapter, and I recommend it to readers still confused by the basic topic. A more in-depth coverage of the topic may be found in Kloeden and Platen [140]. Those looking for depth on the theory behind these methods may also refer to Protter [141]. That text is, unfortunately, likely to be incomprehensible to readers without a strong mathematics background. I have yet to find a textbook on the topic that gives a clear explanation aimed at practitioners in the applied sciences. Stochastic differential equations are probably most widely used today in computational finance, and readers interested in a text aimed more at the practitioner than the theorist may therefore look to the mathematical finance literature to find more practical discussion than appears to be available for scientists. In that regard, readers may consider Lamberton and Lapeyre [142] or Øksendal [143].

17 Case Study: Simulating Cellular Biochemistry

In the preceding chapters, we have seen several methods for simulating the time evolution of complex systems with many components. Today, many researchers are working to create predictive simulations of one particular very complex system: a living cell. In particular, they are working to describe the complicated chain of coupled chemical reactions that collectively produce the essential processes of life. Several of the modeling methods we have seen, for both discrete and continuous simulation, are today being used in these efforts to model biochemistry at the cellular scale. In this chapter, we will see how the general techniques we covered in the preceding chapters are contributing to current research practice in whole-cell modeling and get an idea of some of the added complications one needs to worry about in using them in an extremely complicated real system. We cannot hope to consider every attempt to model any biochemical reaction that occurs in a cell; rather, we will examine a few recent attempts to create a single simulation system that can collectively encompass the large biochemical networks found in a cell. These examples collectively illustrate some of the many options available for whole-cell modeling.

17.1 Differential Equation Models

The simplest way to model biochemistry in a cell is to treat it as a very large system of coupled ordinary differential equations (ODEs). In such a model, one assumes that the cell can be treated as a homogeneous, well-mixed solution containing all of the reactants needed for biochemistry. The biochemical pathways in the cell or any given subsystem of the cell can then be described as a set of coupled reaction equations, which we can convert into a system of ordinary differential equations, as we have seen in chapter 14. For example, a Michaelis–Menten enzyme-mediated catalysis reaction might be represented by the following reaction equations:

$$S + E \underset{k_2}{\overset{k_1}{\rightleftharpoons}} I$$

$$I \to^{k_3} I^*$$

$$I^* \to^{k_4} P + E.$$

These reaction equations then imply a set of coupled differential equations:

$$\frac{d[S]}{dt} = k_2[I] - k_1[S][E]$$

$$\frac{d[E]}{dt} = k_2[I] + k_4[I^*] - k_1[S][E]$$

$$\frac{d[I]}{dt} = k_1[S][E] - (k_2 + k_3)[I]$$

$$\frac{d[I^*]}{dt} = k_3[I] - k_4[I^*]$$

$$\frac{d[P]}{dt} = k_4[I^*].$$

This ODE approach can be expanded to arbitrarily complex reaction networks at the cost of adding one additional equation per reactant and a small number of additional terms per reaction. A straightforward ODE model was used in some of the first attempts at developing whole-cell biochemical simulators. The GEPASI program [144], one of the first widely used programs for whole-cell modeling, adopted essentially this ODE approach in its initial version. The original version of the popular E-Cell system [145] likewise used this ODE approach for general biochemical simulation.

A cell is far from homogeneous, and most recent approaches have incorporated some means of representing the spatial configuration of a cell. The simplest variant of a spatial configuration is to represent the cell as a finite set of compartments and allow transfer between compartments in addition to reaction events. Conceptually, a compartment model is not different from a pure ODE model. One can treat each reactant as a separate species for each localization it takes on, treat transfer between compartments as a new kind of reaction, and instantiate a separate set of reaction equations for each compartment. In the enzyme-mediated catalysis reaction above, for instance, we may imagine that the enzyme is permanently anchored in the Golgi apparatus, the substrate transitions between Golgi and endoplasmic reticulum (ER), and the product between Golgi and cytoplasm. If we denote these different locations by subscripts G (Golgi), E (ER), and C (cytoplasm), we can create a model of the compartmented version of the system with the reactions

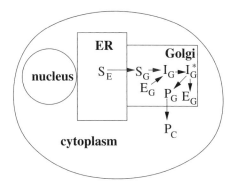

Figure 17.1
A compartmented cell model for an enzyme-mediated reaction. In this simple version, the cell is divided into four compartments, three of which are relevant to the reaction being modeled.

$$S_G + E_G \rightleftharpoons_{k_2}^{k_1} I_G$$

$$I_G \to^{k_3} I_G^*$$

$$I_G^* \to^{k_4} P_G + E_G$$

$$S_E \to^{k_5} S_G$$

$$P_G \to^{k_6} P_C.$$

We can then convert this system of effectively seven reactants and six reactions into a mass-action ODE model just as we did with the five-reactant, four-reaction original model. Figure 17.1 illustrates the model. The GEPASI program evolved in this direction, adding a compartment model in its version 3 release [146]. Note that one can, with a bit of effort, implement a compartment model in a standard ODE simulator by manually adding the compartment labeling and reaction changes in the manner described above.

Even the compartment approach simplifies quite a bit, though. Distributions of molecular components can be quite heterogeneous even within a defined compartment because of variations in where these components are created, transported, sequestered, or degraded. One can take a further step toward realistic modeling of spatial heterogeneity through a partial differential equation (PDE) model. One example of this approach was taken by the Virtual Cell project [147]. The virtual cell uses a spatial model by which detailed cell geometries can be derived from microscopy images, discretized, and divided into compartments. Diffusion occurs between neighboring discrete spatial elements within the same compartment, and models of membrane transport establish fluxes between elements separated by compartment boundaries. In addition to reaction equations, this system also includes a detailed model of electrostatic potential across the cell volume. The global set of PDEs is numerically integrated using an implicit method called the *line-by-line* method. GEPASI also continued to evolve in this direction through an extension called the Model Extender for GEPASI (MEG) [148], which allowed users to define grids of many pseudo compartments (essentially a PDE spatial discretization), which would then be automatically converted into a GEPASI compartment definition.

All of these differential equation approaches share the assumptions that the system is continuous and that it is deterministic. Neither assumption is exactly true, though, as cells have finite numbers of any given reactant and undergo reactions randomly as these individual reactants find one another in the cell. At the cellular scale, then, reactions exhibit a discretization and stochasticity that would be missing from any of the above models. One may assume that stochastic differential equations (SDEs) will provide a natural way to extend ODE models to noisy reaction systems. As far as I know, though, no attempt at whole-cell modeling has actually gone the SDE route. Approaches to modeling stochastic reactions on the whole-cell level have so far favored abandoning the differential equation framework in favor of Markov models, as we will see in the next section.

17.2   Markov Models Methods

One popular approach to modeling stochastic reaction chemistry uses a variant of the continuous-time Markov model (CTMM). This approach is known as the stochastic simulation algorithm (SSA) and is commonly called a Gillespie model after its inventor [149]. In this approach, we keep track of discrete counts of a finite set of reactants and create transitions between states when those states differ from one another by a single reaction event. We saw a special case of this method when looking at a trimer assembly system in chapter 11. We can generalize this approach to any arbitrary reaction system. Given a set of reactants $X_1, X_2, \ldots, X_n$ and a reaction

$$a_1 X_1 + a_2 X_2 + \cdots a_n X_n \rightarrow^k b_1 X_1 + b_2 X_2 + \cdots b_n X_n,$$

we can establish a transition from any state with sufficiently many input reactants to a corresponding output state. We can define the state of the system by a vector of counts of all the reactants. Thus, a system with N_1 copies of reactant X_1, N_2 copies of reactant X_2, N_3 copies of reactant X_3, and so on, will have state

$$(N_1, N_2, N_3, \ldots, N_n).$$

Then, provided $N_1 \geq a_1, N_2 \geq a_2, \ldots$, we can establish the transition

$$(N_1, N_2, N_3, \ldots, N_n) \rightarrow (N_1 - a_1 + b_1, N_2 - a_2 + b_2, N_3 - a_3 + b_3, \ldots, N_n - a_n + b_n).$$

The rate of this transition depends on the intrinsic rate of the reaction, k, as well as on the number of ways of picking the reactants that participate in the reaction. For example, there are $\binom{N_1}{a_1}$ ways to pick the a_1 copies of X_1 that we need from among the N_1 copies available. The overall rate of the transition, then, is

$$\lambda = k \times \binom{N_1}{a_1} \times \binom{N_1}{a_2} \times \cdots \times \binom{N_n}{a_n}.$$

If the system contains many reactions, we add a set of transitions in this way to correspond to each of the possible reactions. This Gillespie approach was incorporated into the first release of a program called StochSim [150], although StochSim soon evolved to a spatial model described below.

When implementing such a method in practice, one must give a considerable amount of attention to the algorithms one uses to implement it. Unless we are modeling a very small number of copies of just a few reactants, it will not be feasible to explicitly construct the entire CTMM graph. Our goal usually will be to explicitly model only the current state of the graph plus a small amount of additional state, and efficiently sample among the possible next states of the system without modeling any of the graph outside the immediate neighborhood of the current state. When Gillespie originally proposed the use of CTMMs for modeling systems of reactions, he suggested two algorithms for this purpose. The first approach, called the *direct method*, on each simulation step determines the sum of all transition rates out of the current state, samples an exponential variable representing the time to leave the current state, and chooses the reaction by which the state is exited among all possible next reactions weighted by their individual rates. Figure 17.2(a) provides pseudocode for the direct method. The second approach, called the *first reaction method*, on each event samples a separate waiting time for each possible next reaction and chooses the one with shortest time as the next reaction event. Figure 17.2(b) provides pseudocode for the first reaction method. These two methods correspond to the two views of CTMMs we saw in chapter 11. These have been largely displaced since then by a technique called the *next reaction method* [151]. The next reaction method is a discrete-event implementation of CTMM simulation in which one stores an event queue containing all possible next events, and on each step implements the event with shortest time and updates only those pending events that correspond to species affected by this most recent reaction. The method therefore should require in practice much less runtime than is needed to sample all possible events, as in the first reaction method. Figure 17.2(c) provides pseudocode for the next reaction method. Although

(a)

1. Identify all transitions $e_1, \ldots, e_k$ out of the current state (i.e., all possible reactions).
2. Calculate the sum $\lambda = \sum_i \lambda_i$ of rates of all reactions.
3. Sample a time τ as an $\text{Exp}(\lambda)$ random variable.
4. Sample a transition e_i from the discrete distribution $(\lambda_1/\lambda, \lambda_2/\lambda, \ldots, \lambda_k/\lambda)$.
5. Update species counts to reflect transition e_i.
6. Go to step 1.

(b)

1. Identify all transitions $e_1, \ldots, e_k$ out of the current state (i.e., all possible reactions).
2. For each transition e_i with rate λ_i
   A. Sample a time τ_i as an $\text{Exp}(\lambda_i)$ random variable.
3. Choose the reaction i for which τ_i is the minimum.
4. Update species counts to reflect reaction e_i.
5. Go to step 1.

(c)

1. Identify all transitions $e_1, \ldots, e_k$ out of the current state (i.e., all possible reactions).
2. For each transition e_i with rate λ_i
   A. Sample a time τ_i as an $\text{Exp}(\lambda_i)$ random variable.
   B. Place an event for transition e_i in the event queue.
3. Choose the transition e_{min} with minimum time from the queue.
4. Update species counts to reflect reaction e_{min}.
5. For all reactions e_i containing a species r that was a reactant or product of e_{min}
   A. Invalidate any existing event for e_i.
   B. Sample a time τ_i as an $\text{Exp}(\lambda_i)$ random variable.
   C. Place an event for the updated transition e_i in the event queue.
6. Go to step 3.

Figure 17.2
Pseudocode for algorithms for implementing SSA simulations. (a) Direct method. (b) First reaction method. (c) Next reaction method.

the next reaction method is currently the preferred method in practice, the overhead of the queue can make it less efficient than Gillespie's direct method when the number of reactants is not too large [152]. Furthermore, there are specialized methods for more efficient modeling of cases where the number of reactants becomes problematic even for the the next reaction method [153].

Markov models, like ODEs, can be extended to cover spatial effects. The simplest method for this purpose is to assume that the space is described by a regular lattice similar to the spatial discretization used in the spatial PDE models we considered previously. We can then treat "jumps" between neighboring lattice points as a special kind of reaction event that can occur side by side with chemical reactions among the reactants within each lattice point. One way of modeling this effect is to use a fixed time step and select the probability of a given reactant's jumping between positions so as to produce a realistic diffusion rate for the particle. This approach is used by the MCell simulation system [154] [155], which uses a very fine lattice to represent cell geometries with high precision. MCell has particularly used this capability for studies of neuronal signaling, where complex geometries of cells and ion stores within them make spatial effects critical to realistic models. We can also combine a lattice model with a CTMM representation.

By assuming a Brownian diffusion process, we can determine exponential jump rates that allow us to track particles in continuous time as they move along the grid. Particle jumps can then occur in continuous time in parallel with reaction events in continuous time, giving us all the advantages of the Gillespie method without sacrificing spatial detail. This "spatial Gillespie" algorithm was developed by Stundzia and Lumsden [156] and extended to use the next reaction algorithm [151] by Elf et al. [157]. A grid-based Gillespie approach was added to the StochSim simulator [150]. It is also used by the SmartCell [158] whole-cell simulation system, which allows for highly detailed spatial models of cells similar to those used by the discrete-time MCell system.

17.3 Hybrid Models

All of the methods described above have advantages and disadvantages for various kinds of systems. Continuous, deterministic models perform well if there are few kinds of reactants but many copies of each. They are, however, unrealistic if copy numbers are low and inefficient if numbers of distinct reactant species are large. Gillespie models work well when few copies of any given reactant are available, but are inefficient when many copies of some reactants are available. These sorts of trade-offs present a problem when one wants to model large numbers of coupled biochemical reactions because no one method is likely to be appropriate for all of the components of the system. Efforts at whole-cell modeling are therefore largely turning to hybrid approaches, which typically combine different algorithms for different kinds of reactants or for different conditions.

One approach to hybrid modeling is to start with an inherently discrete model, but allow automatic adjustment to continuous models as problem characteristics allow. This idea is the basis of a method called the *Tau leap algorithm* [159]. The Tau leap algorithm is an approximate method for simulating a Gillespie model that automatically substitutes continuous approximations for the discrete ones for well-populated reaction intermediates. The idea behind the approach is that species with large copy numbers do not need to be updated one reaction at a time. The small random fluctuations induced by counting individual reaction events will average out to a continuous, deterministic change with a relatively small error. One can exploit this observation to speed up a simulation by taking continuous "leaps" that update the well-populated species alongside discrete single-reaction events that update the poorly populated species. The complication in such an approach is finding the amount of time one can leap (a time Gillespie refers to as τ) without introducing a significant error into any species counts. Gillespie suggests a method by which one examines the distribution of changes as a function of time in each well-populated species of the reaction system, and chooses τ such that the expected change in any species is a small fraction of its preleap population. The resulting method can be orders of magnitude faster than the standard SSA when dealing with systems having many copies of some or all reactants.

The GEPASI program has evolved in the direction of a similar automated hybrid model, resulting in a new program called COPASI [160]. COPASI allows users to define a single reaction network model and then simulate that model using ODE integration, CTMM simulation using the next reaction method, or a hybrid approach combining the next reaction method with an ODE solver. The COPASI hybrid method uses a dynamic, automated partitioning of reactants into continuous and discrete sets based on copy numbers. Continuous ODE steps and discrete Gillespie reaction steps are then handled serially by a common discrete event loop. Thus, the queue on a given step may signal either the next step of a numerical integrator or a single discrete reaction.

One can, alternatively, allow a user to explicitly separate a model into discrete and continuous components. For example, a user of one of these systems may note that the calcium concentration in a cell is so high that it can always be treated as a continuous variable, whereas a particular calcium channel may occur in such small numbers that it must always be treated as a discrete variable. A variant of this manual approach was developed by Takahashi et al. [161], and has been incorporated into the E-Cell system since its third release. Like the COPASI hybrid method, the E-Cell composite approach maintains a single discrete event queue that handles both discrete reaction events and numerical integration time steps. In the case of numerical integration, the method chooses a new time step based on an embedded Runge–Kutta method. On a discrete event, continuous values are inferred between time steps using an interpolation method, a topic we will consider in chapter 22.

17.4 Handling Very Large Reaction Networks

One of the problems that has emerged from models of more complicated systems is that some systems may have too many reactions to explicitly represent them all. This issue is a problem particularly for reactions involving the formation of macromolecular complexes, where an enormous number of possible partially formed structures may in principle occur in a reaction system, even if few are ever present at any given time. Figure 17.3 shows an example of a hypothetical hexameric complex with two kinds of binding interactions, along with the ten partially formed species for which

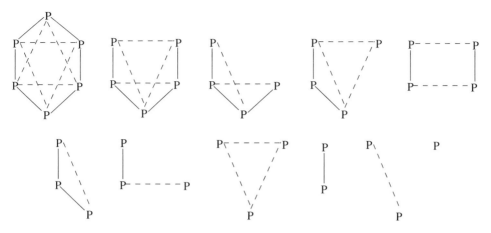

Figure 17.3
A hypothetical macromolecular complex and its assembly intermediates. The complex is formed of six protein copies (P) connected by two types of binding interaction (solid lines and dashed lines). The upper-left shows the complete complex. The other structures are the various partially assembled species implied by the complete complex.

one would need to account in simulating its assembly. The number of these partially formed reactants will in general grow exponentially with the size of the complete complex, making it infeasible to explicitly create a model of the reaction for complexes of tens or hundreds of subunits, such as ribosomes, viral capsids, or cytoskeletal structures.

A few attempts have been made to simulate these processes using implicitly specified reaction networks. In any such model, one establishes a set of components and a set of rules for how components bind to one another. In figure 17.3, for example, the rule may specify the particular pattern of neighbors of each copy of protein P, from which one can derive all of the possible ways of forming a complex of Ps. A simulator will then implicitly model the set of reactions implied by those rules, either by automatically creating the full reaction network or by sampling reaction trajectories from the network without ever explicitly creating it. The BioNetGen program [162] adopted the former approach, establishing a language for specifying interaction patterns and binding constraints for macromolecular complexes, which are then used to automatically generate reaction equations. This tool has been incorporated into the Virtual Cell [147] since version 4.3.

The latter approach, where the reaction network is simulated without ever explicitly creating it, was used by the Moleculizer program [163]. Moleculizer implements a CTMM model of the reaction network, using the next reaction method without ever creating the full set of reaction equations. The DESSA program [164] similarly establishes implicit stochastic reaction networks from subunit–subunit binding rules, but with an explicit model of geometries of complexes that allows modeling of systems where even the final assembled structures may be unknown (e.g., an amorphous aggregate structure or a misassembled complex).

17.5   The Future of Whole-Cell Models

Though whole-cell biochemical modeling has a relatively short history, we can make some conjectures about the future of the field based on its progress to date. Figure 17.4 breaks down the methods we have examined based on how they handle time progress (continuous/deterministic, discrete/stochastic, or hybrid) and how they model space (no spatial model, compartmented, or continuous space). Though this grid does not capture all the nuances between the models, it does provide a rough

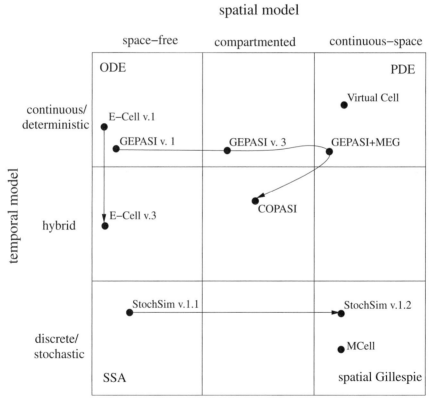

Figure 17.4
Grouping of whole-cell simulators by temporal and spatial models. Filled circles mark individual simulators or different releases of a single simulator when accompanied by major changes of model. Arrows follow progression of particular research efforts.

picture of how the field has been evolving over the past few years. Two general trends we can observe are movements toward explicit modeling of spatial features in cells and toward hybrid discrete/continuous models. The shift toward spatial models is a natural trend toward greater realism as the importance of space to cellular reaction chemistry has become better appreciated. The shift toward hybrids reflects a compromise between realism and tractability, as much of the field has come to appreciate the importance of stochastic effects for realistic cellular models, but has found some continuous component necessary for modeling large networks in reasonable times.

It is worth noting, though, that the field is still quite far from being able to build a truly comprehensive simulation of biochemistry in even the simplest cells. Papers are constantly appearing on new advances in methods or extensions to new systems, and a current literature search is likely to turn them up. There are also numerous examples of particularly interesting or difficult subsystems of cellular biochemistry for which specialized methods have been developed, which we will not attempt to survey here. If we had to predict which algorithms will power the successful whole-cell simulators of the future, we might guess that they will use a hybrid of PDE and spatial Gillespie models. There are, however, alternative approaches we have not considered here because they are so far too computationally demanding to be used for complex reaction networks or long time scales. Examples include Brownian dynamics models [165], which explicitly model diffusion of reactants in three-dimensional space, and a recent approach called Green's function reaction dynamics (GFRD) [166], which attempts to speed up Brownian dynamics models by stepping between events with an event queue. It may be, though, that predictive whole-cell models will eventually require these sorts of more realistic models or even something wholly different and as yet unanticipated.

17.6   An Aside on Standards and Interfaces

One of the major trends in whole-cell modeling has been increased attention to standardization and interface design. Though this topic is somewhat tangential to our focus on algorithms and numerical methods, it is worth noting for anyone interested in working in this area. These standards are aimed at creating a language in which one can describe any kind of cellular reaction system independently of the methods with which it has been or will be simulated. Such a standard will facilitate interconnection of the many kinds of programs now available for creating, simulating, and analyzing such models. One may, for example, wish to define a model with one tool, fit parameters to the model from some experimental data with another, simulate the model with a third, and analyze the results with a fourth. By building a common language familiar to every tool builder, one makes it possible to interconnect many sorts of components. Likewise, common standards make it possible for experimental scientists to publish models in a common format that others can interpret in a uniform way or use in many kinds of models. Both users of whole-cell modeling and analysis tools and developers of the tools therefore need to know something about the standards available.

Unfortunately, as Andrew Tannenbaum once remarked, "The best thing about standards is that there are so many to choose from." Though there have been many attempts to create standards for biochemical modeling, two choices seem to have emerged as the de facto standard standards of the field: SBML [167] [168] and CellML [169] [170]. Both create extremely detailed frameworks for modeling reaction networks and reactant distributions, either in compartments or across heterogeneous spatial regions. Both also support many extensions beyond reaction networks, for example, by allowing electrophysiological or mechanical components in their models. One or both of these standards are likely to be supported by most recent codes for reaction network simulation, optimization, or inference, and one is likely to have little trouble converting among them. Anyone planning to work in using or developing such tools would therefore be well advised to familiarize himself or herself with these efforts.

References and Further Study

This chapter has covered several of the leading efforts to model biochemistry on a cellular scale. The various primary references cited in the text provide much more detail than we can cover here on the precise algorithms and numerical methods they use, other practical implementation details, and further design issues that are beyond the scope of this text. Several of the programs described are available free of charge to researchers. Interested users can check out the Web sites for the E-cell [171], Virtual Cell [172], COPASI [173], and MCell [174] projects to get information about obtaining software. Others may be tracked down through the primary references cited in the text. Readers may also refer to these Web sites for citations to publications describing applications of the aforementioned simulation tools to various complex systems in biology. For more information on the general methods used in this field, as well as much more advanced material on whole-cell modeling, readers may refer to Wilkinson [110] or to Fall et al. [175]. The references in the preceding chapters of this text on numerical integration and Markov models will also provide more information on the use of these techniques in general.

III PARAMETER-TUNING

18 Parameter-Tuning as Optimization

We are now starting the third and final segment of this book: parameter-tuning. We have examined many kinds of models over the course of this text, and almost all of them have required specifying some unknown parameters. For example, if we want to simulate a biochemical model

$$A + B \rightleftharpoons C,$$

we need to know forward and reverse rate constants of the reaction, k_+ and k_-. If we want to implement a lattice protein-folding model, like those we saw in chapter 1, we need to know the contact potentials between different amino acid types. Sometimes we can look up the parameters we need in the literature or measure them directly by experiment, as with the rate constants. In other cases, we can derive parameters from first principles, such as by using detailed molecular dynamics models to estimate contact energies. Very often, though, we need to infer parameter values more indirectly. For example, for the biochemistry model, we may have experimental measurements of the concentrations [A], [B], [C] of the system at various time points in a series of experiments, and need to infer k_+ and k_- from those data points. For the lattice protein model, we may have some solved protein structures, and we need to infer E_{AL} and E_{RK} from the frequencies of contacts in the structures. Over the course of this section of the text, we will see several specialized methods for doing this sort of parameter tuning.

Before we get to those more specialized methods, though, it is worth noting that we covered some excellent tools for parameter-tuning when we looked at optimization problems in the first part of the text. Virtually any sort of parameter-tuning problem can be posed as an optimization problem. If we can figure out what function we want to optimize when fitting our parameters and can evaluate that function for any given choice of parameters, we can often directly apply the kinds of optimization tools we have already seen to the parameter-tuning problem. There may be other, better methods we can use for certain problem types, examples of which we will see later. But if we are not sure what tool to use, we can almost always revert to the general optimization tools we saw in the first section of the text, and make at least some progress on fitting our model parameters.

18.1 General Optimization

It may be easiest to see the relationship between parameter-tuning and general optimization by looking at an example. Suppose we are interested in expression of some protein in a cell. We have created a green fluorescent protein (GFP) construct of some low-copy-number protein, which allows us to see when copies of that protein appear in the cell. We then watch a single cell and record each time a new copy of the protein is formed. We start observing the system at some time t_0 and get a set of time points $t_1, t_2, \ldots, t_n$ at which we observe new proteins appearing. We want to use these points to build a model of that protein's expression. We may then use a model of the protein's expression events as one component of some more complex model of the behavior of the cell.

To build a model of the protein's expression events, we first need to specify a general class of models from which it will be constructed. We may, for example, decide that we believe the times between translation events—$t_1 - t_0, t_2 - t_1, t_3 - t_2, \ldots,$ $t_n - t_{n-1}$—are described by a gamma distribution, $Pr\{\Delta t = \tau\} = \lambda^2 \tau e^{-\lambda \tau}$, for some unknown rate parameter λ.

We then need to decide how to judge the quality of our model, which will specify the objective function for parameter-tuning. For example, we may decide that we want to maximize the probability of producing all of our observed time points over all possible λ values. That is, we want to find

$$\max_{\lambda} \prod_{i=1}^{n} \Pr\{\Delta t = (t_i - t_{i-1})\}.$$

This is called a *maximum likelihood model*, since we are seeking to find a model M, given a data set D that maximizes $\Pr\{D \mid M\}$, the *likelihood* of M, given D. Such a model is often optimized relative to some prior distribution on possible parameter sets M, meaning that we will seek M maximizing $\Pr\{D \mid M\} \Pr\{M\}$. For now, though, we will assume all λs are equally likely. Our objective is then the following:

$$C(\lambda) = \prod_{i=1}^{n} \lambda^2 (t_i - t_{i-1}) e^{-\lambda (t_i - t_{i-1})}.$$

Our goal in the parameter-tuning problem is to maximize this objective function $C(\lambda)$ over $\lambda > 0$. In practice it is often easier, with this kind of product of probabilities, to perform an equivalent optimization using the log of the objective:

$$\ln C(\lambda) = 2n \ln \lambda - \sum_{i=1}^{n} \lambda(t_i - t_{i-1})^2.$$

We now have a well-defined objective function, for which we can optimize by any of our general optimization methods. If we want to find the optimal λ by the Newton–Raphson method, for example, we first note that we want to find a zero of

$$y = \frac{d(\ln C)}{d\lambda} = \frac{2n}{\lambda} - \sum_{i=1}^{n} (t_i - t_{i-1})^2.$$

In this case, we can in fact analytically solve for λ. If we cannot figure out the analytical solution, though, we then proceed to find the second derivative:

$$y' = \frac{d^2(\ln C)}{d\lambda^2} = -\frac{2n}{\lambda^2}.$$

We then pick some initial guess for λ and make successive updates with the usual Newton–Raphson formula:

$$\lambda \leftarrow \lambda - \frac{y}{y'}.$$

The same method generalizes easily to multiparameter optimization. For example, suppose we want to allow the degree of the gamma distribution to vary as well, maximizing likelihood over distributions of the form $p(t) = h(\lambda, \nu)t^{\nu-1}e^{-\lambda t}$ for any $\lambda > 0$ and $\nu \geq 1$. We can simultaneously optimize over λ and ν by finding the gradient and hessian of $D(\lambda, \nu) = \ln \prod_{i=1}^{n} p(t_i - t_{i-1})$ with respect to λ and ν, and applying multidimensional Newton–Raphson or any of our other multidimensional optimization methods.

18.2   Constrained Optimization

In many cases, the constrained optimization techniques we have covered will be more appropriate formulations of parameter tuning problems than the techniques for general unconstrained optimization. Imagine, for example, that we want to build a Markov model to simulate the evolution of a DNA base through successive generations of an organism. In contrast to chapter 12, though, we will assume there are selective pressures making some particular bases more likely than others. We decide to model this evolution with selection by proposing that there is some intrinsic probability p that the base is preserved from one generation to the next, independent of its current value. If it changes values, though, it selects the new base from some distribution (p_A, p_C, p_G, p_T) that is independent of its prior value. So, for example, if it is A in generation n, then the probability it is still A in generation $n + 1$ is $(1 - p) + pp_A$. That is, either the selective pressures preserve A with probability $1 - p$ or they do not preserve A with probability p but A is selected anyway by chance, with probability p_A. The probability that A in generation n is T in generation $n + 1$ is pp_T, and so on.

Assume we can observe a string of bases in successive generations, $s_0, s_1, \ldots, s_n$, and we count the number of occurrences of each possible transition: n_{AA}, n_{AT}, n_{AC}, and so on. We want to fit our model to this data by finding our five parameters: p, p_A, p_C, p_G, and p_T.

As above, we will need to formulate an objective function. A common "default" objective function for this kind of problem is the sum of the squares of the differences between the observed values and the model predictions of those values. In this case, that sum-of-squares objective will be the following function:

$$\Phi = (n_{AA} - ((1 - p) + pp_A)(n_{AA} + n_{AC} + n_{AG} + n_{AT}))^2$$

$$+ (n_{AC} - (pp_C)(n_{AA} + n_{AC} + n_{AG} + n_{AT}))^2$$

$$+ (n_{AG} - (pp_G)(n_{AA} + n_{AC} + n_{AG} + n_{AT}))^2$$

$$+ (n_{AT} - (pp_T)(n_{AA} + n_{AC} + n_{AG} + n_{AT}))^2$$

$$+ (n_{TT} - ((1 - p) + pp_T)(n_{TA} + n_{TC} + n_{TG} + n_{TT}))^2 + \cdots$$

We can try to solve this as a general optimization problem. That method is likely to fail, though, by fitting to values of the parameters that do not make sense for the problem. For example, our parameters are all probabilities, so we do not want to fit any parameter values that are not between zero and 1. We can resolve this problem by adding the following constraints to the optimization:

$$p \geq 0, \quad p \leq 1$$

$$p_A \geq 0, \quad p_A \leq 1$$

$$p_C \geq 0, \quad p_C \leq 1$$

$$p_G \geq 0, \quad p_G \leq 1$$

$$p_T \geq 0, \quad p_T \leq 1.$$

We can further note that p_A, p_C, p_G, and p_T form a multinomial distribution, so they must sum to 1. We therefore need one additional constraint:

$p_A + p_C + p_G + p_T = 1.$

Alternatively, we can remove one variable, say p_T, and make the substitution $p_T = 1 - (p_A + p_C + p_G)$ in the objective. Nonetheless, we now have some simple linear constraints. The objective is not linear, though, so we will want to use an interior point method. And it apparently is not convex, so we may not be able to get a globally optimal solution. But we can at least get some locally optimal parameter set that is likely to do a reasonably good job of fitting the data in practice.

18.3   Evaluating an Implicitly Specified Function

When working on parameter-tuning for simulation problems, we will often be unable to create a closed-form expression for our optimization objective since our objective function is likely to depend on the output of a simulation. We can still handle such cases by using the simulation as a black box to evaluate the objective function, then apply the black-box techniques we covered in chapter 5.

Suppose, for example, we want to fit parameters to a biochemical reaction model. Let us imagine we are studying a system in which a set of proteins, A and B, forms a complex consisting of two copies of A and one copy of B. We can model this as a single reaction,

$2A + B \rightleftharpoons C,$

with some forward rate k_1 and some reverse rate k_2. Further suppose we are given some time points at which we have measured the quantities of the different reactants, very likely with some experimental noise:

(t_0, A_0, B_0, C_0)

(t_1, A_1, B_1, C_1)

(t_2, A_2, B_2, C_2)

$\vdots$

$(t_n, A_n, B_n, C_n).$

If we want to fit k_1 and k_2 to that data, what can we do?

We first need to specify an error model, just as in the previous examples. We can again assume that we want to minimize the sum of the squares of the errors in each of our reactant concentrations for the model relative to the data. That is, we want to derive k_1 and k_2 such that we minimize

$$E(k_1, k_2) = \sum_{i=0}^{n} (A(t_i) - A_i)^2 + (B(t_i) - B_i)^2 + (C(t_i) - C_i)^2,$$

where $A(t)$, $B(t)$, and $C(t)$ are functions that we cannot explicitly evaluate but that are implied by our model for any given guess as to our parameters.

We can then create code to evaluate $E(k_1, k_2)$, using any of the numerical integration methods we have already seen. For example, we can evaluate E using forward Euler by the pseudocode of figure 18.1(a). The pseudocode performs the standard forward Euler iteration, except that each time it passes some measured time points, it linearly interpolates back to those points to estimate the reactant concentrations there. Note that if we were using a higher-order scheme, we would need a more ac-

(a)
```
1. t ← 0
2. A ← A(0)
3. B ← B(0)
4. C ← C(0)
5. i ← 0
6. E ← (A − A_i)² + (B − B_i)² + (C − C_i)²
7. while t < t_n
    A. t_next ← t + Δt
    B. A_next ← A + 2Δt(−k₁A²B + k₂C)
    C. B_next ← B + Δt(−k₁A²B + k₂C)
    D. C_next ← C + Δt(k₁A²B − k₂C)
    E. while t_next > t_i
        i. E ← E + ((A + (A_next − A)·(t_i−t)/(t_next−t)) − A_i)² + ((B + (B_next − B)·(t_i−t)/(t_next−t)) − B_i)² +
              ((C + (C_next − C)·(t_i−t)/(t_next−t)) − C_i)²
        ii. i ← i + 1
    F. t ← t_next, A ← A_next, B ← B_next, C ← C_next
```

(b)
```
1. dE_i/dk₁ ← (E(k₁+Δk, k₂) − E(k₁, k₂))/Δk
2. dE_i/dk₂ ← (E(k₁, k₂+Δk) − E(k₁, k₂))/Δk
3. find r minimizing E(k₁ + r·dE_i/dk₁, k₂ + r·dE_i/dk₂)
4. k₁ ← k₁ + r·dE_i/dk₁
5. k₂ ← k₂ + r·dE_i/dk₂
```

Figure 18.1
Pseudocodes for fitting rate parameters to an implicitly specified reaction system model. (a) Pseudocode for using forward Euler integration to evaluate the least-squares error for a given set of rate constants, k_1 and k_2, for the example reaction system. (b) Pseudocode for using steepest descent to minimize the least-squares deviation between the model and the observed data, using the forward Euler evaluation.

curate interpolation method as well, an issue we will consider in chapter 22. Once we have code for evaluating $E(k_1, k_2)$, we can treat that as a black box for our general optimization methods. For example, given some initial guess as to k_1 and k_2, we could use steepest descent to find improved guesses for the rate constants, with finite difference approximations to estimate the gradient, as in the pseudocode of figure 18.1(b).

This approach may be very slow if our simulations take a long time, so we probably want to use an optimization method that keeps evaluations of E to a minimum. In practice, therefore, Newton–Raphson or Levenberg–Marquardt may be a better choice than steepest descent. We may also have to come up with problem-specific solutions to some other implementation issues we have seen before, such as how to get a good starting guess. But here we have at least the core of a workable method to optimally fit k_1 and k_2 at least locally for our metric.

The problem is complicated quite a bit if our model is stochastic. Suppose, for example, we use the same system but assume the system itself is noisy and is described by a system of stochastic differential equations, such as

$$dA = 2(-k_1 A^2 B + k_2 C) \, dt + 2\kappa AB \, dW(t)$$

$$dB = (-k_1 A^2 B + k_2 C) \, dt + \kappa AB \, dW(t)$$

$$dC = (k_1 A^2 B - k_2 C) \, dt - \kappa AB \, dW(t).$$

How do we optimize for E if E is a random variable? One way is to average E over a sufficiently large number of runs that the noise is effectively eliminated. This will slow down the process even more, but may be our only choice in practice. We need to be very careful with such uses of stochastic models when doing numerical optimization, however, because the optimization methods require derivatives of E and derivatives are very sensitive to noise. We can take derivatives from many runs and average them until they converge, or we may need to apply some kind of smoothing to our curve to suppress the noise enough to get accurate derivatives. In this particular case, it may be best to fit to the deterministic version of the model, even if we are working with the stochastic version, and then treat the noise fitting as a separate problem.

In same cases, we may be able to solve for the distribution of E even if E is a random variable. For example, if we are trying to infer rates for a continuous-time Markov model, we may numerically integrate the Kolmogorov equations and match the state distribution as a function of time to some observed data points rather than directly simulating the model.

References and Further Study

The same references we cited when we examined optimization problems will be relevant here, particularly Press et al. [82] and Nash and Sofer [94]. Any of the many other references we considered for optimization may be relevant to parameter-tuning, though. For particular well-studied applications, there may be much better approaches available than the general methods we have considered in this chapter. For instance, readers looking to fit models of reaction systems can refer to Wilkinson [110] for a much deeper treatment of the sorts of data generally available for that problem, effective fitting methods, and practical considerations in their application.

19 Expectation Maximization

We will now begin to look at more specialized methods for parameter-tuning, starting with a broadly useful and general algorithm design method called expectation maximization (EM) that is widely used for parameter-tuning problems. The theory behind EM can be confusing, but the method is generally very easy to use. We will start out by considering the theory, then see how to apply it in practice. Our goal in a parameter-tuning problem is generally to take a set of observations x and infer a set of parameters λ that is the "best fit" by some measure. EM is well suited to cases where the "best fit" is defined in terms of a probability. There are two main variants of these probability-based parameter-tuning problems, one of which we have seen before.

Case 1   Find λ maximizing $Pr\{x \mid \lambda\}$ (called *maximum likelihood*).

Case 2   Find λ maximizing $Pr\{\lambda \mid x\}$ (called *maximum a posteriori probability* (MAP)).

Maximum likelihood models are more common in practice, mainly because it is generally easier to develop a formula expressing the probability of the observations, given the model, than the other way around. We will therefore assume we are solving a maximum likelihood problem, although it is not too hard to generalize to an MAP problem, given the MAP objective function.

Suppose, for example, that we have an organism with some set of unknown DNA base frequencies: p_A, p_C, p_T, and p_G. We would like to infer these frequencies based on observation of some string of DNA. Suppose we observe n_A As, n_C Cs, n_G Gs, and n_T Ts. Then we can cast this as a maximum likelihood problem as follows:

- x is our set of observations: n_A, n_C, n_G, and n_T.
- λ is our parameter set: p_A, p_C, p_T, and p_G.

Our likelihood of making any given observations x from a parameter set λ is

$$Pr\{x \mid \lambda\} = Pr\{n_A, n_C, n_G, n_T \mid p_A, p_C, p_T, p_G\} = p_A^{n_A} p_C^{n_C} p_G^{n_G} p_T^{n_T}.$$

So the solution to our maximum likelihood problem is the set of probabilities $\lambda = \{p_A, p_C, p_T, p_G\}$ maximizing $p_A^{n_A} p_C^{n_C} p_G^{n_G} p_T^{n_T}$.

In this case, we can analytically solve for the maximum likelihood parameter set:

$$p_A = \frac{n_A}{n_A + n_C + n_G + n_T}$$

$$p_C = \frac{n_C}{n_A + n_C + n_G + n_T}$$

$$p_G = \frac{n_G}{n_A + n_C + n_G + n_T}$$

$$p_T = \frac{n_T}{n_A + n_C + n_G + n_T}.$$

Often, though, the likelihood is defined by some more complicated probabilistic model, and we will not be able to maximize for it directly. For example, suppose we have some complicated Markov model of gene structure, with different parameter distributions for different sequence types, as in figure 19.1. In such cases, there may be no way to solve analytically for these problems. EM is an iterative approach to these hard parameter tuning problems by which we successively refine a guess as to the parameters, often converging to the true optimal parameter set.

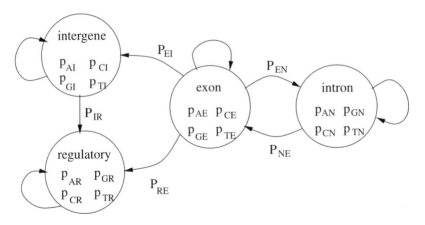

Figure 19.1
Possible Markov model for generating bases for a simulated gene sequence with differential probabilities for different sequence types.

19.1 The "Expectation Maximization Algorithm"

EM is often referred to as "the EM algorithm" but it is really an algorithm design method, not a specific algorithm. The basic principle behind the method is fairly simple. If we cannot figure out how to solve for λ maximizing $Pr\{x \mid \lambda\}$, then we will create some new variables y, called "latent variables," that represent a hypothesis about some unobserved aspect of our system. We can then iteratively improve estimates of λ by finding an expected value of y, given the current guess for λ, and then optimizing λ over this distribution of possible values of y. We choose y to make it easy to evaluate $Pr\{x, y \mid \lambda\}$ and $Pr\{y \mid x, \lambda\}$. If y is well chosen, then both steps of the iterative process will be easy to perform.

From here on, the theory gets a little tougher. We need to understand how the method works, though, if we are going to be able to pick an appropriate set of latent variables for any given problem. We will begin by defining a quantity called the "expected log," $Q(\lambda, \bar{\lambda})$:

$$Q(\lambda, \bar{\lambda}) = \sum_y Pr\{y \mid x, \bar{\lambda}\} \log Pr\{x, y \mid \lambda\}.$$

This expected log is the expectation of the log of the likelihood over the distribution of the latent variables y. In performing the EM optimization, we will implicitly evaluate Q for λ and $\bar{\lambda}$ where λ is a variable representing an arbitrary parameter set, which we do not know, and $\bar{\lambda}$ is our current best guess as to that optimal parameter set.

EM is an iterative method by which to find a good parameter set by repeatedly applying two steps. These are generally formally stated as follows:

1. Find $Q(\lambda, \bar{\lambda})$ in terms of y.
2. Find λ maximizing $Q(\lambda, \bar{\lambda})$.

Step 1 is called the *expectation step* (or E-step) since we compute the expectation of log $Pr\{x, y \mid \lambda\}$ over the possible values of y, where y has a distribution $Pr\{y \mid x, \bar{\lambda}\}$. At least, this is how it is usually presented in the literature. I consider this statement of the problem to be extremely confusing, since we almost never actually compute Q. Q is a concept that is useful in proving the correctness of the EM method, but it is not usually something we really need to know. What we really want to know is the distribution of y, given x and the current guess as to λ. And in fact we often really need only the expected value of y over its distribution, since for many problems, log $Pr\{x, y \mid \lambda\}$ evaluated at this expected y will be the same as Q. This whole issue should be much easier to understand after seeing some examples.

Step 2 is called the *maximization step* (or M-step) since we maximize Q over possible values of λ. We will generally choose our y so as to make sure this is an easy optimization to perform, given x and y.

Repeatedly cycling between the two steps gives the full expectation-maximization method.

19.2 EM Theory

In this section we will explore the theory behind why the method works. Since our goal is to improve $Pr\{x\,|\,\lambda\}$ iteratively, we will want to use some $Pr\{x\,|\,\lambda_i\}$ to find $Pr\{x\,|\,\lambda_{i+1}\}$, such that

$$Pr\{x\,|\,\lambda_{i+1}\} > Pr\{x\,|\,\lambda_i\}$$

or, equivalently,

$$\log Pr\{x\,|\,\lambda_{i+1}\} > \log Pr\{x\,|\,\lambda_i\}.$$

The EM method gives us a guarantee that at least

$$\log Pr\{x\,|\,\lambda_{i+1}\} \geq \log Pr\{x\,|\,\lambda_i\}.$$

To see why this is so, assume that λ is some arbitrary parameter set and λ_i is our current best guess. Then:

$$Pr\{x, y\,|\,\lambda\} = Pr\{y\,|\,x, \lambda\}Pr\{x\,|\,\lambda\}$$

$$Pr\{x\,|\,\lambda\} = \frac{Pr\{x, y\,|\,\lambda\}}{Pr\{y\,|\,x, \lambda\}}$$

$$\log Pr\{x\,|\,\lambda\} = \log Pr\{x, y\,|\,\lambda\} - \log Pr\{y\,|\,x, \lambda\}$$

$$Pr\{y\,|\,x, \lambda_i\} \log Pr\{x\,|\,\lambda\} = Pr\{y\,|\,x, \lambda_i\} \log Pr\{x, y\,|\,\lambda\}$$
$$- Pr\{y\,|\,x, \lambda_i\} \log Pr\{y\,|\,x, \lambda\}$$

$$\sum_y Pr\{y\,|\,x, \lambda_i\} \log Pr\{x\,|\,\lambda\} = \sum_y Pr\{y\,|\,x, \lambda_i\} \log Pr\{x, y\,|\,\lambda\}$$
$$- \sum_y Pr\{y\,|\,x, \lambda_i\} \log Pr\{y\,|\,x, \lambda\}$$

Since $\log Pr\{x\,|\,\lambda\}$ does not depend on y, $\sum_y Pr\{y\,|\,x, \lambda_i\} \log Pr\{x\,|\,\lambda\} = \log Pr\{x\,|\,\lambda\}$. Therefore:

$$\log Pr\{x \mid \lambda\} = \sum_y Pr\{y \mid x, \lambda_i\} \log Pr\{x, y \mid \lambda\} - \sum_y Pr\{y \mid x, \lambda_i\} \log Pr\{y \mid x, \lambda\}$$

$$\log Pr\{x \mid \lambda\} = Q(\lambda, \lambda_i) - \sum_y Pr\{y \mid x, \lambda_i\} \log Pr\{y \mid x, \lambda\}.$$

Plugging in λ_i for λ in the preceding equation gives us

$$\log Pr\{x \mid \lambda_i\} = Q(\lambda_i, \lambda_i) - \sum_y Pr\{y \mid x, \lambda_i\} \log Pr\{y \mid x, \lambda_i\}.$$

Subtracting the previous equation from the one before it then gives us

$$\log Pr\{x \mid \lambda\} - \log Pr\{x \mid \lambda_i\}$$

$$= Q(\lambda, \lambda_i) - Q(\lambda_i, \lambda_i) + \sum_y Pr\{y \mid x, \lambda_i\}(\log Pr\{y \mid x, \lambda_i\} - \log Pr\{Y \mid x, \lambda\}).$$

$\sum_y Pr\{y \mid x, \lambda_i\} \log Pr\{y \mid x, \lambda\}$ is maximized over λ when $\lambda = \lambda_i$, so

$$\sum_y Pr\{y \mid x, \lambda_i\}(\log Pr\{y \mid x, \lambda_i\} - \log Pr\{y \mid x, \lambda\}) \geq 0.$$

Therefore,

$$\log Pr\{x \mid \lambda\} - \log Pr\{x \mid \lambda_i\} \geq Q(\lambda, \lambda_i) - Q(\lambda_i, \lambda_i).$$

We find λ_{i+1} by maximizing $Q(\lambda, \lambda_i)$ over possible values of λ. Therefore

$$Q(\lambda_{i+1}, \lambda_i) \geq Q(\lambda_i, \lambda_i).$$

Thus,

$$Q(\lambda_{i+1}, \lambda_i) - Q(\lambda_i, \lambda_i) \geq 0$$

$$\log Pr\{x \mid \lambda_{i+1}\} - \log Pr\{x \mid \lambda_i\} \geq 0$$

$$\log Pr\{x \mid \lambda_{i+1}\} \geq \log Pr\{x \mid \lambda_i\}.$$

In summary, we have shown that if we follow the EM method, we are guaranteed to find a new parameter set λ_{i+1} that gives us a log likelihood at least as good as the log likelihood of λ_i. Furthermore, the likelihoods will be the same only if λ_i maximizes $Pr\{x \mid \lambda\}$. That does not in itself guarantee that we will find the global optimum λ, since we could converge toward a local optimum. But for a broad class of problems, expectation maximization will find the optimal λ. And it will often yield very good answers even when it cannot optimize globally.

I think EM is easiest to understand intuitively if we ignore most of the theory and just think of it in terms of optimizing a function f. We are given some x and want to choose λ maximizing $f(x, \lambda)$. We first make up some other variables y whose distributions we can determine from x and λ. We then repeatedly do the following:

1. Estimate the distribution of y, given the current x and λ.
2. Optimize λ, given x and our current guess as to the distribution of y.

One may note from the above theory that we do not actually need an optimum λ in step 2, just a local improvement, for the proof of the EM method to go through. In some cases, one does in fact use a local optimizer in the M-step, such as steepest descent, when an analytical maximum likelihood value cannot be found. We refer to the resulting method as *weak EM*, as opposed to *strong EM*, where we globally optimize λ in the M-step.

19.3 Examples

We can get a better understanding of EM by examining two examples inspired by population genetics, derived in part from work by Niu et al. [176].

Example 1 Haplotype frequency estimation. Haplotype frequency estimation is a problem that arises in population genetics when studying how different people's DNA sequences differ. We each have more or less the same DNA, but with slight variations, as illustrated in the following hypothetical collection of four pieces of the same DNA region in different individuals:

```
    *                           *

A C T T G G A C T G T T A C A

A C T T G G A C T G T T A A A

A C G T G G A C T G T T A A A

A C T T G G A C T G T T A C A

    *                           *
```

A string of possible bases on a single chromosome in a region is known as a *haplotype*. We want to know how common each possible haplotype is in the population. A problem arises, though, because the methods for determining someone's DNA sequence cannot distinguish the two versions on the two homologous copies of a chromosome. For each variable site, we would see a pair of bases but we would not know how to connect pairs. So if we sequence a particular person's DNA in this region, we might see

AA CC TG TT GG GG AA CC TT GG TT TT AA CA AA,

and we would not know if the real pair of sequences was

```
    *                       *

A C T T G G A C T G T T A C A

A C G T G G A C T G T T A A A

    *                       *
```

or

```
    *                       *

A C T T G G A C T G T T A A A

A C G T G G A C T G T T A C A

    *                       *
```

More generally, if we look at n biallelic sites (sites with two possible base values), there can be 2^{n-1} possible resolutions into haplotypes. We want some way to find the most likely set of frequencies of the possible haplotypes, given a set of observed sequences. These frequencies can be used in various kinds of models for studying, for example, the history of this population. This is a hard problem to solve optimally, but we can use EM to estimate these frequency parameters.

To do this, we first need to formalize the problem a bit. We can ignore all sites that do not vary, since they are not informative. We can also arbitrarily label one version of each variant site 0 and the other 1. For example, suppose we look at a two-base version of this problem. We might say that at the first site G=0 and T=1, and at the second site, A=0 and C=1. Then our possible resolutions (haplotypes) are 00 (G at the first site and A at the second), 01 (G at the first site and C at the second), 10 (T at the first site and A at the second), and 11 (T at the first site and C at the second). The frequencies of these four haplotypes are the parameters we wish to infer: $\lambda = \{f_{00}, f_{01}, f_{10}, f_{11}\}$.

Our input is the data in which we cannot distinguish 01 and 10 pairs. At each site, then, we have three possible observable pairs: a pair of 0s, a pair of 1s, or a 0 and a 1. We can encode this input concisely by denoting a pair of 0s at a given site by the character 0, a pair of 1s by the character 1, and a 0 and a 1 by the character 2.

Then we can denote each full input sequence as a string from the alphabet $\{0,1,2\}$. We call these $\{0,1,2\}$ strings *genotypes*. Our input then will consist of the counts of each of the genotypes in our observed population:

$$x = \{n_{00}, n_{01}, n_{02}, n_{10}, n_{11}, n_{12}, n_{20}, n_{21}, n_{22}\}.$$

We still need a probability model to tell us how to choose among possible outputs. We will use a maximum likelihood model based on the assumption of *Hardy–Weinberg equilibrium*. Hardy–Weinberg equilibrium means that we assume there is a global f_{ij} for each haplotype ij and that each haplotype of each person is sampled independently of the other from this global haplotype distribution. So, for example:

$$Pr\{20\} = Pr\left\{\begin{matrix} 0 & 0 \\ 1 & 0 \end{matrix}\right\} + Pr\left\{\begin{matrix} 1 & 0 \\ 0 & 0 \end{matrix}\right\} = 2f_{00}f_{10}$$

$$Pr\{01\} = Pr\left\{\begin{matrix} 0 & 1 \\ 0 & 1 \end{matrix}\right\} = f_{01}^2.$$

Then the likelihood of a particular parameter set

$$\lambda = \{f_{00}, f_{01}, f_{10}, f_{11}\}$$

for an input set

$$x = \{n_{00}, n_{01}, n_{02}, n_{10}, n_{11}, n_{12}, n_{20}, n_{21}, n_{22}\}$$

is given by

$$Pr\{x \mid \lambda\} = \left(f_{00}^2\right)^{n_{00}} \times \left(f_{01}^2\right)^{n_{01}} \times \left(f_{01}f_{00} + f_{00}f_{01}\right)^{n_{02}} \times \left(f_{10}^2\right)^{n_{10}} \times \left(f_{11}^2\right)^{n_{11}}$$

$$\times \left(f_{10}f_{11} + f_{11}f_{10}\right)^{n_{12}} \times \left(f_{00}f_{10} + f_{10}f_{00}\right)^{n_{20}} \times \left(f_{01}f_{11} + f_{11}f_{01}\right)^{n_{21}}$$

$$\times \left(f_{00}f_{11} + f_{01}f_{10} + f_{10}f_{01} + f_{11}f_{00}\right)^{n_{22}}.$$

Or, more concisely,

$$Pr\{x \mid \lambda\} = \prod_{\substack{\text{genotypes } g_i}} \sum_{\substack{(ab,cd) \text{ consistent} \\ \text{with } g_i}} f_{ab}f_{cd}$$

That is, for each observed genotype, we find its likelihood by taking the sum of the probabilities of all ways of generating it from two haplotypes. We then take the product of these sums over all genotypes. This full product gives us the total maximum likelihood of the observations (genotype counts), given the parameters (estimated haplotype frequencies).

Many of the components of the maximum likelihood parameter set are easy to determine analytically. If we see the genotype 00, we know we have two instances of haplotype 00. If we see genotype 02, we know we have one haplotype 00 and one haplotype 01. The complication is the 22 case. We cannot tell how frequent haplotypes are because we do not know how many 22s are from haplotype pair

$$\begin{matrix} 0 & 0 \\ 1 & 1 \end{matrix}$$

and how many are from haplotype pair

$$\begin{matrix} 0 & 1 \\ 1 & 0 \end{matrix}.$$

More generally, if we had k bases, we would not be able to resolve any genotype with more than a single 2 in its string. EM gives us a way to work around this problem.

The first thing we need for any EM approach is a set of latent variables y. We try to choose y to be a piece of information that is not known to us but that will make our parameter estimation problem easy if we know it. We can accomplish this for the current example by choosing y to be a guess as to the counts of our haplotype pairs:

$$y = \{g_{00,00}, g_{01,00}, g_{10,00}, g_{11,00}, g_{00,01}, \ldots\}.$$

For example, $g_{01,10}$ is the number of individuals who have the 01 haplotype on the first chromosome and the 10 haplotype on the second chromosome.

Given our inputs x and some current guess as to our parameters λ, we can find an expectation of y as follows:

$$g_{00,00} = n_{00}$$

$$g_{01,00} = \frac{1}{2}n_{02}$$

$$\vdots$$

$$g_{01,10} = n_{22}\left(\frac{f_{01}f_{10}}{f_{00}f_{11} + f_{01}f_{10} + f_{10}f_{01} + f_{11}f_{00}}\right)$$

$$g_{00,11} = n_{22}\left(\frac{f_{00}f_{11}}{f_{00}f_{11} + f_{01}f_{10} + f_{10}f_{01} + f_{11}f_{00}}\right).$$

Furthermore, given any x and our expected value of y, we can find an optimal value of λ:

$$f_{ij} \approx \frac{g_{ij,ij} + \sum_{xy} g_{ij,xy}}{\sum_{ab} n_{ab}}.$$

Normally, we will need to use the distribution over the latent values to estimate the parameters. In this case, though, the maximum likelihood estimate of each parameter is a linear function of the latent variables, and we can therefore use the expectations of the latent variables in place of a sum over their distribution.

And that gives us our EM iteration:

1. Estimate each $g_{ij,kl}$ from the inputs n_{ij} and the current parameter estimates f_{ij}.
2. Optimize the parameter estimates f_{ij} from the $g_{ij,kl}$s and n_{ij}s.

There is a small complication, in that we need a first guess at the fs to get started. One way to get this is to enumerate over possible resolutions of each genotype and assume that all are equally likely. We then apportion the contribution from each ambiguous genotype equally to all of its possible resolutions.

We can see how this method works by running through a few steps with a sample input:

$$n_{00} = 1 \quad n_{01} = 5 \quad n_{02} = 4$$

$$n_{10} = 8 \quad n_{11} = 17 \quad n_{12} = 22$$

$$n_{20} = 2 \quad n_{20} = 19 \quad n_{22} = 22.$$

Let N be the total count, $n_{00} + n_{01} + n_{02} + n_{10} + \cdots = 100$.

We first need to get an initial guess as to our haplotype frequencies $\lambda_i = (f_{00}, f_{01}, f_{10}, f_{11})$:

$$f_{00} = \frac{2n_{00} + n_{02} + n_{20} + \frac{1}{2}n_{22}}{2N} = \frac{19}{200} = 0.095$$

$$f_{01} = \frac{2n_{01} + n_{02} + n_{21} + \frac{1}{2}n_{22}}{2N} = \frac{44}{200} = 0.22$$

$$f_{10} = \frac{2n_{10} + n_{12} + n_{20} + \frac{1}{2}n_{22}}{2N} = \frac{51}{200} = 0.255$$

$$f_{11} = \frac{2n_{11} + n_{12} + n_{21} + \frac{1}{2}n_{22}}{2N} = \frac{86}{200} = 0.43.$$

Now we perform an E-step by estimating $y = (g_{00,00}, g_{01,00}, \ldots)$ from x and our current λ. We already know most of these terms. For instance:

$$g_{00,00} = n_{00} = 1$$

$$g_{00,01} = g_{01,00} = \frac{1}{2}n_{02} = 2$$

$$\vdots$$

The only unknowns are $g_{00,11} = g_{11,00}$ and $g_{01,10} = g_{10,01}$. We estimate these given our current f values:

$$g_{00,11} = g_{11,00} = \frac{f_{00}f_{11}n_{22}}{2(f_{00}f_{11} + f_{10}f_{01})} = 4.64$$

$$g_{10,01} = g_{01,10} = \frac{f_{01}f_{10}n_{22}}{2(f_{00}f_{11} + f_{10}f_{01})} = 6.37.$$

Now we perform an M-step by reestimating our set of f values, λ:

$$f_{00} = \frac{2(g_{00,00} + g_{00,01} + g_{00,10} + g_{00,11})}{2N} = \frac{2\left(1 + \frac{1}{2}(4) + \frac{1}{2}(2) + 4.64\right)}{200} = 0.0864$$

$$f_{01} = \frac{2(g_{01,01} + g_{01,00} + g_{01,10} + g_{01,11})}{2N} = \frac{2\left(5 + \frac{1}{2}(4) + 6.37 + \frac{1}{2}(19)\right)}{200} = 0.2287$$

$$f_{10} = \frac{2(g_{10,10} + g_{10,00} + g_{10,01} + g_{10,11})}{2N} = \frac{2\left(8 + \frac{1}{2}(2) + 6.37 + \frac{1}{2}(22)\right)}{200} = 0.2637$$

$$f_{11} = \frac{2(g_{11,11} + g_{11,00} + g_{11,01} + g_{11,10})}{2N} = \frac{2\left(17 + 4.64 + \frac{1}{2}(19) + \frac{1}{2}(22)\right)}{200} = 0.4214.$$

We can then perform another E-step by estimating the gs, given x and the new λ:

$$g_{00,11} = g_{11,00} = \frac{f_{00}f_{11}n_{22}}{2(f_{00}f_{11} + f_{10}f_{01})} = 4.14$$

$$g_{10,01} = g_{01,10} = \frac{f_{01}f_{10}n_{22}}{2(f_{00}f_{11} + f_{10}f_{01})} = 6.86.$$

Then we apply another M-step:

$$f_{00} = \frac{2(g_{00,00} + g_{00,01} + g_{00,10} + g_{00,11})}{2N} = 0.0814$$

$$f_{01} = \frac{2(g_{01,01} + g_{01,00} + g_{01,10} + g_{01,11})}{2N} = 0.2336$$

$$f_{10} = \frac{2(g_{10,10} + g_{10,00} + g_{10,01} + g_{10,11})}{2N} = 0.2686$$

$$f_{11} = \frac{2(g_{11,11} + g_{11,00} + g_{11,01} + g_{11,10})}{2N} = 0.4164.$$

Then another E-step:

$$g_{00,11} = g_{11,00} = \frac{f_{00}f_{11}n_{22}}{2(f_{00}f_{11} + f_{10}f_{01})} = 3.86$$

$$g_{10,01} = g_{01,10} = \frac{f_{01}f_{10}n_{22}}{2(f_{00}f_{11} + f_{10}f_{01})} = 7.14.$$

Then another M-step:

$$f_{00} = \frac{2(g_{00,00} + g_{00,01} + g_{00,10} + g_{00,11})}{2N} = 0.0786$$

$$f_{01} = \frac{2(g_{01,01} + g_{01,00} + g_{01,10} + g_{01,11})}{2N} = 0.2364$$

$$f_{10} = \frac{2(g_{10,10} + g_{10,00} + g_{10,01} + g_{10,11})}{2N} = 0.2714$$

$$f_{11} = \frac{2(g_{11,11} + g_{11,00} + g_{11,01} + g_{11,10})}{2N} = 0.4136,$$

and so on. Our estimates seem to be converging, so we could probably stop in a few steps. Normally, we would need some kind of stopping criterion, such as $\|\lambda_{i+1} - \lambda_i\| < \varepsilon$, to determine when to terminate the iteration.

Example 2   Inferring haplotypes from noisy data. We will look at another example involving haplotype data that illustrates a common use of EM methods: making inferences in the presence of missing or erroneous data. To illustrate the problem, suppose we have the following set of haplotype sequences:

0 1 0 × 100 copies

0 1 1 × 1 copy

1 1 0 × 100 copies

1 * 0 × 1 copy,

where * means an unknown value. We may reasonably infer that 1 * 0 should really be 1 1 0, since we know 1 1 0 is a common haplotype. If we know our methods for experimentally determining haplotypes are noisy, we may also infer that 0 1 1 is likely a mistake, and probably is really supposed to be 0 1 0. EM will give us a formal way to make those sorts of inferences.

We begin by defining our parameters λ to be the haplotype frequencies, as in the prior example:

$$\lambda = \{f_{000}, f_{001}, f_{010}, f_{011}, f_{100}, f_{101}, f_{110}, f_{111}\}.$$

We now assume that our observed data set x contains haplotypes, not genotypes, which will make our problem easier. However, we also must consider that we might observe data including missing values:

$$x = \{n_{000}, n_{001}, n_{00*}, n_{010}, n_{011}, n_{01*}, \ldots\}.$$

To use EM, we then need to create a probability model to say how likely any given observation is, given a parameter set. We can deal with missing data by saying we have some probability μ of outputting a $*$ in place of a 0 or 1. Then, for example:

$$Pr\{\text{observe } 1*0\} = (f_{100} + f_{110})(1 - \mu)\mu(1 - \mu).$$

We derive this formula from the fact that there are 2 true haplotypes that could give the observation $1*0$: 100 and 110. For either one, we have probability $1 - \mu$ of observing the first or third base, and probability μ of not observing the second base.

We can similarly handle erroneous data by saying there is a probability ε of flipping 0 to 1 or vice versa. This makes things somewhat more complicated, since now in theory any observation could derive from any true haplotype with some probability:

$$\begin{aligned}
Pr\{\text{observe } 1*0\} = p_{1*0} = & (f_{100} + f_{110})(1 - \mu - \varepsilon)\mu(1 - \mu - \varepsilon) \\
& + (f_{000} + f_{010})\varepsilon\mu(1 - \mu - \varepsilon) + (f_{101} + f_{111})(1 - \mu - \varepsilon)\mu\varepsilon \\
& + (f_{001} + f_{011})\varepsilon\mu\varepsilon.
\end{aligned}$$

For each base in the actual sequence, we now have probability μ of observing $*$, ε of observing the wrong value, and $1 - \mu - \varepsilon$ of observing the right value. For simplicity, we will refer to this function as p_{1*0}, defining other p_{ijk} values analogously. Then we can define the overall likelihood of our parameter set $Pr\{x \mid \lambda\}$ to be

$$p_{000}^{n_{000}} p_{001}^{n_{001}} p_{00*}^{n_{00*}} \times \cdots = \prod_{ijk \in \{0,1,*\}^3} p_{ijk}^{n_{ijk}}.$$

Our goal is then to find λ maximizing this likelihood $Pr\{x \mid \lambda\}$ for the given x.

We now need to choose our latent data set, y. A reasonable choice for y is the true count of each possible haplotype (i.e., what the input would be if we had no noise in the data):

$$y = \{N_{000}, N_{001}, N_{010}, N_{011}, N_{100}, N_{101}, N_{110}, N_{111}\}.$$

We also need a way of establishing an initial guess λ_0. We can guess that all 0s and 1s are correct, and that all $*$s are equally likely to be 0 or 1. Then, with our initial data set given above, we will make the initial guess:

$$f_{010} = \frac{100}{202} \approx 0.495$$

$$f_{011} = \frac{1}{202} \approx 0.00495$$

$$f_{110} = \frac{100.5}{202} \approx 0.4975$$

$$f_{100} = \frac{0.5}{202} \approx 0.00248$$

$$f_{ijk} = 0 \quad \text{otherwise.}$$

We then perform our EM algorithm. For our E-step, we find the expected values of our latent variables. The formula for this gets pretty complicated to write down, but it is not too hard to understand. To estimate the true count for a given haplotype, we need to consider all possible observations and then, for each, evaluate the fraction of times that observation should correspond to the given true haplotype. For instance,

$$N_{000} =$$

$$n_{000} \frac{f_{000}(1 - \mu - \varepsilon)^3}{f_{000}(1 - \mu - \varepsilon)^3 + (f_{001} + f_{010} + f_{100})\varepsilon(1 - \mu - \varepsilon)^2 + (f_{011} + f_{110} + f_{101})\varepsilon^2(1 - \mu - \varepsilon) + f_{111}\varepsilon^3}$$

$$+ n_{001} \frac{f_{000}\mu(1 - \mu - \varepsilon)^2}{f_{001}(1 - \mu - \varepsilon)^3 + (f_{000} + f_{011} + f_{101})\varepsilon(1 - \mu - \varepsilon)^2 + (f_{010} + f_{111} + f_{100})\varepsilon^2(1 - \mu - \varepsilon) + f_{110}\varepsilon^3}$$

$$+ \cdots$$

The M-step is easy for this choice of latent variables:

$$f_{ijk} = \frac{N_{ijk}}{N},$$

where *ijk* is any true haplotype and N is the total number of data points observed.

If we repeatedly apply these two steps, then we should converge on some estimate of the parameters λ. We will also, as a side benefit, get the most likely assignment for any missing or possibly incorrect values. EM is often a very good choice for these kinds of parameter-tuning problems in the presence of noisy data because it is usually so easy to handle missing or incorrect data: define latent variables corresponding to the true values of all observed variables. It can also be useful for cleaning up noisy data to make it more amenable for some other inference algorithm that does not deal well with noise in the data.

References and Further Study

EM is a topic often covered in statistical contexts, and good coverage is likely to be found in many introductory-to-intermediate general statistics texts in the context of statistical inference. Wasserman [112] is one example. The general topic of statistical inference is also a source of many other methods suitable for parameter tuning of similar maximum likelihood and MAP models. EM is similarly an important tool of the field of machine learning, and introductory texts on that topic are likely to cover EM, as well as other methods for fitting probabilistic models that we will not cover here. Mitchell [177], for example, provides a good introduction to EM in the context of machine learning.

The basic EM method is generally credited to two papers: Baum et al. [178], which first developed the method for the specific application of learning hidden Markov models (HMMs), and Dempster et al. [179], which first formally described it as a general method for maximum likelihood problems. Either of these papers also does a fine job as a tutorial on the method. The method appears to have been independently derived many times in the literature with reference to specific problems, though, and dedicated readers may find earlier examples of its use. The haplotype examples are derived from a technique for inferring haplotypes from genotypes due to Niu et al. [176]. EM is widely used in statistical genetics applications, and examples too numerous to cite can be found throughout that literature.

We will now examine a particularly important use of expectation maximization in biology: fitting data sets to hidden Markov models (HMMs). HMMs are a special kind of Markov model widely used in biological sequence analysis problems. Recall that a standard Markov model is defined by three elements:

- A set of states $Q = \{q_1, q_2, \ldots, q_n\}$
- An initial state distribution $\Pi = \{\pi_1, \pi_2, \ldots, \pi_n\}$
- A set of transition probabilities $P = \begin{bmatrix} p_{11} & \cdots & p_{1n} \\ \vdots & \ddots & \vdots \\ p_{n1} & \cdots & p_{nn} \end{bmatrix}$

An HMM is called "hidden" because in an HMM we usually assume that we do not see the states of the model, but rather a set of outputs influenced by them. Each time the model makes a state transition, it emits one output chosen from a distribution that depends on the new state. To make an HMM, we extend our standard Markov model with the following two features:

- A set of possible outputs $\Sigma = \{\sigma_1, \sigma_2, \ldots, \sigma_m\}$
- A set of output probabilities $B = \begin{bmatrix} b_{11} & \cdots & b_{1m} \\ \vdots & \ddots & \vdots \\ b_{n1} & \cdots & b_{nm} \end{bmatrix}$, where b_{ij} is the probability of emitting output j in state i.

For example, suppose we define the following HMM:

- $Q = \{q_1, q_2, q_3\}$
- $\Pi = \{\frac{1}{3}, \frac{1}{3}, \frac{1}{3}\}$
- $P = \begin{bmatrix} 0 & \frac{1}{2} & \frac{1}{2} \\ \frac{1}{2} & 0 & \frac{1}{2} \\ \frac{1}{2} & \frac{1}{2} & 0 \end{bmatrix}$

- $\Sigma = \{a, b, c\}$
- $B = \begin{bmatrix} 1 & 0 & 0 \\ 0.1 & 0.8 & 0.1 \\ 0.2 & 0.2 & 0.6 \end{bmatrix}$

Then, as we run through the model, we will get a sequence of states just as with a regular Markov model, such as the following:

$$q_1 \rightarrow q_2 \rightarrow q_1 \rightarrow q_3 \rightarrow q_2 \rightarrow q_3 \rightarrow q_2 \rightarrow q_3.$$

Simultaneously, though, we will generate a sequence of observations:

a   b   a   c   c   a   b   c

Normally, we will assume the outputs are something we actually observe and the states are something we can only infer. Here, for example, we may guess that when we see an "a," it is likely that the state is q_1, but it may be in q_2 or q_3. A "b" means we are likely to be in state q_2, and a "c" means we are likely to be in state q_3. From these rules, we may guess that our states are

$$q_1 \rightarrow q_2 \rightarrow q_1 \rightarrow q_3 \rightarrow q_3 \rightarrow q_1 \rightarrow q_2 \rightarrow q_3,$$

which will be mostly, but not completely, correct. In this chapter, we will cover several more rigorous methods for making this sort of inference.

20.1   Applications of HMMs

Example 1   Gene structure. As mentioned above, HMMs are often used for problems in sequence analysis. For example, we may generate a simple gene structure model such as that of figure 20.1(a). The states will correspond to different regions within or near a gene, and the outputs will be DNA bases. Each state will then have a different base distribution to reflect biases for certain bases in certain regions of a gene. For instance, introns usually start with the bases GT and end with the bases AG, so we will give high probabilities to those bases in the four intron start and end states. We can use such a model, for example, to simulate gene sequences to provide test data for sequence analysis algorithms. More often, our real interest in building the model is to help us infer the structure of actual genes by figuring out how a given true DNA sequence is likely to map to the the model states. For instance, if we see the sequence

C A A T C A G G T G T C C T A G G A T C A G,

we may infer that it is best explained by the states

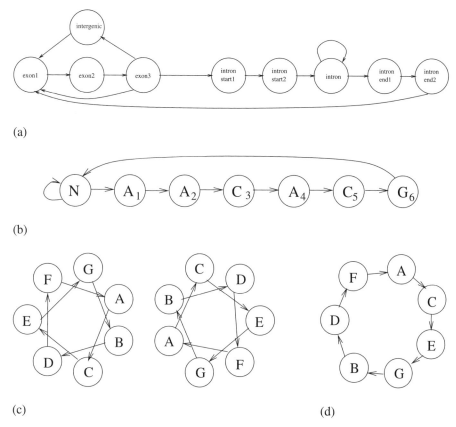

Figure 20.1
Examples of hypothetical HMMs for sequence analysis problems. (a) An HMM of gene structure. (b) An HMM for a DNA binding motif. (c,d) A model of a coiled-coil domain in proteins and an HMM for recognizing that domain type.

$$E_1 E_2 E_3 E_1 E_2 E_3 E_1 E_2 E_3 IS_1 IS_2 IIIIIIIE_1 IE_2 E_1 E_2 E_3,$$

which will show us where the introns and exons are likely to occur in the sequence.

Example 2 Transcription factor binding motif. HMMs are also used for identifying binding motifs and other short but noisy patterns in nucleotide and amino acid sequences. We may, for example, have some canonical DNA binding motif, AACACG, for which we wish to search. Usually, such a motif will be only a consensus sequence for a binding protein that can bind not just to that sequence but also to many similar ones. Some bases are likely to be strongly conserved, and others to be very flexible. To create an HMM that models DNA containing the motif, we may create one state N representing any base outside the motif and then one state for each position in the motif, as in figure 20.1(b).

The output probability matrix will then tell us how strongly conserved each base is. If state A_1 is highly conserved, then we may have an output distribution where $p_A \approx 1$ for state A_1. If state C_3 is weakly conserved, then maybe $p_C = 0.6$ for state C_3. Given this kind of knowledge, we may be able to infer that the DNA sequence

A C A A T A A G A C G G T G A

probably contains the motif and is explained by the state set

$NNNNNA_1A_2C_3A_4C_5G_6NNNN,$

whereas the sequence

T T C C A C A C G T G

probably does not contain the motif even though it also has a subsequence that matches in five of six motif positions (CACACG).

Example 3　Protein domain recognition. HMMs are also useful for problems related to protein sequences, such as detecting conserved domains in proteins. One example of this application is the detection of coiled-coil domains, which are produced by two (or more) alpha helices that wrap around one another to produce a helical twist made of the component alpha helices. Examined end-on, coiled coils have a structure approximately defined by seven repeating positions around the axis of each helix. In the case of a two-stranded coiled coil, for instance, we have the structure of figure 20.1(c). Because the A and B states form the interface packed between the two helices, those states tend to be hydrophobic. The D, E, and F states are exposed to the solvent and are therefore more likely to be polar. We can encode the sequence of states into the HMM of figure 20.1(d).

If we have output probabilities chosen so that the A and B states are likely to emit hydrophobic residues and the D, E, and F states are likely to emit polar residues, then this HMM will tend to emit sequences like the following:

$\underline{L}$–C–P–D–$\underline{V}$–Q–R–$\underline{I}$–P–K–A–$\underline{A}$–H–K–$\underline{V}$–M–C–Q–$\underline{F}$–K–V.

This pattern of hydrophobic residues at alternating 3-residue and 4-residue separations is the hallmark of coiled coils. We can detect this pattern by fitting the sequence to the HMM, leading to a state set like the following:

A–C–E–G–B–D–F–A–C–E–G–B–D–F–A–C–E–G–B–D–F.

Although this is a relatively easy pattern to fit to an HMM because of its simple repeat structure, HMMs can be used to fit many different protein domain types. For example, the Pfam [180] protein domain family database uses HMMs to identify a library of known domains from protein sequences.

20.2   Algorithms for HMMs

Though there are many things we can do with HMMs, there are three specific problems that are at the core of working with them. These problems, drawn from a widely read tutorial by Rabiner [181], can be stated as follows:

1. Given an observed set of outputs x and an HMM $\lambda = \{Q, P, \Pi, \Sigma, B\}$, find the best state string S to produce x from λ.
2. Given x and λ, find the probability of generating x from λ. (This is useful for evaluating different possible models as the source of x.)
3. Given the observations x and the geometry (graph structure) of λ, find the parameters of λ (P, B, and Π) that maximize the probability of producing x from λ (i.e., the maximum likelihood parameter set for generating x).

We will now proceed through each of these problems in turn. Our presentation of these methods closely follows that of Rabiner [181].

20.2.1   Problem 1: Optimizing State Assignments
If we have a sequence of observations $x = x_1, x_2, \ldots, x_T$ and an HMM λ and we want to find the best sequence of states $S = S_1, S_2, \ldots, S_T$ to explain x, given λ, we first need to be a bit more precise about what we mean by "best." There are different ways we can specify this, but a good one is to ask for the complete state set S maximizing the total likelihood of the outputs and states, given the model

$$\max_{S} \ Pr\{x, S \,|\, \lambda\}.$$

We can make this problem tractable by noting that by the rules of conditional probability,

$$Pr\{x, S \,|\, \lambda\} = Pr\{x \,|\, S, \lambda\} \ Pr\{S \,|\, \lambda\}.$$

If we know S, then both terms of this conditional probability will be easy to evaluate. $Pr\{x \,|\, S, \lambda\}$ is the product over all states of the probability of emitting the ith output x_i from the ith state S_i:

$$\prod_{i=1}^{T} b_{S_i, x_i}.$$

$Pr\{S \,|\, \lambda\}$ is the probability of transitioning among the consecutive states of S from λ:

$$\pi_{S_1} \prod_{i=2}^{T} p_{S_{i-1}, S_i}.$$

We can therefore easily evaluate the probability of the observations, given the model for any given S.

Choosing S to optimize for these probabilities is not so simple, but it turns out that we can do it efficiently by a dynamic programming algorithm called the *Viterbi algorithm*. As is usual with dynamic programming algorithms, we can visualize how this one works by imagining that we are filling in a table. In this case, the table will consist of the state set in one dimension and the sequence of steps during a run of the model in the other dimension.

	1	2	3		$j-1$	j		$T-2$	$T-1$	T
1										
2										
$\vdots$				$\cdots$			$\cdots$			
$n-1$										
n										

Each element of the table, M_{ij}, will record the best possible sequence of states for the first j observations such that the jth state is q_i. That is:

$$M_{ij} = \max_{S_1, \ldots, S_j} Pr\{x_1, \ldots, x_j \wedge S_1, \ldots, S_j \mid \lambda \wedge S_j = q_i\}$$

Figure 20.2(a) illustrates the overall process of filling in the table. Figure 20.2(b) provides pseudocode for the full algorithm. Line 1 of figure 20.2(b) is the initialization step. It fills in the first column of the matrix by defining a probability for each possible initial state of the model. Line 2 is the induction step. It fills in the remainder of the matrix by maximizing for each element over all states k the probability of having been in state q_k at the previous step, times the probability of transitioning from q_k to q_i, times the probability of outputting x_j from state q_i. The A matrix records which state q_k from the prior step led to the maximum for each state q_i in the next step. Lines 3 and 4 are the termination step. This step finds the most likely final state of the model by maximizing among probabilities in the final column of the matrix. Line 5, finally, is the backtracking step. It reconstructs the optimal path through the M matrix, using the A matrix to trace backward from the optimal ending state found in the termination step. At the end of the backtracking process, $q_1^*, \ldots, q_T^*$ will be our

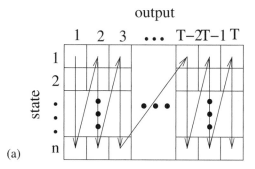

(a)

(b)
1. for $i = 1$ to n
   A. $M_{i1} = Pr\{x_1 \wedge S_1 = q_i | \lambda\} = Pr\{x_1 | S_1 = q_i \wedge \lambda\} Pr\{S_1 = q_i | \lambda\} = b_{i,x_1} \times \pi_i$
2. for $j = 2$ to T
   A. for $i = 1$ to n
      i. $M_{ij} = \max_k Pr\{x_1, \ldots, x_j | \lambda, S_j = q_i, S_{j-1} = q_k\} = \max_k M_{k,j-1} p_{ki} b_{i,x_j}$
      ii. $A_{ij} = \operatorname{argmax}_k M_{k,j-1} p_{ki} b_{i,x_j}$
3. $M^* = \max_i M_{iT}$
4. $q_T^* = \operatorname{argmax}_i M_{iT}$
5. for $j = T - 1$ to 1
   A. $q_j^* = A_{q_{j+1}^*, j+1}$

Figure 20.2
The Viterbi algorithm. (a) Illustration of the algorithm, showing how we fill in table elements during the dynamic programming process. (b) Pseudocode for the algorithm.

single best explanation of the observations, given the model, where the best sequence is defined to be the one maximizing the likelihood of the outputs and states, given the model.

20.2.2 Problem 2: Evaluating Output Probability

Our second problem is to measure the goodness of fit of a given model for a given output sequence. That is, given x and λ, we want to find $Pr\{x | \lambda\}$. This probability of the output, given the model, will be computed by summing over all possible assignments of states to the outputs.

Before we consider how we will do this, it is worth considering why we might do it. We can illustrate this with a real-world use of HMMs we mentioned briefly above: protein domain recognition. Suppose we have a hidden Markov model λ_1 that generates sequences of residues corresponding to some conserved protein domain D_1, and we also have a second HMM λ_2 that generates residues corresponding to a different domain D_2. If we are provided a real amino acid sequence x, having a method for evaluating the probability of an output set, given a model, will provide us a way to say whether our sequence x is more likely to have been generated by λ_1 or λ_2, and thus whether it is more likely to contain D_1 or D_2. More generally, if we have a library of D domains, each with its own HMM λ_i, as well as a background HMM λ_B representing the lack of any known motif, then we can identify the most likely domain for a real sequence by finding

$$\max_{\lambda_i} \ Pr\{x \,|\, \lambda_i\} \ Pr\{\lambda_i\},$$

where $Pr\{\lambda_i\}$ represents a *prior probability* of λ_i, essentially a guess as to how frequent that domain is across all protein sequences.

If runtime is not an issue, then problem 2 will be easy to solve. We will just enumerate over all possible sequences of states S and add up the likelihoods over all of these state sequences:

$$\sum_S Pr\{x, S \,|\, \lambda\} = \sum_S Pr\{x \,|\, S, \lambda\} \ Pr\{S \,|\, \lambda\}$$

$$= \sum_S (b_{S_1 x_1} \times b_{S_2 x_2} \times \cdots \times b_{S_T x_T})(\pi_{S_1} \times p_{S_1 S_2} \times p_{S_2 S_3} \times \cdots \times p_{S_{T-1} S_T}).$$

This generally is not practically possible, though, since the there are N^T possible state sequences, a number which may be quite large. We can do much better with something called the *forward algorithm*. The forward algorithm computes a set of values α_{ij}, defined as follows:

$$\alpha_{ij} = Pr\{x = x_1, \ldots, x_j \wedge s_j = q_i \,|\, \lambda\}.$$

The definition of α_{ij} is very similar to the definition of M_{ij} from problem 1, and it can be found by a similar dynamic programming algorithm. The forward algorithm, shown in figure 20.3(a), is basically the same as the Viterbi algorithm, except that during the induction (step 2) we sum over possible prior states instead of maximizing over them. We also do not need to backtrack, since we are no longer looking for one best sequence of states.

One might reasonably guess that if there is a forward algorithm, there is also a backward algorithm. The backward algorithm solves for the following related quantities:

$$\beta_{ij} = Pr\{x_{j+1}, x_{j+2}, \ldots, x_T \,|\, S_j = q_i \wedge \lambda\}.$$

That is, it finds the probability of generating the suffix of x from $j+1$ to T, given that the model is in state q_i at step j. It is also solved by an algorithm similar to the

(a)

1. for $i = 1$ to n
   A. $\alpha_{i1} = Pr\{x_1 \wedge S_1 = q_i | \lambda\} = Pr\{x_1 | S_1 = q_i \wedge \lambda\} Pr\{S_1 = q_i | \lambda\} = b_{i,x_1} \times \pi_i$
2. for $j = 2$ to T
   A. for $i = 1$ to n
      i. $\alpha_{ij} = \sum_{k=1}^{n} (\alpha_{k,j-1} p_{ki}) b_{i,x_j}$
3. $Pr\{x | \lambda\} = \sum_{i=1}^{n} \alpha_{iT}$

(b)

1. for $i = 1$ to n
   A. $\beta_{iT} = 1$
2. for $j = T - 1$ downto 1
   A. for $i = 1$ to n
      i. $\beta_{ij} = \sum_{k=1}^{n} \left(p_{i,k} b_{k,x_j} \right) \beta_{k,j+1}$

Figure 20.3
Pseudocode for the forward and backward algorithms. (a) Forward algorithm. (b) Backward algorithm.

Viterbi and forward algorithms, although it lacks a termination step. Figure 20.3(b) provides pseudocode for the backward algorithm.

Note that the forward and backward algorithms put together give us an alternative method for solving problem 1. Instead of finding the one complete sequence that maximizes the likelihood, we can find the one maximum likelihood state S_j for each step j. This is not necessarily the same sequence as the Viterbi method would give us. This new formulation will maximize the expected number of correct states over the full sequence. Yet the complete sequence of states derived by this definition may have very low or even zero probability. We solve for this new sequence by filling a table Γ with entries γ_{ij}, where

$$\gamma_{ij} = \frac{\alpha_{ij}\beta_{ij}}{Pr\{x | \lambda\}} = \frac{\alpha_{ij}\beta_{ij}}{\sum_k \alpha_{kj}\beta_{kj}}.$$

γ_{ij} is then $Pr\{S_j = q_i | x_1, \ldots, x_t, \lambda\}$, the probability of being in state q_i at step j, given the observations and the model. The overall best sequence by this alternative definition is then the sequence for which each S_j is the state i maximizing γ_{ij}.

20.2.3 Problem 3: Training the Model

Problem 3 is the problem of inferring the HMM parameters from a set of training data. Training the model can be relatively easy if we have *labeled* training data (i.e., data in which we know the true assignment of states). For example, if we are training a gene structure model like that of figure 20.1(a), we may use a data set in which someone has experimentally determined the true intron–exon structure of some genes and labeled the state assignment of each base in those genes. In that case, we can directly formulate a maximum likelihood estimate of each parameter. For example, if we observe that state I_1 outputs A 90 percent of the time, then our best estimate is that $b_{I_1,A} = 0.9$. If we observe that state E_3 is followed by E_1 98 percent of the time, then our best estimate is that $p_{E_3,E_1} = 0.98$.

As we have phrased it, though, problem 3 is the hardest of the three problems. We get to see only the outputs, and have to decide what parameter set would be most likely to have yielded those outputs, given that we do not know the state assignment. This problem is generally solved using an EM algorithm called the Baum–Welch algorithm. To show how to do this, we will start by casting our problem in terms of the formalisms of EM:

1. *Observations* x Our observables for the EM algorithm are the outputs of the HMM model, which we have also called x. Often, we will have multiple training sequences, such as some $x_1 = \{x_{11}, x_{12}, \ldots\}, x_2 = \{x_{21}, x_{22}, \ldots\}, \ldots$. Then our observation vector is the union of all of these input sequences:

$$x = x_{11}, x_{12}, \ldots, x_{1T_1}, x_{21}, x_{22}, \ldots, x_{2T_2}, \ldots$$

2. *Parameters* λ We will assume that the structure of the Markov model graph is given to us, but that we do not know the values of the probability parameters. These parameters, which we will try to infer, are the following:

$$\lambda = \{\Pi, P, B\},$$

where again Π is the vector of starting state probabilities, P is the matrix of transition probabilities, and B is the matrix of output probabilities. Note that in many cases we may have additional problem-specific constraints on these. For example, we may assume that some probabilities are known to us and others need to be inferred. Some of these constraints will be easy to handle within the framework of the Baum–Welch algorithm, but others will require significant revisions of the algorithm.

3. *Latent variables* y We want our latent variables to make it easy to infer the parameters. As the discussion of labeled data above shows us, it is easy to infer the parameters if we know the state assignments. Therefore, it may make sense to choose the state assignments as our latent variables. It turns out to be convenient to choose a slightly different set of latent variables that give us almost the same information as the state assignments. Specifically, we will choose the following:

• $C_t(i, j)$ We define this to be the probability that we go from state q_i to state q_j during the transition from step t to step $t + 1$.

• γ_{it} This is the same γ we used earlier, the probability we are in state q_i at step t.

These two quantities in effect define a probability distribution over possible state assignments. Finding this distribution will constitute the E-step of the Baum–Welch EM algorithm.

We can compute both of these quantities using the tables constructed by the forward and backward algorithms. First, we can compute the $C_t(i, j)$ terms by the following formula:

$$C_t(i, j) = \frac{\alpha_{it} \times p_{ij} \times \beta_{j,t+1} \times b_{j,x_{t+1}}}{Pr\{x \,|\, \lambda\}} = \frac{\alpha_{it} \times p_{ij} \times \beta_{j,t+1} \times b_{j,x_{t+1}}}{\sum_{i'=1}^{n} \sum_{j'=1}^{n} \alpha_{i't} \times p_{i'j} \times \beta_{j',t+1} \times b_{j',x_{t+1}}}.$$

The numerator consists of four components: the probability of generating the prefix of the observables up to state t, the probability of transitioning from q_i to q_j at step t, the probability of generating the suffix of the observables after state $t + 1$, and the probability of giving the observed output from state $t + 1$. Together, these give the probability of a particular state transition from q_i to q_j at this one step, given the complete observed x. The denominator is a scaling factor which normalizes the whole thing by the probability of getting our observed x from λ.

We already know how to compute γ_{it}, although we can derive a simpler expression for it once we know the C values:

$$\gamma_{it} = \sum_{j=1}^{n} C_t(i, j)$$

for $t < T$, or

$$\gamma_{it} = \sum_{j=1}^{n} C_{t-1}(j, i)$$

for $t > 1$. That is, the probability of being in state q_i at time t is the sum over all possible next states q_j of the probability of transitioning from q_i to q_j at step t, or the sum over prior q_js of the probability of transitioning from q_j to q_i at step $t - 1$.

The M-step of the Baum–Welch algorithm will then consist of finding maximum likelihood estimators for Π, P, and B, given the state distribution implied by the C and γ values. Estimating $\Pi = \{\pi_1, \pi_2, \ldots, \pi_n\}$ is the easiest:

$$\pi_i \approx \gamma_{i1}.$$

That is, the best estimate of the starting probability of state q_i is the estimated probability that we are in q_i at step 1.

We can also estimate P by observing that the probability of taking a transition from q_i to q_j is best approximated by the fraction of the times the model is in state q_i and next goes to state q_j. This fraction is given by

$$p_{ij} \approx \frac{\sum_{t=1}^{T-1} C_t(i,j)}{\sum_{t=1}^{T-1} \gamma_{it}}.$$

Finally, we can estimate B by observing that the best estimate of the probability of emitting output k from state j is the fraction of times that output k is produced by the model when it is in state j. This fraction is given by

$$b_{jk} \approx \frac{\sum_{t=1, x_t=\sigma_k}^{T} \gamma_{jt}}{\sum_{t=1}^{T} \gamma_{jt}}.$$

As with any EM algorithm, we have to repeatedly apply our E-step and our M-step until they converge on some solution. We establish our distributions of state and transition usages by calculating the Cs and γs (E-step), then estimating Π, P, and B from those distributions (M-step), repeating until convergence. Note that because EM is a local optimization method, our starting conditions can be important. Prior knowledge about the system that is helpful in making an initial best guess can make a big difference in the quality of the final output. For example, if we have many unlabeled data but few labeled data, we may use the labeled data to get a reasonable best guess as to the parameters, then train by Baum–Welch from there to get a better fit to the full data set.

An Aside on Fitting to HMMs   Baum–Welch is one optimization method that is widely useful and generally easy to implement for EM problems, but it is not the only one we may use. In general, we have a well-defined objective function ($Pr\{x \mid \lambda\}$) that we wish to optimize for a set of variables:

- n π_i variables
- n^2 p_{ij} variables
- $n|\Sigma|$ $b_{j\sigma}$ variables.

We can therefore treat this as a generic multivariate optimization problem and use any general-purpose optimization method, such as Newton–Raphson, steepest descent, constraint satisfaction, or others we have not covered. For some kinds of constraints we might put on our HMMs, Baum–Welch might prove impossible to implement, and it is therefore useful to be able to conceptualize this kind of parameter tuning as just another optimization problem.

We may also want to use a hybrid approach. For example, some constraints on the probabilities may allow us to keep the standard Baum–Welch E-step but require us to substitute a specialized constrained optimizer for the M-step. That kind of complication arises particularly if probabilities from different states or transitions are pre-

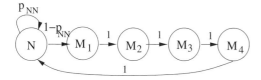

Figure 20.4
Simple HMM model for finding a motif of four bases. State N represents a background base frequency, and states M_1, M_2, M_3, and M_4 represent four motif positions. We assume here that the motif cannot contain insertions or deletions.

sumed to be dependent on one another or to be constrained by some restrictive prior distribution.

20.3   Parameter-Tuning Example: Motif-Finding by HMM

We can better understand how these methods work by going through a simple example in more detail. We will use a simplified variant of the motif-finding HMM we saw in figure 20.1(b). This variant is illustrated in figure 20.4. Here we have a motif of four bases, each with one state in the model, and also an extra background state N representing bases that are not part of the motif. To make the math a bit easier, we will assume that there are no insertions or deletions possible in the motif, making most of the transition probabilities either 1 or 0. This is generally a reasonable assumption for a DNA-binding motif, although one might more easily use a position-specific scoring matrix instead of an HMM in such cases. We will use this system to see how we may train an HMM model like this in practice.

We will generally need a place to get started, so we will suppose that our real motif has some general consensus pattern, say ACTG, but that we know it can vary from this pattern. We will also assume that we have some training data containing examples of the motif, although we do not know exactly how many examples the training data contain or where they are. For example, we may have the following sequence which, unknown to us, contains two real examples of the motif:

GACACTGCCTACGGT.

Initializing   The first thing we need to do is set some initial probabilities to get the EM method started. We would ordinarily set the probability of starting the motif, p_{NM_1}, to be a *prior estimate* of the frequency of the motif in the data. So, for example, if we believe this motif occurs about once every thousand bases in our data, we may initially say $p_{NM_1} = 0.001$.

Note that we have to be careful here because the parameter will depend on what kind of data we are using. If we choose a parameter based on the frequency of the motif in gene regions, the model is unlikely to work well if we use a whole genome as training data; the model will likely predict far more motifs than are really there. Likewise, if we choose our prior estimate based on the frequency of the motif in the whole genome, then the model is likely to perform poorly if we train it using only data from promoter regions of genes, since the model will expect the motif to be much less frequent than it actually is in that data set.

For the sake of this example, let us assume that we believe the motif occurs on average once every five bases in the training data. Then we will guess:

$$p_{NM_1} = \frac{1}{5}$$

$$p_{NN} = 1 - p_{NM_1} = \frac{4}{5}.$$

All other values of P are fixed in this model.

We also need to set Π, which we can also do using our prior estimate of the motif frequency. If the motif occurs once every five bases, then each motif state should have probability $\frac{1}{5}$. That means the background state N also has probability $\frac{1}{5}$. Therefore, we guess

$$\pi_i = \frac{1}{5} \quad \forall i.$$

Finally, we need some initial values for the elements of the B matrix. We need a reasonable guess here if we are to have any hope of solving the problem. If we guess that the consensus is perfectly preserved, then we have no hope of learning nonconsensus motifs. If we do not use an initial guess that approximately matches the consensus, we are unlikely to be able to detect the motif at all. And if we allow too much flexibility in the definition, the model may converge to some different but more frequent motif. We can try to avoid all of these possible problems by using a starting distribution which represents a noisy guess that the consensus is correct. We will suppose that all bases are equally likely in state N and that the consensus base has probability 0.7, and the others 0.1 each, in the motif states. Then our initial estimate of B will be

$$B = \begin{array}{c} N \\ M_1 \\ M_2 \\ M_3 \\ M_4 \end{array} \begin{bmatrix} .25 & .25 & .25 & .25 \\ .7 & .1 & .1 & .1 \\ .1 & .7 & .1 & .1 \\ .1 & .1 & .7 & .1 \\ .1 & .1 & .1 & .7 \end{bmatrix}$$
$$\quad\quad\quad A \quad C \quad G \quad T$$

E-Step   Once we have our initial values, we have to perform our first E-step. We ultimately need the C and γ values required by Baum–Welch. To get those, though, we first need to run the forward and backward algorithms to get the α and β tables.

Suppose our training data is the sequence GACACGTCCTACGCT. We begin by filling in the first column of the α table by the formula

$$\alpha_{i1} = b_{iG} \times \pi_i.$$

We then fill in subsequent columns by the formula

$$\alpha_{ij} = \left(\sum_k \alpha_{k,j-1} p_{ki} \right) b_{ix_j}.$$

This should be fairly easy for this example, since most of the p_{ij} values are either 0 or 1. We will get something like the following:

α	G	A	C	A	C	G	$\cdots$
N	0.05	0.015	0.0065	0.00135	0.000275	6.725×10^{-5}	
M_1	0.02	0.0070	0.0003	0.00091	0.00027	5.5×10^{-6}	
M_2	0.02	0.002	0.0049	0.00003	0.000637	2.7×10^{-6}	$\cdots$
M_3	0.14	0.002	0.0002	0.00049	3.0×10^{-6}	4.459×10^{-4}	
M_4	0.02	0.014	0.0002	0.00002	4.9×10^{-5}	3×10^{-7}	

The columns of the matrix containing the substring ACGT, which is a match to our consensus motif, show high probabilities for the corresponding motif states. This observation suggests that our initial guess is pretty good at finding the consensus motif.

We next need to find the β_{ij} values by the backward algorithm. We now initialize the rightmost column to all 1s and fill in toward the left, using the formula

$$\beta_{ij} = \sum_{k=1}^{n} p_{ik} b_{kx_{j+1}} \beta_{k,j+1}.$$

We will see something like the following:

β	$\cdots$	A	C	G	C	T
N		.0178	.084	.07	.3	1
M_1		.0049	.001	.07	.1	1
M_2	$\cdots$	.00075	.007	.01	.1	1
M_3		.00175	.0075	.01	.1	1
M_4		.021	.0175	.075	.1	1

Once we have the α and β tables, we can compute the C values. For example:

$$C_{T-1}(1,1) = \frac{\alpha_{1,T-1}p_{11}\beta_{1,T}b_{1T}}{Pr\{x\,|\,\lambda\}}.$$

Most of the C values will be zero for this example, since most transitions are presumed to have zero probability. We can also get the γ values using our formulas from above. For example:

$$\gamma_T(1) = \sum_{j=1}^{n} C_T(j,1).$$

We will not go through all of the math in detail. Filling in the rest should be an easy, if tedious, exercise for the reader.

M-Step One we have computed the C and γ values, we need to reestimate the Π, P, and B values. We can easily estimate Π values as follows:

$$(\pi_N, \pi_{M_1}, \pi_{M_2}, \pi_{M_3}, \pi_{M_4}) = (\gamma_{1N}, \gamma_{1M_1}, \gamma_{1M_2}, \gamma_{1M_3}, \gamma_{1M_4}).$$

We get the B values by finding all occurrences of a given output and then counting the fraction of times the model is predicted to be in a given state and emit the given output. For example, if we want to find b_{NA}, we will first note that the data show output A at steps 2, 4, and 11. Therefore, the probability of emitting A from state N will be estimated by

$$b_{NA} = \frac{\gamma_{N,2} + \gamma_{N,4} + \gamma_{N,11}}{\gamma_{N,1} + \gamma_{N,2} + \gamma_{N,3} + \cdots + \gamma_{N,14} + \gamma_{N,15}},$$

that is, the expected number of occurrences of state N that coincide with output A, divided by the total expected number of occurrences of state N.

Finally, to get maximum likelihood estimates of the transition probabilities, we can use the C values to estimate how often the model transitions from state q_i to q_j as a fraction of how often it transitions from state q_i to any state. For this example, we only need to estimate p_{NN}, since $p_{NM_1} = 1 - p_{NN}$ and all other probabilities are fixed at 0 or 1. We can estimate p_{NN} as follows:

$$p_{NN} \approx (C_1(N, N) + C_2(N, N) + C_3(N, N) + \cdots + C_{T-1}(N, N))/$$

$$\left(\begin{array}{c} C_1(N, N) + C_1(N, M_1) + C_1(N, M_2) + C_1(N, M_3) + C_1(N, M_4) \\ + C_2(N, N) + C_2(N, M_1) + C_2(N, M_2) + C_2(N, M_3) + C_2(N, M_4) + \cdots \\ + C_{T-1}(N, N) + C_{T-1}(N, M_1) + C_{T-1}(N, M_2) + C_{T-1}(N, M_3) \\ + C_{T-1}(N, M_4) \end{array} \right).$$

That completes our M-step. If we repeatedly apply the E-step and M-step above, we should eventually converge on some final model fit. If the initial estimates were well chosen, the final fit is likely to be a good model for the motif that one can then use to detect occurrences of the motif in new data sets.

References and Further Study

Hidden Markov models have become such a widely used technique in computational molecular biology that one can find coverage of them in many introductory and intermediate-level computational biology or bioinformatics texts. One may refer, for example, to Durbin et al. [182] or Mount [183] for additional material on HMMs in the context of various biological problems. HMMs are also widely used in other fields, such as language analysis and speech recognition, and readers looking for more extensive coverage may turn to those fields for texts on the topic. The most comprehensible tutorial on the topic of which I am aware is the review article by Rabiner [181], which focuses on speech-recognition applications.

Those interested in referring to the primary literature on HMMs can look up the first descriptions of the Viterbi [184], forward-backward [185], and Baum–Welch [178] algorithms. Baum et al. [178] is also one of our primary references for expectation maximization in general. There are many applications of HMMs in the computational biology literature. One particularly influential example is the Genscan gene finder [186], the first reasonably accurate gene prediction method for eukaryotic DNA. Genscan is at its heart based on HMMs, but with some nonstandard modifications that make it worth a look for those interested in more depth on practical matters in adapting HMMs to real-world data sources. Interested readers may also want to look up Delorenzi and Speed [187] to read about HMMs in the context of coiled-coil recognition, Sonnhammer et al. [180] for information about the use of HMMs for more general domain recognition by the Pfam database, or Krogh et al. [188] for the earliest application of HMMs to protein-fold recognition. For a more general survey of the use of HMMs in biological applications, readers may refer to several review and tutorial articles by Eddy on the topic [189] [190] [191]. These reviews also provide several examples of the use of HMMs for various forms of motif-finding.

21 Linear System-Solving

In this chapter, we will cover some basic tools for solving linear systems. Recall that a linear system is described by the equation

$$A\vec{x} = \vec{b},$$

where A is an $m \times n$ matrix, x is an $n \times 1$ vector, and b is an $m \times 1$ vector. Usually, when we talk about solving linear systems, we assume that A and b are known to us and we want to solve for x. We have already referred to linear system-solving in the context of solving various other problems. For example, solving linear systems is a key step in multidimensional Newton–Raphson optimization, as we saw in chapter 5, and also comes up in linear programming (chapter 6), state distributions of Markov models (chapters 8 and 11), integrating systems of differential equations (chapters 14 and 15), and many other topics in numerical optimization and simulation. As we will see, linear system-solving is also a core part of many parameter tuning problems. Linear system-solving is therefore one of those topics about which one needs to know in order to be prepared to work in any kind of scientific computing.

In simpler applications, we do not really need to know much about how linear systems are solved because there are highly optimized solvers freely available for work in the area. When one deals with harder linear systems, though, an understanding of the underlying algorithms can be critical in achieving reasonable performance in practice. Readers planning to go on to more advanced numerical computing would therefore be well advised to learn this topic, and numerical linear algebra in general, in much greater depth than we have room for here.

For now, we will look at linear systems specifically in the context of parameter-tuning. For example, suppose we want to know the expression level of a gene X, which we know is regulated by genes $A_1, A_2, \ldots, A_k$. Figure 21.1 illustrates this system. We can measure the expression of X in response to different levels of the regulating genes, and want to be able to predict how X will respond under unobserved conditions. One way we can approach this problem is to assume there is a linear

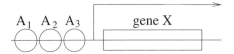

Figure 21.1
A gene X regulated by three transcription factors—A_1, A_2, and A_3—that bind to its promoter.

relationship between X and the A_i genes. Suppose we conduct a series of $k+1$ experiments. We will define the following variables:

- A_{ij} is the concentration of A_i in experiment j.
- X_j is the concentration of X in experiment j.
- c_i is the inferred contribution of $[A_i]$ to $[X]$.
- c is an extra inferred "baseline" value of $[X]$.

Then we can pose our inference problem, using the following linear system:

$$\begin{bmatrix} A_{11} & A_{21} & \cdots & A_{k1} & 1 \\ A_{12} & A_{22} & \cdots & A_{k2} & 1 \\ \vdots & & \ddots & & \vdots \\ A_{1k} & A_{2k} & \cdots & A_{kk} & 1 \\ A_{1,k+1} & A_{2,k+1} & \cdots & A_{k,k+1} & 1 \end{bmatrix} \begin{bmatrix} c_1 \\ c_2 \\ \vdots \\ c_k \\ c \end{bmatrix} = \begin{bmatrix} X_1 \\ X_2 \\ \vdots \\ X_k \\ X_{k+1} \end{bmatrix}$$

To find the c_i parameters, we can perform the experiment $k+1$ times and then solve for $c_1, \ldots, c_k, c$. Then, if we want to estimate the value of $[X]$ for some new $[A_1], \ldots, [A_k]$, we can guess that it is

$$c_1[A_1] + c_2[A_2] + \cdots + c_k[A_k] + c.$$

This process is called *linear regression*. To fit our linear model, we need to know how to solve $A\vec{x} = \vec{b}$. Solving this problem is what we mean by solving a linear system. We will initially assume we have an $n \times n$ matrix of full rank, and therefore the systems will have a unique solution, but later we will see how we may deal with over-determined and underdetermined systems.

21.1   Gaussian Elimination

The classic method for solving linear systems is *Gaussian elimination*. The basic idea behind the method is to repeatedly subtract multiples of one row of the matrix A from another row until we transform the system into something equivalent but easy to solve. Specifically, we will use a series of transformations to try to create an equiv-

alent linear system where the matrix is the identity matrix. To see how the method
works, we can take an example system

$$\begin{bmatrix} 2 & 4 & 2 \\ 3 & 2 & 3 \\ -1 & 0 & 2 \end{bmatrix} x = \begin{bmatrix} 16 \\ 16 \\ 5 \end{bmatrix}$$

To make it a little more readable, we will write $A\vec{x} = \vec{b}$ as $[A \,|\, \vec{b}]$, so our system
becomes

$$\begin{bmatrix} 2 & 4 & 2 & | & 8 \\ 3 & 2 & 3 & | & 16 \\ -1 & 0 & 2 & | & 5 \end{bmatrix}$$

To transform the system into one with the identity matrix, we will first try to fix
the lower triangular portion by filling the diagonal with ones and everything below
it with zeros. The first step is fixing the upper-left entry to 1. We accomplish this by
scaling the first row of the matrix by 1/2, yielding the following:

$$\begin{bmatrix} 1 & 2 & 1 & | & 8 \\ 3 & 2 & 3 & | & 16 \\ -1 & 0 & 2 & | & 5 \end{bmatrix}$$

We then will then use the first row to try to put zeros in the first column of the
second and third rows. We do this by subtracting 3 times the first row from the second row and -1 times the first row from the third:

$$\begin{bmatrix} 1 & 2 & 1 & | & 8 \\ 0 & -4 & 0 & | & -8 \\ 0 & 2 & 3 & | & 13 \end{bmatrix}$$

We will then normalize the second row to try to put a 1 in the second diagonal entry. We do this by dividing the second row by its diagonal entry, -4:

$$\begin{bmatrix} 1 & 2 & 1 & | & 8 \\ 0 & 1 & 0 & | & 2 \\ 0 & 2 & 3 & | & 13 \end{bmatrix}$$

We will then use the second row to eliminate the second column entry from the
third row. We do this by subtracting twice the second row from the third row:

$$\begin{bmatrix} 1 & 2 & 1 & | & 8 \\ 0 & 1 & 0 & | & 2 \\ 0 & 0 & 3 & | & 9 \end{bmatrix}$$

We will then put a 1 on the diagonal in the third row by dividing it by its diagonal entry, 3:

$$\begin{bmatrix} 1 & 2 & 1 & | & 8 \\ 0 & 1 & 0 & | & 2 \\ 0 & 0 & 1 & | & 3 \end{bmatrix}$$

We will now reverse direction and try to place zeros in the upper triangular portion of the matrix. We first use the bottom row to eliminate entries in the third column above the diagonal. We do this by subtracting the third row from the first:

$$\begin{bmatrix} 1 & 2 & 0 & | & 5 \\ 0 & 1 & 0 & | & 2 \\ 0 & 0 & 1 & | & 3 \end{bmatrix}$$

We then use the second row to eliminate entries above the diagonal in the second column by subtracting twice the second row from the first:

$$\begin{bmatrix} 1 & 0 & 0 & | & 1 \\ 0 & 1 & 0 & | & 2 \\ 0 & 0 & 1 & | & 3 \end{bmatrix}$$

We have, finally, reduced this to an easy system to solve. We have converted A to the identity matrix, so the solution to $A\vec{x} = \vec{b}$ is $\vec{x} = \vec{b}$, giving us

$$x = \begin{bmatrix} 1 \\ 2 \\ 3 \end{bmatrix}$$

Figure 21.2(a) provides pseudocode summarizing the basic Gaussian elimination algorithm for transforming an arbitrary linear system $A\vec{x} = \vec{b}$ into an equivalent system $I\vec{x} = \vec{b}$. When we are done running it, A should be the identity matrix and $\vec{b}$ will contain the solution vector $\vec{x}$ to the original system.

21.1.1 Pivoting

There is a problem with the preceding pseudocode, though. What if, as we are running through the algorithm, we end up with an a_{ii} entry that is zero? Because we have to divide by a_{ii}, the algorithm will fail with a division by zero error. Even a nonzero but small a_{ii} is a problem, since a_{ii} close to zero can lead to large numerical errors or, if it occurs repeatedly, to numerical overflows.

This problem is generally solved in practice by a method called *pivoting*. When we want to subtract out all elements below the diagonal in some column i, we ordinarily subtract multiples of row i from the rows below i. With pivoting, we instead first find

1. for $i = 1$ to n
   A. $b_i \leftarrow b_i / a_{ii}$
   B. for $k = n$ downto i
      i. $a_{ik} \leftarrow a_{ik} / a_{ii}$
   C. for $j = i + 1$ to n
      i. $b_j \leftarrow b_j - b_i a_{ji} / a_{ii}$
      ii. for $k = n$ downto i
         a. $a_{jk} \leftarrow a_{jk} - a_{ik} a_{ji} / a_{ii}$
2. for $i = n$ downto 2
   A. for $j = i - 1$ downto 1
      i. $b_j \leftarrow b_j - b_i a_{ji}$
      ii. $a_{ji} \leftarrow 0$

---

1. for $i = 1$ to n
   α. for $j = i + 1$ to n
      i. if $|a_{ij}| > |a_{ii}|$
         a. for $k = 1$ to n
            - $tmp \leftarrow a_{ik}$
            - $a_{ik} \leftarrow a_{jk}$
            - $a_{jk} \leftarrow tmp$
   A. $b_i \leftarrow b_i / a_{ii}$
   B. for $k = n$ downto i
      i. $a_{ik} \leftarrow a_{ik} / a_{ii}$
   C. for $j = i + 1$ to n
      i. $b_j \leftarrow b_j - b_i a_{ji} / a_{ii}$
      ii. for $k = n$ downto i
         a. $a_{jk} \leftarrow a_{jk} - a_{ik} a_{ji} / a_{ii}$
2. for $i = n$ downto 2
   A. for $j = i - 1$ downto 1
      i. $b_j \leftarrow b_j - b_i a_{ji} / a_{ii}$
      ii. $a_{ji} \leftarrow 0$

(a)　　　　　　　　　　　　　　　　(b)

Figure 21.2
Pseudocodes for the Gaussian elimination algorithm. (a) Standard Gaussian elimination. (b) Gaussian elimination with partial pivoting.

the row $k \in [i, n]$ that has the largest absolute value in column i. We then permute the rows to swap rows i and k. Then we use the new row i (the old row k) to eliminate all entries below the diagonal in column i.

To illustrate pivoting, we can use our example matrix from above:

$$\begin{bmatrix} 1 & 2 & 1 & | & 8 \\ 3 & 2 & 3 & | & 16 \\ -1 & 0 & 2 & | & 5 \end{bmatrix}$$

Instead of using row 1 to zero out lower triangular elements in column 1, we will instead begin with a pivoting step. We look for the element with largest absolute value in column 1, which occurs in row 2. We then pivot to swap rows 1 and 2:

$$\begin{bmatrix} 3 & 2 & 3 & | & 16 \\ 1 & 2 & 1 & | & 8 \\ -1 & 0 & 2 & | & 5 \end{bmatrix}$$

Note that this transformation is equivalent to multiplying both sides of the equation by a *permutation matrix*:

$$P_{12} = \begin{bmatrix} 0 & 1 & 0 \\ 1 & 0 & 0 \\ 0 & 0 & 1 \end{bmatrix}$$

That is, we convert the problem $A\vec{x} = \vec{b}$ into the equivalent problem $P_{12}A\vec{x} = P_{12}\vec{b}$.

We can now proceed with the Gaussian elimination by scaling the first row to put a 1 on the diagonal:

$$\begin{bmatrix} 1 & 2/3 & 1 & | & 16/3 \\ 1 & 2 & 1 & | & 8 \\ -1 & 0 & 2 & | & 5 \end{bmatrix}$$

We then subtract out multiples of the first row from the others to place zeros below the diagonal:

$$\begin{bmatrix} 1 & 2/3 & 1 & | & 16/3 \\ 0 & 4/3 & 0 & | & 8/3 \\ 0 & 2/3 & 3 & | & 31/3 \end{bmatrix}$$

We do not need to pivot again for the second step, since the largest value in column 2 is already in row 2, so we normalize the second row:

$$\begin{bmatrix} 1 & 2/3 & 1 & | & 16/3 \\ 0 & 1 & 0 & | & 2 \\ 0 & 2/3 & 3 & | & 31/3 \end{bmatrix}$$

From this point on, the algorithm will proceed exactly as in the nonpivoting version.

The swapping of rows to keep small values off the diagonal is technically known as *partial pivoting*. Partial pivoting generally works very well at controlling numerical errors, but there are cases where it performs poorly. We can avoid these by using *full pivoting*. With full pivoting, when we are trying to put a 1 on the jth diagonal entry, we find the largest element in the submatrix below and to the right of (j, j), then permute both rows and columns to swap that largest entry with the one initially in position (j, j). Full pivoting can handle some pathological cases for which partial pivoting fails, but full pivoting is not usually done in practice.

We can amend the pseudocode in figure 21.2(a) to incorporate partial pivoting by adding a step 1.α to find the row with largest absolute value in a given column and swap it into the diagonal position. Figure 21.2(b) shows the amended pseudocode for Gaussian elimination with partial pivoting. It should be straightforward to see how to amend the code to incorporate full pivoting if that is desired.

Note Gaussian elimination itself is not generally used for demanding problems in practice, since it is inefficient ($O(N^3)$ for an $N \times N$ matrix). In some cases, it can be made efficient for sparse matrices, defined as those that are mostly zeros. The techniques for efficient solution of large, sparse matrices are more advanced than we can cover here. Gaussian elimination can still be useful as an easy-to-code method if one ever needs to write a basic linear system solver. Gaussian elimination is also still important in scientific computing practice because it is equivalent to a widely useful method called *LU decomposition*. In LU decomposition, we convert a matrix A into a product of two matrices L and U, where L is lower triangular and U is upper triangular.

Although it may not be obvious, the Gaussian elimination process actually computes L and U for us. To see how this works, we can note that every step of Gaussian elimination can be rewritten as a matrix multiplication. When we scale a row to place a 1 in its diagonal entry a_{ii}, what we are doing is equivalent to transforming the system $A\vec{x} = \vec{b}$ into the system

$$\begin{bmatrix} 1 & 0 & 0 \\ & \ddots & \\ 0 & (a_{ii})^{-1} & 0 \\ & & \ddots \\ 0 & 0 & 1 \end{bmatrix} A\vec{x} = \begin{bmatrix} 1 & 0 & 0 \\ & \ddots & \\ 0 & (a_{ii})^{-1} & 0 \\ & & \ddots \\ 0 & 0 & 1 \end{bmatrix} \vec{b}$$

Similarly, eliminating the entries below a diagonal entry a_{ii} that has already been scaled to 1 can be represented as the matrix multiplication

$$\begin{bmatrix} 1 & 0 & 0 \\ & \ddots & \\ 0 & 1 & 0 \\ 0 & -a_{i+1,i} & 0 \\ & & \ddots \\ 0 & -a_{n,i} & 1 \end{bmatrix} A\vec{x} = \begin{bmatrix} 1 & 0 & 0 \\ & \ddots & \\ 0 & 1 & 0 \\ 0 & -a_{i+1,i} & 0 \\ & & \ddots \\ 0 & -a_{n,i} & 1 \end{bmatrix} \vec{b}$$

Eliminating entries above the diagonal has a similar interpretation as matrix multiplication:

$$
\begin{bmatrix} 1 & & -a_{1,i} & & 0 \\ & \ddots & & & \\ 0 & & -a_{i-1,i} & & 0 \\ 0 & & 1 & & 0 \\ & & & \ddots & \\ 0 & & 0 & & 1 \end{bmatrix} A\vec{x} = \begin{bmatrix} 1 & & -a_{1,i} & & 0 \\ & \ddots & & & \\ 0 & & -a_{i-1,i} & & 0 \\ 0 & & 1 & & 0 \\ & & & \ddots & \\ 0 & & 0 & & 1 \end{bmatrix} \vec{b}
$$

Furthermore, the first phase of Gaussian elimination produces only lower triangular matrices (including the diagonal scaling matrices), whereas the second phase produces only upper triangular matrices. The result is that as we go through the Gaussian elimination process, we are really accumulating a series of transformations of the form

$$
U_n U_{n-1} \cdots U_1 L_n L_{n-1} \cdots L_1 A\vec{x} = U_n U_{n-1} \cdots U_1 L_n L_{n-1} \cdots L_1 \vec{b},
$$

where $U_n U_{n-1} \cdots U_1 L_n L_{n-1} \cdots L_1 = A^{-1}$. The product of all of the upper triangular matrices will itself be an upper triangular matrix whose inverse is lower triangular. The product of the lower triangular matrices is a lower triangular matrix whose inverse is upper triangular. By accumulating the two upper triangular and lower triangular products and inverting them, which can be done efficiently for an upper or lower triangular matrix, we end up with the L and U matrices for the decomposition. Note that we can also accumulate the permutation matrices used for pivoting in the course of this transformation if necessary.

Once we find the L and U matrices by Gaussian elimination, we can efficiently solve $A\vec{x} = \vec{b}$ by solving two linear systems, $L\vec{y} = \vec{b}$ and $U\vec{x} = \vec{y}$, each of which requires only $O(N^2)$ time. Therefore, LU decomposition can be very useful if we have to repeatedly solve linear systems for one A matrix but different $\vec{b}$ vectors. That kind of situation can come up in many practical contexts. For example, implementation of an implicit numerical integration method for a linear system of ODEs is likely to lead to repeated solution of $A\vec{x} = \vec{b}$ for one A but many $\vec{b}$s.

21.2 Iterative Methods

Historically, Gaussian elimination was largely displaced in practice by a set of methods known as *iterative methods*. With an iterative method, we guess an initial solution and then apply an iterator that will successively improve the solution, as long as it is close enough to the right answer to begin with. Iterative methods are generally superior to Gaussian elimination in practice once we have a good initial guess. Typically, they require quadratic time per iteration (equivalent to a matrix multiplication) and potentially linear time for sparse matrices. They will therefore outperform

Gaussian elimination, provided they converge to their solution in a sublinear number of steps, something that is generally achieved in practice. There are three principal iterative methods:

- Jacobi
- Gauss–Seidel
- Successive overrelaxation.

Each method has a simple intuition behind it. In the Jacobi method, we improve our guess as to the value of $\vec{x}$ by solving for each variable in $\vec{x}$ in turn as if all of the other x_i values are fixed. Once we have a complete set of new x_i values, we repeat the process, solving for each x_i and again assuming all the others are correct in the new estimate. Suppose we denote the upper triangular portion of A by U, the lower triangular portion by L, and the diagonal portion by D. (Note that these are not the same L and U as in LU decomposition.) Then we can represent the Jacobi iteration by the linear operation

$$\vec{x}_{i+1} \leftarrow D^{-1}((L + U)\vec{x}_i + \vec{b}).$$

Gauss–Seidel follows nearly the same process, except that entries are updated one at a time, using the most recent updates when computing the next one rather than the entries from the prior round. This is equivalent to the update operation

$$\vec{x}_{i+1} \leftarrow (D - L)^{-1}(U\vec{x}_i + b).$$

Successive overrelaxation is the same as Gauss–Seidel except that it deliberately "overcorrects" when doing updates. This practice can often accelerate the convergence rate beyond that of Gauss–Seidel. It is described by the iteration

$$\vec{x}_{i+1} \leftarrow (D - \omega L)^{-1}((\omega U + (1 - \omega)D)\vec{x}_i + \omega\vec{b})$$

for some $\omega > 0$. Though these methods were historically significant, and are still important in some contexts, such as parallel linear system-solving, they have largely been superseded in practice by the next class of methods we will examine. We will therefore skip any detailed coverage of these classic iterative methods.

21.3 Krylov Subspace Methods

Large linear systems are generally solved today by a class of methods known as *Krylov subspace methods*. A Krylov subspace method depends on having a matrix M and a vector $\vec{v}$, which may be the A and $\vec{b}$ from our $A\vec{x} = \vec{b}$ linear system. The method iteratively solves the linear system by repeatedly finding a best-fit solution within a restricted solution space. It initially finds the best solution to $A\vec{x} = \vec{b}$ such that $\vec{x}$ is a linear multiple of $\vec{v}$. It then finds the best solution in the plane defined by $\vec{v}$ and $M\vec{v}$. Next, it finds the best solution in the three-dimensional space defined by $\vec{v}$, $M\vec{v}$, and $M^2\vec{v}$; and so on. Eventually, the method will find the optimal solution in the space defined by $\vec{v}, M\vec{v}, \ldots, M^{n-1}\vec{v}$, which is the full space of an $n \times n$ matrix A and thus contains the true solution to the problem. In practice, though, the method usually will have an almost exact answer in just a few steps. If the method is well designed, each new step requires only a small amount of additional work beyond that already performed for the previous step. Krylov subspace methods can thus be far more efficient in practice than Gaussian elimination or the classic iterative methods. Different Krylov subspace methods are distinguished by how we choose M and $\vec{v}$ and how we define the "best" solution.

We briefly saw one Krylov subspace method, the conjugate gradient method, in the context of nonlinear optimization. We will reconsider the method here in its original context, linear system-solving. The conjugate gradient method was the first Krylov subspace method, and is still widely used in practice. It requires that the matrix A is symmetric and positive definite. Recall that positive definite matrices are matrices in which all eigenvalues are positive. An alternative definition of positive definite is that if we run Gaussian elimination without pivoting, the a_{ii} values by which we divide (the pivots) will always be positive. Here we will see a slightly less general formulation of the method than we saw in its general optimization guise. Figure 21.3(a) provides the pseudocode.

1. $\vec{x}_0 = \vec{b}$
2. $\vec{r}_0 = \vec{b} - A\vec{x}_0$
3. $\vec{p}_0 = \vec{r}_0$
4. for $i \leftarrow 0$ to $n - 1$
   A. if $||\vec{r}_i|| < \epsilon$ then stop
   B. else
     i. $\alpha_i = \frac{\vec{r}_i^T \vec{r}_i}{\vec{p}_i^T A \vec{p}_i}$
     ii. $\vec{x}_{i+1} = \vec{x}_i + \alpha_i \vec{p}_i$
     iii. $\vec{r}_{i+1} = \vec{r}_i - \alpha_i A \vec{p}_i$
     iv. $\beta_i = \frac{\vec{r}_{i+1}^T \vec{r}_{i+1}}{\vec{r}_i^T \vec{r}_i}$
     v. $\vec{p}_{i+1} = \vec{r}_{i+1} + \beta_i \vec{p}_i$

1. $\vec{q}_i = \vec{b}/||\vec{b}||$
2. for i=1 to n
   A. $\vec{v} = A\vec{q}_i$
   B. for j= 1 to i
     i. $h_{ji} = \vec{q}_i^T \vec{v}$
     ii. $\vec{v} \leftarrow \vec{v} - h_{ji}\vec{q}_j$
   C. $h_{i+1,i} = ||\vec{v}||$
   D. $\vec{q}_{i+1} = \vec{v}/h_{i+1,i}$
   E. find $\vec{y}$ minimizing $||AQ_n\vec{y} - \vec{b}||$
     (we will see how to do this later in the chapter)
   F. $\vec{x}_i = Q_i\vec{y}$

(a)   (b)

Figure 21.3
Pseudocode for Krylov subspace methods for linear system-solving. (a) Conjugate gradient method. (b) Generalized minimal residual (GMRES) method.

For the most part, the method should remind us of steepest descent. At each step, we pick a vector $\vec{p}_i$ by which we will improve $\vec{x}_i$, find a distance on $\vec{p}_i$ along which to move, and make the update. Here, however, $\vec{p}_i$ is essentially a "corrected" version of the gradient, which we create by subtracting off components of the previous $\vec{p}$ vector from the gradient. This correction prevents $\vec{p}_i$ from undoing work done by $\vec{p}_{i-1}$ on the prior step. Lines 4.B.iii–v identify a $\vec{p}_i$ that is *conjugate* to the previous ones with respect to A, meaning that $\vec{p}_i A \vec{p}_j = 0$ for $i \neq j$. This choice has the effect of choosing movement vectors that are all at right angles to each other, guaranteeing that work which is done in one step is not undone in subsequent steps.

If we need a method that works for matrices that are not positive definite, a simple one is *biconjugate gradient*. To solve $A\vec{x} = \vec{b}$ by the biconjugate gradient method, we first convert it to the problem $A^T A \vec{x} = A^T \vec{b}$. $A^T A$ is symmetric positive definite for any matrix A, so we can run conjugate gradient on $A^T A$. In other words, biconjugate gradient means solving $A^T A \vec{x} = A^T \vec{b}$ by conjugate gradient. The method tends not to work as well as standard conjugate gradient if the matrix is symmetric positive definite to begin with, for reasons that are too advanced for this text. The technical explanation is that the runtime of any of these methods depends on a quantity called the *condition number* of the matrix A, defined as $(\|A\|)(\|A^{-1}\|)$, which is always at least 1. Converting from A to $A^T A$ squares the condition number, so it will roughly square the number of steps needed to achieve convergence.

Another Krylov subspace method is called generalized minimal residual (GMRES), which works by successively minimizing the *residual vector* $\vec{r}_n = \vec{b} - A\vec{x}_n$ over the nth Krylov subspace of A and $\vec{b}$. Figure 21.3(b) provides pseudocode for the method. This method is slower than conjugate gradient because it has to orthogonalize the update vector by all prior update vectors rather than just the most recent one. This added work allows it to drop the requirement for a symmetric positive definite input matrix.

21.3.1 Preconditioners

In practice, we are rarely likely to need to implement a Krylov subspace method because there are highly optimized implementations already available for almost any computer. Therefore, we do not really need to know how to implement them. We only strictly need to be aware that they exist and understand the restrictions on different Krylov subspace methods. But there are some additional things we do need to know in order to use Krylov subspace methods effectively. Most important, we need to be aware of something called a *preconditioner*.

We generally do not run our linear system-solving methods directly on the system $A\vec{x} = \vec{b}$ that we are interested in solving. Instead, we solve the system $(M^{-1}A)x = (M^{-1}\vec{b})$, where M is a matrix called the preconditioner. A good M is one that satisfies two properties:

• It is easy to compute $M^{-1}y$ for any vector y.
• $M^{-1}A$ is "well-behaved."

A well-behaved matrix $M^{-1}A$ is intuitively "close" to the identity matrix, and $(M^{-1}A)\vec{x} = (M^{-1}\vec{b})$ is therefore easy to solve. This concept again relates to the condition number mentioned above. The identity matrix has condition number 1, and a well-behaved matrix is really any matrix with a condition number close to 1.

Picking a good preconditioner is not trivial. If we choose $M = A$, we satisfy our condition that $M^{-1}A \approx I$, but solving $M^{-1}\vec{y}$ is as hard as solving our original system. If we choose $M = I$, then solving $M^{-1}\vec{y}$ is easy, but $M^{-1}A$ is just A and is therefore probably not close to I. Choosing a really good preconditioner is a difficult engineering issue that requires more knowledge than we can cover in a single chapter. A reasonable general-purpose choice is $M = diag(A)$, the matrix that has the same diagonal as A but has zeros everywhere else. There will usually be better choices, though, depending on the specific A under consideration.

21.4 Overdetermined and Underdetermined Systems

So far we have been considering how to solve systems defined by full-rank $n \times n$ matrices. Very often, though, we will overdetermine or underdetermine matrices. Recall that an overdetermined system is one in which the number of rows of the matrix exceeds its rank, essentially meaning that there are more constraints than variables. An underdetermined system is one in which the number of columns exceeds the rank, meaning that there are more variables than constraints. We will close with a consideration of what to do with these kinds of systems.

There is generally no exact solution to an overdetermined system. Such systems come up very often in practice, though. For example, in our regulated gene example at the beginning of the chapter, we will get an overdetermined system if we do more than $k + 1$ experiments. Very often, we want to gather more data points (constraints) than we need to get a full-rank system because experiments can be noisy and having an excess of data lets us reduce the effects of the noise. The noise also means, though, that we will almost certainly have no parameter set that exactly fits all experiments.

A widely used way of dealing with the problem of overdetermined systems is to find an $\vec{x}$ that is "close" to a solution to $A\vec{x} = \vec{b}$. "Close" is often defined in terms of the *least-squares* measure. This means that if we call row i of our matrix $\vec{a}_i$, then we want to find $\vec{x}$ minimizing

$$\sum (\vec{a}_i \cdot \vec{x} - b_i)^2.$$

An alternative way of saying that is in terms of the residual $\vec{r} = A\vec{x} - b$. We want to find $\vec{x}$ minimizing $\vec{r}^T\vec{r} = (A\vec{x} - \vec{b})^T(A\vec{x} - \vec{b})$. Given an $m \times n$ matrix A, the least-squares solution to $A\vec{x} = \vec{b}$ is the solution to

$$A^T A \vec{x} = A^T \vec{b},$$

which will never be overdetermined, since $A^T A$ is $n \times n$. So to find the least-squares solution to $A\vec{x} = \vec{b}$, we solve the linear system

$$(A^T A)\vec{x} = (A^T \vec{b}).$$

Note that in the GMRES algorithm that we saw earlier, we had one step for which we needed to find $\vec{y}$ minimizing $\|AQ_n\vec{y} - \vec{b}\|$. This is a least-squares problem, which we now know how to solve.

When our system is underdetermined, how we handle it is somewhat arbitrary, since an underdetermined system usually has an infinite number of solutions. We can in principle select any of these as the answer. One possibility is to define an objective function on possible solutions and then solve for the solution with optimal objective value. If our objective function is linear or convex, then this becomes a linear programming or convex programming problem, which we already know how to solve. One other common solution is to use something called a *pseudoinverse*. To find the pseudoinverse, we first find the singular value decomposition of matrix A, a common linear algebra operation that transforms A into the product

$$A = Q_1 \Sigma Q_2^T,$$

where Σ is a diagonal matrix of the singular values of A (i.e., the eigenvalues of $A^T A$). We then find a new matrix $\bar{\Sigma}$, which is Σ with the nonzero values inverted. In other words, we perform the following transformation:

$$\Sigma = \begin{bmatrix} \sigma_1 & 0 & \cdots & 0 & 0 \\ 0 & \sigma_2 & \cdots & 0 & 0 \\ \vdots & \vdots & \ddots & \vdots & \vdots \\ 0 & 0 & \cdots & 0 & 0 \\ 0 & 0 & \cdots & 0 & 0 \end{bmatrix} \Leftrightarrow \bar{\Sigma} = \begin{bmatrix} \sigma_1^{-1} & 0 & \cdots & 0 & 0 \\ 0 & \sigma_2^{-1} & \cdots & 0 & 0 \\ \vdots & \vdots & \ddots & \vdots & \vdots \\ 0 & 0 & \cdots & 0 & 0 \\ 0 & 0 & \cdots & 0 & 0 \end{bmatrix}$$

Then the pseudoinverse of A is defined as

$$\bar{A} = Q_2 \bar{\Sigma} Q_1^T.$$

The vector $\bar{A}\vec{b}$ will give us a solution to $A\vec{x} = \vec{b}$ that sets as many values as possible to zero, which can be a very good choice from a modeling point of view. Effectively, the pseudoinverse gives us the simplest possible linear model satisfying all of

our constraints. It thus is a good option when we have no other basis on which to set
an objective function for the solution.

References and Further Study

Linear system-solving is one of the core problems of numerical linear algebra, and
any general numerical linear algebra text is likely to cover the methods we have
seen here. Trefethen and Bau [86] is a highly regarded text for this topic. The Numerical Recipes books [82] and Stoer and Bulirsch [132] also provide coverage. A general
introductory linear algebra text, such as Strang [192], will often have more basic coverage of at least some of these methods in the context of general linear algebra, and
may therefore be a more suitable reference for beginners.

Many of the matrices one will encounter in practice will be sparse, and it is therefore well worth learning special methods for sparse systems. Often, the advantages of
a sparse matrix will come for free because one generally uses a linear system solver to
solve $A\vec{x} = \vec{b}$ by supplying it with a function for multiplying by A rather than the
actual matrix A. If $A\vec{x}$ can be evaluated quickly, then the linear system solver will
run quickly. It can nonetheless be quite helpful to learn about some specialized methods for sparse systems. Saad [192] provides coverage of this topic.

As mentioned above, readers interested in doing more advanced work in scientific
computing would do well to learn about some of the other core numerical linear algebra problems, such as finding eigenvalues and eigenvectors. The references covered
above will serve one well in beginning more advanced study on these topics.

22 Interpolation and Extrapolation

A common problem when dealing with real-world data sources is translating discrete samples into a continuous model. Raw experimental data are almost always discretized at some level, yet many of our models assume continuous data. Furthermore, we often are required to impute values that fall between our discrete observed data points. We therefore need methods for fitting broad classes of continuous functions to discrete data sets. Variants of this basic problem come up in many guises in the context of modeling and simulation in biology. We can begin by considering some examples.

Example 1   We may be interested in creating a model of a reaction network in which we know that the activity a of one reactant is affected by some external signal s, such as a sensory stimulus. We want to a create a differential equation model of the full reaction network so we can see how it will respond to different patterns of stimulus, but we do not know how the external signal affects the reactant activity. We may address this problem by observing a few discrete data points of reactant activity versus signal strength. We can then fit a continuous function $a(s)$ to the observations, as in figure 22.1(a), and plug $a(s)$ into our differential equation model. We then have a model of the network on which we can test the effects of different driving stimuli $s(t)$.

Example 2   We may want to predict the behavior of a model at a limit we cannot simulate. For example, suppose we monitor a protein expression model at several discrete time points and observe the data in figure 22.1(b). We may wish to know from this model the equilibrium to which it is converging. If the simulation is computationally demanding, there may be no way to reach that limit directly. We can, however, try to infer the limit from the points already seen.

Example 3   Biological data sets are often very noisy, and we may wish to fit a curve to a set of data points as a way of smoothing the noisy data. For example, suppose we are interested in monitoring the expression of some gene in a microorganism throughout its cell cycle. We can in principle accomplish this by synchronizing the

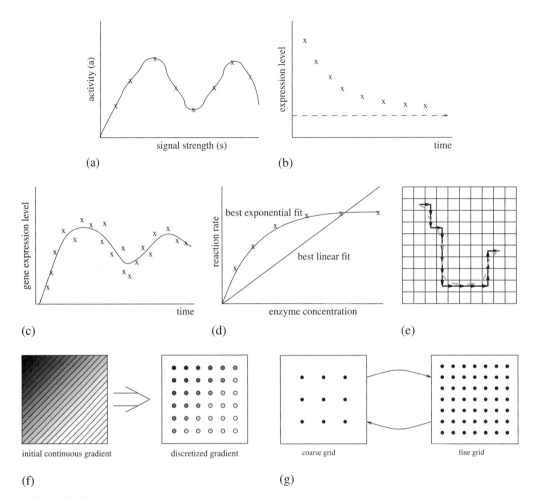

Figure 22.1
Possible applications of interpolation and extrapolation in biological modeling. (a) Model of an enzyme system for which we require a continuous input (solid line) for our numerical methods, but must infer it from discrete experimental data (x's). (b) Discrete data points (x's) from which we wish to infer a long-term limit (dashed line). (c) A model of noisy gene expression data we wish to denoise by fitting a continuous curve. (d) Two possible curve classes, linear and exponential, that we might fit to a set of reaction data to evaluate possible reaction mechanisms. (e) A random walk on a lattice (solid arrows) from which we wish to infer a continuous trajectory (dashed arrows). (f) A continuous chemical gradient that we need to discretize for use in a numerical model. (g) Two grids between which we may need to translate in implementing a multigrid optimization method.

cell cycles of a set of the microorganisms and then measuring gene expression in subsets of the organisms at various time points. In practice, the process of synchronizing cell cycles will be imperfect and the measurements somewhat inaccurate, so we may observe something like the set of xs in figure 22.1(c). By fitting a curve with only a few degrees of freedom to the data, we are likely to get a relatively smooth representation of the data, such as the line in the figure, that may be more amenable to model computation.

Example 4   We may want to infer what general kind of model best describes a particular data set by fitting various kinds of curves to the model. For instance, suppose we observe that the activity of an enzyme depends on its concentration according to a set of data points, as in figure 22.1(d). We may want to know if those points are better described by a linear function or an exponential function, since that is likely to tell us something about the activity of the enzyme. We can attempt to fit the data to both curve types and observe which provides a more reliable fit.

Example 5   We may want to create a continuous model of a system using an inherently discrete modeling method. For example, suppose we want to create a model of a particle diffusing in a two-dimensional space. We may choose to use a lattice model for high efficiency. We can then create a continuous model of the particle's movement by inferring smooth paths between the discrete steps, as in figure 22.1(e).

Example 6   We sometimes need to convert between different discretizations of a problem. For example, if we are trying to find the spatial equilibrium of a chemical diffusion system starting at some initial state, we may begin by converting the input into a set of discrete points such as we might use for the spatial grid of a reaction–diffusion model, as in figure 22.1(f). Even if our input is already discrete, we may need to change its discretization to one more suitable for our numerical methods. Furthermore, some advanced numerical methods we have not seen can solve such problems more quickly by jumping between different grid spacings during the solution of the problem. Such methods are known as *multigrid methods*. To use a multigrid method, we generally need a way to take a discretized state, convert it to a continuous model, and then convert that continuous model back into a discretized model on a different grid, as in figure 22.1(g). Similar problems may arise in applications related to image analysis, as we need to move between different levels of resolution of an image.

What all of these problems have in common is that they can be described in terms of two basic operations:

- Interpolation: inferring missing data within the range of a set of known data points
- Extrapolation: inferring missing data outside the range of the observed data points.

In this chapter, we will learn some methods for performing interpolation and extrapolation from discrete data sets.

22.1   Polynomial Interpolation

The most common kind of interpolation is *polynomial interpolation*: fitting data points to a polynomial, or possibly several polynomials stitched together. The simplest way to specify a polynomial interpolant is to require one passing through all of our observed data points. Given n data points $(x_1, y_1), (x_2, y_2), \ldots, (x_n, y_n)$, we can uniquely specify an $n-1$ degree polynomial passing through all of those points as follows:

$$P(x) = \frac{(x - x_2)(x - x_3) \cdots (x - x_n)}{(x_1 - x_2)(x_1 - x_3) \cdots (x_1 - x_n)} y_1 + \frac{(x - x_1)(x - x_2) \cdots (x - x_n)}{(x_2 - x_1)(x_2 - x_3) \cdots (x_2 - x_n)} y_2 + \cdots$$

$$+ \frac{(x - x_1)(x - x_2) \cdots (x - x_{n-1})}{(x_n - x_1)(x_n - x_2) \cdots (x_n - x_{n-1})} y_n.$$

To understand why this works, try plugging in any x_i for x. The coefficient of y_i will have the same numerator and denominator, and will therefore be 1. The coefficients of all y_j for $j \neq i$ have $(x - x_i)$ terms in their numerators and will therefore be zero. The polynomial thus passes through each of the (x_i, y_i) input points.

22.1.1   Neville's Algorithm

The formulation above is correct for a polynomial fitting a set of points, but is also cumbersome. There is a better way to compute a polynomial fitting a set of points, called *Neville's algorithm*. Neville's algorithm hierarchically constructs a solution by merging solutions to subsets of the points. Given four points

$$(x_1, y_1), (x_2, y_2), (x_3, y_3), (x_4, y_4),$$

the method will first construct a zero-order polynomial fitting each individual point:

$$P_{11}(x) = y_1, \quad P_{22}(x) = y_2, \quad P_{33}(x) = y_3, \quad P_{44}(x) = y_4.$$

It will then merge pairs of points to produce three first-order curves: $P_{1,2}$ fitting points (x_1, y_1) and (x_2, y_2), $P_{2,3}$ fitting points (x_2, y_2) and (x_3, y_3), and $P_{3,4}$ fitting points (x_3, y_3) and (x_4, y_4). From there, it will construct second-order polynomials $P_{1,3}$ and $P_{2,4}$ matching triplets of points, and finally a third-order polynomial $P_{1,4}$ fitting all four points. Each polynomial of higher order is generated by combining two polynomials of the next lower order, as in figure 22.2.

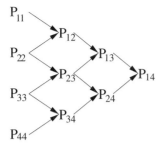

Figure 22.2
Data dependence in Neville's algorithm for polynomial interpolation. Each polynomial $P_{i,j}(x)$ is constructed using two lower-order polynomials $P_{i+1,j}(x)$ and $P_{i,j-1}(x)$.

These merging steps are accomplished by the following formula for deriving $P_{i,i+m}$ from $P_{i,i+m-1}$ and $P_{i+1,i+m}$:

$$P_{i,i+m}(x) = \frac{(x - x_{i+m})P_{i,i+m-1} + (x_i - x)P_{i+1,i+m}}{x_i - x_{i+m}}.$$

We can show that the formula works correctly by induction, assuming as our inductive hypothesis that $P_{i,i+m-1}$ gives correct values for data points $x_i \ldots x_{i+m-1}$ and that $P_{i+1,i+m}$ gives correct values for data points $x_{i+1} \ldots x_{i+m}$. We then need to show that $P_{i,i+m}$ gives correct values for all data points $x_i \ldots x_{i+m}$. We can consider first how it behaves for those points covered by both parent functions. If we evaluate the new function at some $j \in [i+1, i+m-1]$, we will get the following:

$$\frac{(x_j - x_{i+m})y_j + (x_i - x_j)y_j}{x_i - x_{i+m}} = \frac{x_i - x_{i+m}}{x_i - x_{i+m}}y_j = y_j.$$

In other words, the formula will give us the correct value at any point that is included in both parent polynomials.

We can then look at the point x_i, which is covered by $P_{i,i+m-1}$ but not by $P_{i+1,i+m}$. At x_i, the formula will yield the following value:

$$\frac{(x_i - x_{i+m})P_{i,i+m-1}(x_i) + (x_i - x_i)P_{i+1,i+m}(x_i)}{x_i - x_{i+m}}$$

$$= \frac{(x_i - x_{i+m})P_{i,i+m-1}(x_i) + 0 \times P_{i+1,i+m}(x_i)}{x_i - x_{i+m}}$$

$$= \frac{x_i - x_{i+m}}{x_i - x_{i+m}}P_{i,i+m-1}(x_i)$$

$$= P_{i,i+m-1}(x_i)$$

$$= y_i.$$

Similarly, if we plug in x_{i+m}, we get

$$\frac{(x_{i+m} - x_{i+m})P_{i,i+m-1}(x_{i+m}) + (x_i - x_{i+m})P_{i+1,i+m}(x_{i+m})}{x_i - x_{i+m}}$$

$$= \frac{0 \times P_{i,i+m-1}(x_{i+m}) + (x_i - x_{i+m})P_{i+1,i+m}(x_{i+m})}{x_i - x_{i+m}}$$

$$= \frac{x_i - x_{i+m}}{x_i - x_{i+m}} P_{i+1,i+m}(x_{i+m})$$

$$= P_{i+1,i+m}(x_{i+m})$$

$$= y_{i+m}$$

Therefore, if the parent functions of degree $m - 1$ are correct, the child function of degree m will be correct. We then only need to consider the base case, which is the set of degree zero functions $P_{i,i}$. These are correct by design, since $P_{i,i}(x) = y_i$. This completes a proof by induction that Neville's method will produce a polynomial matching all of the input points.

To look at a concrete example, consider the points $(0, -4)$, $(1, -2)$, $(2, 2)$, and $(3, 14)$. We first get a zero-order approximation for each point:

$$P_{11}(x) = -4$$

$$P_{22}(x) = -2$$

$$P_{33}(x) = 2$$

$$P_{44}(x) = 14.$$

We then get the first-order approximations

$$P_{12} = \frac{(x-1)(-4) + (0-x)(-2)}{(0-1)} = \frac{-2x+4}{-1} = 2x - 4$$

$$P_{23} = \frac{(x-2)(-2) + (1-x)(2)}{(1-2)} = \frac{-4x+6}{-1} = 4x - 6$$

$$P_{34} = \frac{(x-3)(2) + (2-x)(14)}{(2-3)} = \frac{-12x+22}{-1} = 12x - 22.$$

Next we get the second-order approximations

$$P_{13} = \frac{(x-2)(2x-4) + (0-x)(4x-6)}{(0-2)} = \frac{-2x^2 - 2x + 8}{-2} = x^2 + x - 4$$

$$P_{24} = \frac{(x-3)(4x-6) + (1-x)(12x-22)}{(1-3)} = \frac{-8x^2 + 16x - 4}{-2} = 4x^2 - 8x + 2.$$

Finally, we get a third-order approximation that fits all four points:

$$P_{14} = \frac{(x-3)(x^2 + x - 4) + (0-x)(4x^2 - 8x + 2)}{(0-3)}$$

$$= \frac{-3x^3 + 6x^2 - 9x + 2}{-3} = x^3 - 2x^2 + 3x - 4.$$

We can try plugging in any of the four data points to verify that this cubic polynomial does in fact pass through all four points.

22.2   Fitting to Lower-Order Polynomials

Although we can always fit n data points to a polynomial of degree $n-1$, it is often better to fit to a lower order. Choosing a lower order can be a good way to smooth out noise in the data and make the fit less sensitive to outliers. Figure 22.3 shows a hypothetical noisy data set for which we may prefer a low-order fit to a high-order fit.

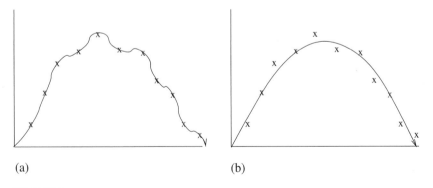

(a)   (b)

Figure 22.3
A noisy data set, shown with xs, for which a more strictly accurate high-order fit (a) may be inferior in practice to a less accurate low-order fit (b).

In general, using a lower-order polynomial means that we will not fit every point exactly. We therefore need to find a best fit by some metric. To see how we may do this, let us pretend at first that we can solve exactly for a kth-order polynomial:

$$c_0 + c_1 x + c_2 x^2 + \cdots + c_k x^k$$

for n data points, where $k < n - 1$. Solving this is equivalent to finding the set of coefficients $c_0, \ldots, c_k$ satisfying the following set of equations:

$$y_1 = c_0 + c_1 x_1 + \cdots + c_k x_1^k$$

$$y_2 = c_0 + c_1 x_2 + \cdots + c_k x_2^k$$

$$\vdots$$

$$y_n = c_0 + c_1 x_n + \cdots + c_k x_n^k.$$

These equations are linear in the coefficients we want to determine, so we can represent this as a linear system-solving problem:

$$
\begin{bmatrix}
1 & x_1 & x_1^2 & \cdots & x_1^k \\
1 & x_2 & x_2^2 & \cdots & x_2^k \\
& & \vdots & & \\
1 & x_n & x_n^2 & \cdots & x_n^k
\end{bmatrix}
\begin{bmatrix}
c_0 \\ c_1 \\ c_2 \\ \vdots \\ c_k
\end{bmatrix}
=
\begin{bmatrix}
y_0 \\ y_1 \\ y_2 \\ \vdots \\ y_k
\end{bmatrix}
$$

If we denote the matrix of x values X, the vector of coefficient values as $\vec{c}$, and the vector of y_i values as $\vec{y}$, then we want to solve the linear system $X\vec{c} = \vec{y}$. If $k < n - 1$, the system is overdetermined and we will not be able to solve for it exactly. In chapter 21 we saw one good way to deal with overdetermined systems: least-squares. We can find the least-squares best fit for $\vec{c}$ by solving for the full-rank linear system

$$(X^T X)\vec{c} = X^T \vec{y}.$$

22.3   Rational Function Interpolation

Though polynomial interpolation is the most commonly used interpolation method, there are others that may be more appropriate for some problem types. Interpolating by Fourier components is one important example that we will omit here since most scientifically educated readers will cover it elsewhere in their studies. Another kind of interpolation one is less likely to see elsewhere is *rational interpolation*. A rational function is a ratio between two polynomials:

$$R(x) = \frac{P(x)}{Q(x)} = \frac{p_0 + p_1 x + p_2 x^2 + \cdots + p_m x^m}{q_0 + q_1 x + q_2 x^2 + \cdots + q_n x^n}.$$

Rational functions may give a better match than polynomials for functions that go to $\pm\infty$ for finite x or approach a constant as x goes to $\pm\infty$. There is an algorithm due to Bulirsch and Stoer [132] to fit data points to a rational function that is similar to the Neville method for polynomials. The Bulirsch–Stoer method uses the following iterator formula to derive a higher-order rational interpolant $R_{i,i+m}$ from the lower-order approximations $R_{i+1,i+m}$ and $R_{i,i+m-1}$:

$$R_{i,i+m} = R_{i+1,i+m}$$

$$+ \frac{R_{i+1,i+m} - R_{i,i+m-1}}{(x - x_i)/(x - x_{i+m})(1 - (R_{i+1,i+m} - R_{i,i+m-1})/(R_{i+1,i+m} - R_{i+1,i+m-1})) - 1}.$$

22.4 Splines

Very often, if we have a lot of data points, we want a function that is a good match locally to every small region of our data points. We therefore do not want to fit a low-order function that will give a good global fit at the expense of poor local fits. At the same time, though, we do not want to have a huge number of parameters in our function because the fit is likely to be too sensitive to noisy data, as well as costly to compute. We can often solve this dilemma by solving the problem with a different simple function in the neighborhood of each data point. We then get a good local fit in each neighborhood as well as a relatively smooth fit overall. One trivial variant of this approach is a piecewise linear approximation, as illustrated in figure 22.4(a), which performs linear interpolation between each pair of consecutive points. A piecewise linear fit provides a match at each observed data point and ensures continuity of the whole curve. It has discontinuities in the derivative of the curve, though, making it poorly suited for use in many kinds of numerical algorithms.

Splines are a generalization of this concept of local approximation, using higher-order polynomials to fit local regions. For example, suppose we want to put a curve between each pair of points in our data set so that we match all of the endpoints, just as we would with piecewise linear functions. But we further want to smooth out the discontinuities we get from the piecewise linear functions by insisting that the derivatives also match where we cross between distinct piecewise polynomial regions. We need piecewise interpolants of higher than first order to accomplish this, and may end up with an approximation like that of figure 22.4(b).

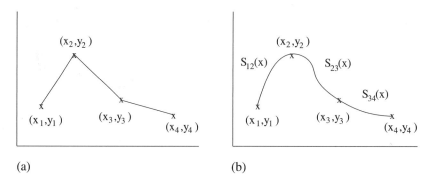

Figure 22.4
Splines for providing piecewise interpolation for a set of data points. (a) A piecewise linear fit, which can be considered a first-order spline. (b) A piecewise quadratic fit to the same data points.

A set of local polynomial interpolants S_{12}, S_{23}, and S_{34} meeting our conditions above would have to satisfy the following constraints:

$$S_{12}(x_1) = y_1$$

$$S_{12}(x_2) = y_2$$

$$\frac{dS_{12}}{dx}(x_2) = \frac{dS_{23}}{dx}(x_2)$$

$$S_{23}(x_2) = y_2$$

$$S_{23}(x_3) = y_3$$

$$\frac{dS_{23}}{dx}(x_3) = \frac{dS_{34}}{dx}(x_3)$$

$$S_{34}(x_3) = y_3$$

$$S_{34}(x_4) = y_4.$$

Suppose we want to satisfy these constraints using some set of piecewise quadratic interpolants

$$S_{i,i+1} \equiv c_{i,0} + c_{i,1}x + c_{i,2}x^2.$$

These interpolants will have derivatives of the form

$$\frac{dS_{i,i+1}}{dx} \equiv c_{i,1} + 2c_{i,2}x.$$

We can then represent the problem of finding the coefficients $c_{i,j}$ to satisfy the above constraints as a linear system-solving problem:

$$
\begin{bmatrix}
1 & x_1 & x_1^2 \\
1 & x_2 & x_2^2 \\
0 & 1 & 2x_2 & 0 & -1 & -2x_2 \\
 & & & 1 & x_2 & x_2^2 \\
 & & & 1 & x_3 & x_3^2 \\
 & & & 0 & 1 & 2x_3 & 0 & -1 & -2x_3 \\
 & & & & & & 1 & x_3 & x_3^2 \\
 & & & & & & 1 & x_4 & x_4^2
\end{bmatrix}
\begin{bmatrix}
c_{1,0} \\
c_{1,1} \\
c_{1,2} \\
c_{2,0} \\
c_{2,1} \\
c_{2,2} \\
c_{3,0} \\
c_{3,1} \\
c_{3,2}
\end{bmatrix}
=
\begin{bmatrix}
y_1 \\
y_2 \\
0 \\
y_2 \\
y_3 \\
0 \\
y_3 \\
y_4
\end{bmatrix}
$$

There is a slight problem here in that the system as specified is underdetermined (eight equations for nine unknowns). If we had n points instead of four, then we would have $3(n-1) - 1$ constraints in $3(n-1)$ unknowns. We therefore need to add one additional constraint in order to fix a unique solution. We may accomplish this by insisting that the derivative is zero at point x_1, adding a row to our linear system and making it full rank.

This system of equations can be solved very efficiently in practice, even if we have a large number of data points, because the matrix is *block diagonal*. This means that entries more than a small distance from the matrix diagonal are all zero. We can solve this sort of matrix in approximately $O(n)$ time, using a Krylov subspace method. For commonly used splines, there are also often analytical formulas into which we can plug any observed data points.

We can easily generalize this spline-fitting procedure to more sophisticated kinds of interpolants. For example, we can use higher-order polynomials and require that more derivatives match at the known data points. Probably the most widely used variant is the *cubic spline*, in which we fit the data to a piecewise cubic function with the constraints that the curves pass through all observed points and that consecutive interpolants match in the first and second derivatives at each of the points. We then need to impose two additional arbitrary boundary conditions to give the system full rank, such as insisting that the first derivatives are zero at both ends of the interval modeled.

In addition to their relevance to biological modeling, splines are commonly used in computer graphics and animation because they allow one to generate curves that are complicated but "smooth-looking." So, for instance, if we wish to show an object moving in space, we can specify a few discrete points the object passes through, fit a spline through those points, and end up with a smooth trajectory through those points.

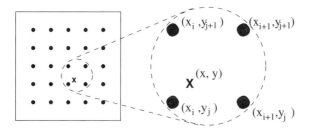

Figure 22.5
A two-dimensional interpolation problem. We are given values on a regular grid, shown by circles, and wish to infer the value at some arbitrary point within the grid, labeled by x.

22.5   Multidimensional Interpolation

Most of the methods we have seen have generalizations into multiple dimensions. They generally require some additional assumptions, though. For example, there is a two-dimensional version of linear interpolation called *bilinear interpolation*. Suppose we have a function of two variables, $f(x, y)$, that is evaluated at a grid of points in the x–y plane, defined by the x coordinates $x_1, x_2, \ldots$ and the y coordinates $y_1, y_2, \ldots$. With bilinear interpolation, we estimate the value of the function at some unobserved point (x, y) by interpolating from the four corners of the grid box in which (x, y) lies. Figure 22.5 illustrates this problem.

Given the values at the four corners, we can compute two quantities measuring the relative distance of (x, y) from the corners of its grid box:

$$t = \frac{x - x_i}{x_{i+1} - x_i}, \quad u = \frac{y - y_j}{y_{j+1} - y_j}.$$

We then estimate the value of the unobserved point as an average of the values at the corners, weighted to favor the closer corners, as follows:

$$f(x, y) = (1 - t)(1 - u)f(x_i, y_j) + t(1 - u)f(x_{i+1}, y_j)$$
$$+ (1 - t)uf(x_i, y_{j+1}) + tuf(x_{i+1}, y_{j+1}).$$

We can come up with all sorts of other ways to define the fit on a grid box, but this one is likely to work well if the function is smooth enough and the grid spacing small enough that we can treat the function as locally linear within each grid box.

22.6   Interpolation with Arbitrary Families of Curves

We have looked at some basic kinds of interpolants which are widely used and for which we have well-developed computational tools, but we may know some other class of curves that is appropriate for a particular problem. For example, many bio-chemical processes are described by decaying exponentials as a system approaches equilibrium. If we want to interpolate between measured data points for biochemical reactions, we may therefore want to insist on exponential interpolants:

$$y(x) = a + be^{-cx},$$

where we must fit a, b, and c to our data. We may not even be able to explicitly state an analytic formula for the curves we are fitting. If we are looking at genetic regula-tory networks, for example, we may know that we want systems described by ordi-nary differential equations, but have no analytical description of the curves those equations will produce. The parameters we are fitting may then be simply rate con-stants of the ODE model. For example, we may have a system of the form

$$\frac{dA}{dt} = a_1 A + b_1 B + c_1 AB + d_1$$

$$\frac{dB}{dt} = a_2 A + b_2 B + c_2 AB + d_2,$$

where we want to find a_1, a_2, b_1, b_2, c_1, c_2, d_1, and d_2 to match some observed data set.

In such cases, we often will not have any convenient exact method to find an opti-mal parameter set. But we can still pose the interpolation problem as a form of optimization problem and then solve it with the continuous optimization tools we have learned about in previous chapters. This is essentially the same idea we saw when we first looked at parameter-tuning as a form of optimization in chapter 18.

Example   Suppose we take our exponential from above,

$$y(x) = a + be^{-cx},$$

and further suppose we have a set of points to fit to that exponential, $(x_1, y_1), \ldots,$ (x_n, y_n). Then we can set up a metric of solution quality by declaring, for example, that we want a least-squares fit to the data. Then, given initial guesses as to a, b, and c, we can evaluate our current solution at each of our observed x values to get the points .

$$\hat{y}_1 = y(x_1), \hat{y}_2 = y(x_2), \ldots, \hat{y}_n = y(x_n).$$

Then our objective function will be

$$\min_{a,b,c}\{(y_1 - \hat{y}_1)^2 + (y_2 - \hat{y}_2)^2 + \cdots + (y_n - \hat{y}_n)^2\}$$

$$= \min_{a,b,c}\{(y_1 - a - be^{-cx_1})^2 + (y_2 - a - be^{-cx_2})^2 + \cdots + (y_n - a - be^{-cx_n})^2\}.$$

We now have a function we can evaluate and differentiate with respect to any of our parameters, so we can treat this as a multidimensional Newton–Raphson or Levenberg–Marquardt optimization problem. We will need to get a reasonable first guess as to the parameters, but if we can do that, it is likely we can get a nearly optimal parameter set, whatever our data points are.

If we have a function we cannot explicitly represent, such as the differential equation model above, we can still apply a similar approach by treating our function as a black box. That is, if we want to generate the points $(x_1, \hat{y}_1), \ldots, (x_n, \hat{y}_n)$ from our current parameter set $a_1, a_2, b_1, b_2, \ldots$, we can integrate the system from $x = 0$ to $x = x_n$, using the methods we learned for numerical integration. We can then solve for our objective function, say

$$\min_{a_1, a_2, \ldots} \{(y_1 - \hat{y}_1)^2 + (y_2 - \hat{y}_2)^2 + \cdots + (y_n - \hat{y}_n)^2\},$$

if we want a least-squares fit again. If we want to find derivatives of the system with respect to various parameters, we can use any of the numerical derivative formulas we have covered. Suppose we call the objective function $\Phi(a_1, a_2, b_1, b_2, \ldots)$. Then, to find, for example, the second derivative of the objective function with respect to a_1, we can evaluate the objective function at $a_1, a_2, \ldots$, evaluate it again at $a_1 + \Delta a$, and again at $a_1 - \Delta a$, and use the estimate

$$\frac{\partial^2 \Phi}{\partial a_1^2} \approx \frac{\Phi(a_1 + \Delta a, a_2, b_1, b_2, \ldots) + \Phi(a_1 - \Delta a, a_2, b_1, b_2, \ldots) - 2\Phi(a_1, a_2, b_1, b_2, \ldots)}{\Delta a^2}$$

Repeating this for all of the first and second derivatives will give us approximations for the gradient and the hessian, so we can again apply Newton–Raphson to find an optimal parameter set. Once again, the method may fail if we do not have a reasonable starting point. But once we get it started well, it is likely to converge very quickly to the optimal parameter set. There are excellent programs available for doing this kind of fitting in certain domains, such as the Dynafit program [194] for fitting rate constants to curves of chemical reactants over time. We should be able to write our own program for any arbitrary family of functions, though.

Once we have a best fit to whatever family of curves we want by whatever metric we want, we can interpolate by evaluating the best-fit curve at any unobserved point. This generic approach will generalize easily to the case of multidimensional curve-fitting, since that just involves expanding the dimensions of what is already a multi-dimensional optimization problem.

22.7 Extrapolation

Extrapolation is similar to interpolation, but with the assumption that we are trying to look beyond our own observed data points to infer values outside their range. In some cases, extrapolation problems can be solved by the same methods as interpolation problems: fit the observed points to a curve and use the curve to predict the function at new data points. Often, though, extrapolation is used to approximate the limit of an infinite series. In such cases, we cannot trust polynomial approximations. Rational approximations may work better, but it is sometimes useful to take a more direct approach to the problem.

22.7.1 Richardson Extrapolation

We mentioned one example of extrapolation very briefly in the context of numerical integration: a technique called Richardson extrapolation. Richardson extrapolation is useful when we have a function g of some variable h where $g(h)$ takes the form

$$g(h) = c_0 + c_1 h + c_2 h^2 + c_3 h^3 + \cdots.$$

We generally require that the c_is themselves converge to zero with increasing i, so that low-order terms are dominant for small h. Our goal with Richardson extrapolation is to determine c_0, which is the limit of g as h goes to zero. This may seem like a very specialized kind of function, but it actually is not. The reason is that this is a description of the error term of a Taylor series for a function $g(x)$ evaluated at the point $x + h$. It therefore describes pretty much any function with well-behaved derivatives if we approximate it in terms of some step size h. Numerical integrals computed with decreasing step sizes are just one example of a class of functions that commonly takes this $g(h)$ form.

With Richardson extrapolation, we use approximations at different step sizes to eliminate low-order terms of $g(h)$. For example, using the two equations

$$g(h) = c_0 + c_1 h + c_2 h^2 + c_3 h^3 + \cdots$$

$$g(h/2) = c_0 + c_1 h/2 + c_2 (h/2)^2 + c_3 (h/2)^3 + \cdots,$$

we can construct a second-order accurate approximation to c_0:

$$\hat{g}(h) = 2g(h/2) - g(h) = c_0 + 0 - c_2(h^2/2) - c_3(7h^2/8) + \cdots.$$

We similarly can use $g(h/2)$ and $g(h/4)$ to construct another second-order approximation:

$$\hat{g}(h/2) = 2g(h/4) - g(h/2) = c_0 + 0 - c_2(h^2/8) - c_3(7h^2/64) + \cdots .$$

We can then combine these two second-order approximations to get a third-order approximation:

$$\hat{\hat{g}}(h) = \frac{4}{3}\hat{g}(h/2) - \frac{1}{3}\hat{g}(h) = c_0 + 0 + 0 + c_3(7h^3/48) + \cdots .$$

By repeatedly applying these kinds of linear combinations to cancel out low-order error terms, we can compute successively higher-order approximations to c_0. As long as the higher-order c values do not blow up, this procedure will converge rapidly on the limit:

$$\lim_{h \to 0} g(h) = c_0.$$

If h is a step size, then this limit will correspond to the value of the function if approximated with infinitely small steps. If our scheme is correctly designed (or, more formally, if it is *consistent*), then this limit will be exactly the value we want to approximate. Note that the practice is not always as clean as the theory. The c_i values may blow up with larger i, especially if there are any discontinuities in the functions we are computing. Even if they do not blow up, using Richardson extrapolation can hurt the stability properties of a numerical method in ways that are difficult to analyze. Nonetheless, Richardson extrapolation can be a powerful technique for accelerating convergence of a broad class of functions.

22.7.2 Aitken's δ^2 Process

There is another general extrapolation technique, called *Aitken's δ^2 process*, that is appropriate for finding the limit of a function that can be expressed as the partial sum of a geometric series:

$$S_i = S_0(1 + \gamma + \gamma^2 + \gamma^3 + \cdots + \gamma^i).$$

Again, this may seem to be a very specialized class of function, but it is actually a format that occurs often, at least approximately, in practice. This formula describes the sum of any sequence in which the change from step i to $i+1$ is a constant fraction smaller than the change from $i-1$ to i. This will be a good approximation to many kinds of convergent processes, such as the behavior of a stable finite difference iteration or the convergence of the bisection method or a similar binary search-type method.

Given a series of these approximations, $S_1, S_2, \ldots, S_n$, we want to infer the limit, $\lim_{i \to \infty} S_i$. We can find this by constructing a new series

$$\hat{S}_1, \hat{S}_2, \ldots, \hat{S}_n,$$

where

$$\hat{S}_i = S_{i+1} - \frac{(S_{i+1} - S_i)^2}{S_{i+1} - 2S_i + S_{i-1}}.$$

If the geometric series approximation is good, then the new sequence $\hat{S}_i$ will converge to the same value as S_i but at a faster rate. We can then find $\hat{S}_i$ to get a series that converges even faster on the same limit, and another, and so on, until we get one that converges within just a few steps.

To understand why this process works, we can imagine that the limit of our series is some value $\overline{S}$. Then, proposing that the error term is a convergent geometric series times a constant is equivalent to claiming

$$S_i \approx \overline{S} + \gamma^i \varepsilon$$

for some unknown γ and an error term ε, where ε is

$$\varepsilon = S(1 + \gamma + \gamma^2 + \gamma^3 + \cdots) = S \sum_{i=0}^{\infty} \gamma^i.$$

Then

$$\hat{S}_i = \overline{S} + \gamma^{i+1} \varepsilon - \frac{(\overline{S} + \gamma^{i+1} \varepsilon - \overline{S} - \gamma^i \varepsilon)^2}{\overline{S} + \gamma^{i+1} \varepsilon - 2\overline{S} - 2\gamma^i \varepsilon + \overline{S} + \gamma^{i-1} \varepsilon}$$

$$= \overline{S} + \gamma^{i+1} \varepsilon - \frac{\gamma^{2i} \varepsilon^2 (\gamma - 1)^2}{\gamma^{i-1} \varepsilon (\gamma^2 - 2\gamma + 1)}$$

$$= \overline{S} + \gamma^{i+1} \varepsilon - \gamma^{i+1} \varepsilon \frac{(\gamma - 1)^2}{(\gamma - 1)^2}$$

$$= \overline{S} + \gamma^{i+1} \varepsilon - \gamma^{i+1} \varepsilon$$

$$= \overline{S}.$$

So if the error term is truly exactly geometric, then one step of this method will convert every term of the sequence into the limit of the original sequence. If the error term is even close to geometric, the method may still work very well in producing a series whose terms are close to the original limit.

References and Further Study

Most of the methods covered in this chapter can be found in the Numerical Recipes books [82], along with others for which we do not have space here. Several other texts on numerical algorithms or numerical analysis also are good references for this topic. Stoer and Bulirsch [132] and Hamming [134] are good choices, covering some of the topics seen here as well as various interpolation and extrapolation methods we have not covered.

The coverage in this chapter is far from exhaustive, and there are several key classes of interpolants we did not cover that are of broad practical use. The most important class of interpolants we did not cover is the Fourier interpolants. Fourier interpolation is a broad enough topic that we cannot hope to do justice to it here. There is also a special class of fitting polynomials called Chebyshev polynomials, which have a variety of very useful mathematical properties. One can refer to Press et al. [82] or to Hamming [134] for the fundamentals and key numerical methods for working with Fourier or Chebyshev approximations. Finally, there is a broad class of functions called *wavelets* that have come into widespread use in recent years, particularly in image- and signal-processing applications. Press et al. [82] provides an introductory coverage of wavelet methods, but they are a broad topic for which one might desire greater depth. For an introductory text on wavelets, one may refer to Walnut [195] or Walker [196], among many others.

Those interested in seeing how data-fitting of the types examined in this chapter applies in a real-world biological context can look at Kuzmic [194] for an example of fitting differential equation models to reaction kinetic data. Wilkinson [110] provides in-depth coverage of issues of data-fitting specific to systems biology reaction models, with much greater depth on the peculiarities of those models and data types, and the specifics of handling them in practice, than we can cover here. I am not aware of any in-depth treatment of the broader topic of interpolation and extrapolation specifically for biological applications.

23 Case Study: Inferring Gene Regulatory Networks

This chapter will cover the last of our case studies, in which we look at how some of the techniques we have seen in the previous chapters have been applied to a real-world biological problem. We will examine the problem of inferring gene regulatory networks from expression data. This is a very active area of research in biology at present, with a rapidly accumulating literature. It is also an excellent subject for the present study because it has been approached with a variety of methods we have seen for parameter-tuning, as well as several others covered in the context of optimization and simulation and sampling.

For the purposes of our discussion, we will assume that our input is a matrix of data representing expression measurements from a series of gene expression microarrays:

$$
A = \begin{bmatrix}
a_{11} & a_{12} & \cdots & a_{1n} \\
a_{21} & \ddots & & \\
\vdots & & & \vdots \\
a_{m1} & a_{m2} & \cdots & a_{mn}
\end{bmatrix}
$$

Element a_{ij} refers to the expression level of gene i under some condition j. Normally, these are not raw expression values, but rather the logarithm of the ratio of expression between the condition and some control. It is further common to normalize these log ratios for each gene, for example, by translating and scaling them so that each gene has mean zero and standard deviation 1 across conditions. The conditions may correspond to different time points if one is studying time-dependent gene expression, to different experimental conditions if one is studying condition-dependent gene regulatory networks, to different patients if one is studying expression changes distinguishing healthy from diseased individuals or distinguishing different disease subtypes, or to many other sets of data. We will typically have many more distinct genes than distinct conditions. A typical experiment, for example, may contain one expression measurement for each known gene in the organism being examined, on the order of 20,000 genes for humans. In some cases, multiple measurements may be available for a given gene, perhaps corresponding to different splice forms, and some noncoding RNA may also be included. The number of distinct experiments may vary from the order of 10 for time-series data, to on the order of a few hundred for clinical studies of individual people.

Our output will be a graph identifying pairs of genes that are presumed to be involved in a regulatory relationship. Many different models have been proposed for gene expression networks, and we will focus on three different levels of abstraction that lend themselves to different kinds of inference techniques: coexpression models, static network models, and kinetic models. Depending on the specific model we are using, the graphs may be directed (implying that the source gene is a regulator of the destination gene) or undirected (implying that the endpoints are correlated with one another). Different variants may also have additional outputs corresponding to the nature or strength of each interaction.

23.1   Coexpression Models

Gene network inference at first glance generally seems to be an impossible problem. The number of edges that may be present in the graph is on the order of the square of the number of genes, and the number of graphs we can potentially choose from is exponential in that number. The number of distinct expression measurements from which to fit the graph, though, is only a few tens or hundreds of times the number of genes. It therefore seems that we cannot hope to distinguish the best among the universe of possible graph models. One approach to the problem, then, is to drastically restrict the space of models we will consider, reducing it to a universe from which we can tractably find the "right" model. Coexpression follows this idea, essentially lowering our sights from trying to find the full network to trying to find a very restricted subclass of the network.

In a coexpression model, we seek to group our genes into "clusters" such that the members of a cluster approximately follow the same expression pattern across the samples. The goal is to figure out how to partition the genes into clusters so as to have a small error by some measure. Coexpression is, then, a degenerate form of network inference in which we are trying to infer an unweighted, undirected graph that consists of a union of cliques. A clique, we may recall from chapter 3, is a subgraph in which there are edges between all pairs of nodes.

23.1.1   Measures of Similarity

One way we may approach this problem is to start with some trivially computable graph identifying pairs of genes with similar expression, then find some "union-of-cliques" graph that is close to this pairwise graph. Several approaches in the literature start by creating a weighted, complete graph on all genes reflecting pairwise similarities. There are many measures one can use to decide how similar different pairs of genes are. Consider two rows of input, $\vec{a}_i$ and $\vec{a}_j$, representing expression levels of genes X_i and X_j across all conditions in the matrix. These two rows will define a set of paired expression values $(a_{i1}, a_{j1}), (a_{i2}, a_{j2}), \ldots, (a_{in}, a_{jn})$. Our goal is to find some metric by which to say how far apart the vectors $\vec{a}_i$ and $\vec{a}_j$ are.

One simple measure we have seen in several prior contexts is the Euclidean distance between the vectors. That is, we compute the length of the difference between the two vectors:

$$\sqrt{(a_{i1} - a_{j1})^2 + (a_{i2} - a_{j2})^2 + \cdots + (a_{in} - a_{jn})^2} = \sum_{k=1}^{n}(a_{ik} - a_{jk})^2.$$

This is a simple, intuitive measure that is easy to compute, making it a popular choice.

An alternative is the use of *correlation coefficients*, defined as follows:

$$r = \frac{COV(\vec{a}_i, \vec{a}_j)}{\sqrt{VAR(\vec{a}_i)VAR(\vec{a}_j)}}$$

$$= \frac{\left(\frac{1}{n}\sum_k a_{ik}a_{jk}\right) - \left(\frac{1}{n}\sum_k a_{ik}\right)^2\left(\frac{1}{n}\sum_k a_{jk}\right)^2}{\sqrt{\left(\frac{1}{n}\sum_k a_{ik}^2 - \left(\frac{1}{n}\sum_k a_{ik}\right)^2\right)\left(\frac{1}{n}\sum_k a_{jk}^2 - \left(\frac{1}{n}\sum_k a_{jk}\right)^2\right)}}.$$

Correlation coefficients have the advantage that we can distinguish statistically significant and insignificant correlations, given an assumption that the relationship between the two genes in question is linear. *P-value calculators*, programs that allow one to find the value of r needed to give us a particular level of confidence in an association, are now freely available online. The above formula defines what is called a *Pearson correlation coefficient*. One can also use the *Spearman correlation coefficient*, in which one ranks the expression levels of the conditions for each gene and substitutes the ranks for the raw values in the formula above. So, for example, if we have the rows $\vec{a}_i = \begin{bmatrix} 0.1 & -0.2 & 1.3 & -2.5 \end{bmatrix}$ and $\vec{a}_j = \begin{bmatrix} 2.7 & 1.8 & 0.2 & -0.1 \end{bmatrix}$, then we will convert them to the rank vectors $\vec{r}_i = \begin{bmatrix} 3 & 2 & 4 & 1 \end{bmatrix}$ and $\vec{r}_j = \begin{bmatrix} 4 & 3 & 2 & 1 \end{bmatrix}$. We will then use $\vec{r}_i$ and $\vec{r}_j$ in the formula for r above.

A third class of approach is based on information theoretic measures. Information theory is based on the concept of *entropy*, the amount of information carried by a signal. In an information theoretic context, two genes are said to be related to one another if the information they carry collectively is not much different from the information they carry individually. Entropy, which literally measures the number of bits of information carried by a signal $\vec{a}_i$, is measured as follows:

$$H(\vec{a}_i) = -\sum_k Pr\{a = a_{ik}\} \log Pr\{a = a_{ik}\},$$

where a is a probability distribution from which we presume elements of $\vec{a}_i$ are drawn. If we are working with continuous, normalized expression values, then we will commonly assume that they are drawn from a unit normal distribution, and therefore

$$Pr\{a = a_{ik}\} = \frac{1}{\sqrt{2\pi}} e^{-(a_{ik})^2/2}.$$

We can also discretize expression values and establish a discrete distribution by counting how often each discrete expression level occurs in $\vec{a}_i$. Whatever definition we use, we can further calculate a *joint entropy* for the pair of vectors $\vec{a}_i$ and $\vec{a}_j$ as follows:

$$H(\vec{a}_i, \vec{a}_j) = -\sum_k Pr\{a = a_{ik} \wedge b = a_{jk}\} \log Pr\{a = a_{ij} \wedge b = a_{jk}\},$$

where we may assume that distributions a and b are independent unit normals:

$$Pr\{a = a_{ik} \cap b = a_{jk}\} = \frac{1}{2\pi} e^{-(a_{ik} + a_{jk})^2/2}.$$

We can, finally, define the *mutual information* to be

$$M(\vec{a}_i, \vec{a}_j) = H(\vec{a}_i) + H(\vec{a}_j) - H(\vec{a}_i, \vec{a}_j).$$

Intuitively, mutual information measures how much of the information contained in the pair of genes is shared by both of them. It therefore provides a third good alternative for establishing edge weights between genes.

23.1.2   Finding a Union-of-Cliques Graph

Once we have an initial graph of pairwise gene similarity, we still have various options for finding a union-of-cliques graph, or *clustering*, from that complete graph. Nearly any meaningful, rigorous definition of the "closest" union-of-cliques graph to the similarity graph will, unfortunately, result in an NP-hard problem. For example, we may propose to split the graph into a set of k subsets that maximize the sum of pairwise similarities within the subsets. We should recognize this as a max-k-cut problem from chapter 3. A variety of heuristics are therefore actually used in practice.

One alternative is to use simple graph heuristics to find some reasonably good set of clusters. An example of this approach is the guilt by association method [196]. In this method, we pick some test gene and then pull out every gene whose similarity to the test gene is above some threshold. These genes form a clique. We can then repeat the process with the unassigned genes and a new test gene. Because measures of similarity are not necessarily transitive, we may end up incorporating into a clique some pairs of genes that are not very similar to one another even if they are both similar to some common third gene. It is easy to come up with various heuristic modifications one may use with this method to make an approach more or less tolerant of some poor pairwise similarity within clusters.

A second heuristic approach is known as "hierarchical clustering," a technique adopted by Eisen et al. [197] for some of the earliest microarray analysis. With hierarchical clustering, we build a tree of similar groups such that the leaves of the tree correspond to individual genes, and the root to the whole data set. The trees are assembled through some greedy operation of either merging or splitting a gene set to produce a parent and usually two children. For example, if we are using a Euclidean distance measure, then we can define the distance between two clusters as the mean distance between their members. We can then create a hierarchical clustering by placing each gene into its own cluster and next repeatedly merging the two closest clusters. Although one would normally continue the approach until all genes are merged into the root, we can arbitrarily halt the approach at any desired distance cutoff to produce a discrete set of clusters. Figure 23.1 provides an illustration of a hierarchical clustering solution. In addition to this "bottom-up" approach to hierarchical clustering, we can use a "top-down" approach in which we start with the cluster of all genes and successively split it into pairs of child clusters [198].

A third broad class of approaches treats cluster assignment as a kind of missing-data inference problem, in which the cluster assignments are latent variables we must infer. As we might expect, expectation maximization provides one way to solve this problem [199]. In an EM approach to the problem, we treat the expression levels, A, as the observed data and create a matrix of cluster assignment variables, C, where c_{ik} is the probability that gene i is assigned to cluster k. The model λ is then described in terms of a set of variables expressing the distribution of values in each cluster. For example, we can define g_{jk} to be the mean expression of cluster k in condition j. We can also optionally consider standard deviations to be part of the model. If we assume unit standard deviations for simplicity, though, we can calculate the likelihood of the model as follows:

$$Pr\{A \mid C, \lambda\} = \prod_i \sum_k c_{ij} \prod_j \frac{1}{\sqrt{2\pi}} e^{(a_{ij}-g_{kj})^2/2}.$$

That is, the probability is the product over all genes of the weighted probability over all possible cluster assignments of generating the expression of that gene for a given cluster assignment. In the E-stage of the algorithm, we evaluate $Pr\{A \mid C, \lambda\}$

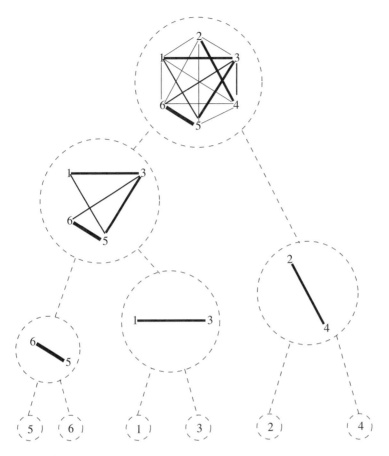

Figure 23.1
Hierarchical clustering. Dashed circles and lines mark the cluster graph. Within each cluster are subsets of the full expression graph (shown in the root node), where thicker edges correspond to stronger similarity. The cluster graph is formed by successively accumulating the two most similar expression subgraphs into a larger subgraph until the complete expression graph is formed at the root.

for each possible cluster assignment of a given gene, and normalize to create a new vector of c_{ik} values for that gene. In the M-stage, we find maximum likelihood values of g_{jk} by averaging all expression values for condition j weighted by the probability that gene i is in cluster k:

$$g_{jk} \leftarrow \sum_i a_{ij} c_{ik}.$$

Several other popular approaches use a similar iterative inference algorithm, although they are not formally EM algorithms. With k-means clustering [200], we perform essentially the same iterations as with the EM approach, except that in the E-phase we assign each gene entirely to its one most plausible cluster rather than establishing a distribution of cluster assignments. That gene then contributes only to the mean of that one cluster in the M-phase. Several other methods use similar EM-like approaches, of which the most popular for expression clustering is the self-organizing map [201] [202].

23.2   Bayesian Graphical Models

Coexpression models provide one possible balance between models too rich to learn robustly and those too simple to be of any practical value, but those models are too crude to capture many kinds of expression patterns of interest to us. For example, if we examine three genes A, B, and C, knowing that they have similar expression does not tell us whether A regulates B and C, or A regulates B and B in turn regulates C, or if all three are regulated by some common fourth factor. Yet this kind of inference about chains of expression is one of the main reasons we are interested in expression. If, for instance, our goal is to find a drug to shut down these genes, then it will help to know which genes lie earliest in the network. It will therefore be very useful in some cases to be able to identify more general classes of networks. Bayesian models have proved to be a powerful approach for learning general graph models from expression data [203], and we will therefore consider them next.

Bayesian models are a broadly used class of models in statistics and machine-learning contexts, and have many uses in computational biology beyond gene expression analysis. These other applications include motif inference and various problems related to comparative genomics. They tend to deal very well with "messy" data of any form, and thus are often useful for inference in which one must work with real-world biological data sets. In a Bayesian model, we must generally separate the inference of graph structure from the inference of probability distributions describing dependence within the graph. We will begin by considering inference of the probabilistic component of the model and then discuss graph inference of the model.

23.2.1   Defining a Probability Function

The basic idea behind a Bayesian model is to assume that each variable in the model is drawn from a probability distribution based on a small set of other variables. In the simplest version, each variable represents one observed gene. Figure 23.2 shows an example of a hypothetical simple network linking four genes. This network implies that gene X_1's expression is drawn from some distribution $Pr\{X_1\}$ independent of the other genes. Genes X_2 and X_3 then have their expression drawn from distributions $Pr\{X_2 \mid X_1\}$ and $Pr\{X_3 \mid X_1\}$ that depend on the previously drawn value of X_1. Finally, X_4's expression level is presumed to be drawn from a distribution

Figure 23.2
A hypothetical regulatory network among four genes.

$Pr\{X_4 \,|\, X_1, X_2\}$ that is conditional on the values of both X_1 and X_2. Given these assumptions, we can define the likelihood of the complete network model, given the data, as the product over all conditions of the probabilities of the observed expression levels for each condition, given the model:

$$Pr\{A \,|\, \lambda\} = \prod_j \prod_i Pr\{X_i = a_{ij}\}$$

$$= \prod_j Pr\{X_1 = a_{1j}\} \times Pr\{X_2 = a_{2j} \,|\, X_1 = a_{1j}\}$$

$$\times Pr\{X_3 = a_{3j} \,|\, X_1 = a_{1j}\} \times Pr\{X_4 = a_{4j} \,|\, X_1 = a_{1j}, X_2 = a_{2j}\}.$$

We can then in principle find the best model by evaluating this function for every possible model and choosing the one that maximizes likelihood.

Before we get to the optimization, though, we need to be a bit more precise about what these probability functions look like. In the simplest case, we can discretize our expression values, in which case the probability functions become discrete binomial or multinomial distributions. We may, for example, declare that each gene is either "on" or "off"; "overexpressed" or "underexpressed"; or "(H)igh," "(M)edium," or "(L)ow," based on various cutoffs. If we assume the H/M/L labeling, we can empirically estimate the probability distributions by counting how often each label is used. For instance, if we observe that X_1 has medium expression in ten conditions and that in three of those conditions X_2 has low expression, then we can estimate

$$Pr\{X_2 = L \,|\, X_1 = M\} = \frac{3}{10}.$$

More often, we want to maintain our use of continuous expression values, in which case we need to presume some class of continuous distributions. A typical class of distribution for gene expression inference is a *Gaussian linear model*, in which we assume that each expression value is drawn from a Gaussian distribution that is a linear function of its input values. For example, we will assume that the distribution for gene X_4 from figure 23.2 takes the form

$$Pr\{X_4 = a_{4j} \mid X_1 = a_{1j}, X_2 = a_{2j}\} = \frac{1}{\sqrt{2\pi\sigma}} e^{(a_{4j} - (\alpha_0 + \alpha_1 a_{1j} + \alpha_2 a_{2j}))^2 / 2\sigma^2}.$$

The α values are regression coefficients we will normally estimate by finding a least-squares best fit model for the observed expression values of X_1, X_2, and X_4, as we saw in chapter 21. σ may be empirically measured for each gene, presumed to be 1, or fit by some other inference procedure. Regardless of the model we assume, we end up with a likelihood function we can evaluate to test the quality of a given network.

It is usually a little more complicated than this, though, because we will we want to assume prior probabilities on various parameters that will bias the model in the direction of our previous expectations. For example, we may wish to penalize complicated models by assuming that an edge is prima facie unlikely to be present. We may then say that a given network structure G has its own prior probability $Pr\{G\} = p^{|E|}$, where $|E|$ is the number of edges in G and p is some small per-edge prior probability. If the genes have been studied before, we may have more specific ideas about exactly how likely each possible edge will be. We may also attach prior distributions to regression parameters or any other model parameter, in which case we may need a more complicated inference procedure to find those parameters optimally. We can even assume that the priors themselves are described by unknown parameters with their own priors, or that those priors have priors, and so on. For the purposes of the rest of this discussion, we will assume that there is some overall graph prior $Pr\{G\}$ and that we know how to evaluate it.

23.2.2 Finding the Network

Our goal from this point on, then, is to find the network structure G, assuming that we can evaluate $Pr\{A \mid G\}Pr\{G\}$ (i.e., the likelihood of the observed expression data, given the network times the probability of the network). There are several ways to do this that draw on many of the techniques we have seen. We will consider some approaches in turn.

A straightforward approach to the problem is to treat the inference as an optimization problem. The input to the problem is the expression matrix A, the output is the graph G, and the metric is the likelihood function $Pr\{A \mid G\}Pr\{G\}$. We therefore have a well-defined discrete optimization problem that we can in principle solve with the tools we covered in chapters 2 and 3. Unfortunately, the objective function is so complicated, and the search space so large, that we can really hope to solve the problem only heuristically. Many standard heuristics have been attempted for this problem. For instance, simulated annealing and genetic algorithms, two techniques we have seen previously, have been applied to heuristic inference of Bayesian expression models [204].

An alternative approach is to treat network inference as a sampling problem. In the sampling approach, we assume that we have a distribution of possible networks rather than a single network, and that our goal is to estimate that distribution. The probability of any given network in the distribution is assumed to be proportional to the likelihood function. In principle, one can evaluate every possible network, compute the likelihood of each, and then directly identify the probability of any given network G_i in the distribution as follows:

$$p_i = \frac{Pr\{A \mid G_i\}Pr\{G_i\}}{\sum_j Pr\{A \mid G_j\}Pr\{G_j\}}.$$

We cannot actually hope to perform this calculation, though, so we have to use somewhat more sophisticated samplers. One popular approach is the use of Markov chain Monte Carlo (MCMC) samplers [205]. Since it is easy to compute a small number of likelihood evaluations, we can determine the relative probabilities of two similar states. Thus, for instance, we can evaluate the likelihood $Pr\{A \mid G_i\}Pr\{G_i\}$ for some graph G_i, try adding or removing a single edge to get a modified graph G_j, and evaluate $Pr\{A \mid G_j\}Pr\{G_j\}$. Although we do not know the absolute probability of either graph, we do know the ratio

$$\frac{p_i}{p_j} = \frac{Pr\{A \mid G_i\}Pr\{G_i\}}{Pr\{A \mid G_j\}Pr\{G_j\}}.$$

We can therefore construct a Metropolis sampler that will sample among graphs with the correct stationary distribution. We can similarly construct a Gibbs sampler that will sample from the same distribution. Once we have a sampler, we can see how often each edge is present in the complete Markov chain and infer that those edges often present in the chain should be present in the inferred network.

This sampling approach has the advantage of giving us not just a graph of well-supported edges but also a confidence estimate on each edge. For example, if we observe that the edge from X_1 to X_3 is present 90 percent of the time, then we can say that we are 90 percent confident that that edge is found in the real network. One can also use a hybrid of the optimization and sampling approaches with a statistical method called *bootstrapping* [206]. In the bootstrapping approach, we infer the network by any method, but use only a random subset of the input data. We then repeat the process for many other random subsets of the data. The result is a sampling of networks generated by the solutions to the optimization problem. We can then keep those edges that show up in a large fraction of the sampled network. Bootstrapping provides an alternative to MCMC methods for generating confidence estimates for edges in gene expression graphs [207].

As is often the case, real-world complications make the actual inference problem more complex than we can cover here. First, as we mentioned when considering coexpression models, the inference problem we have described above is likely to be hopeless if we apply these methods naïvely. The model space is simply too large for the data. One way around this problem is to trim the model space quite a bit. It turns out that the coexpression approaches can help us accomplish that trimming by a method called the sparse candidate algorithm [208]. With the sparse candidate algorithm, we first perform a coexpression clustering to identify possible regulatory relationships. When we then infer the full network, we consider only subgraphs of the union-of-cliques graph from the clustering. This will generally reduce the edge set from approximately quadratic in the number of genes to approximately linear. The alternative to reducing the space of models is expanding the data. Most current applications of genetic network inference merge several other possible sources of data about likely regulatory relationships into the inference. Other sources of data may include sequence-based predictions of nearby transcription-factor binding site motifs, experimental evidence of transcription-factor binding, protein–protein interaction data, or comparative genomic data (see, for example, Tavazoie et al. [209], Ideker et al. [210], and Tu et al. [211]). Adding in these other sources makes for a more complicated inference problem. The Bayesian graphical framework tends to be well suited to these sorts of heterogeneous inferences, though, as it is easy to throw in new kinds of nodes and edges based on the possible dependencies among the various data types.

23.3 Kinetic Models

A third class of models occasionally used for gene network inference applications is kinetic models describing the time progress of a set of genes. These kinetic models are generally represented by systems of ordinary differential equations (ODEs). This approach was first developed under the name network identification by multiple regression (NIR) [212], and was generalized with an approach called mode-of-action by network identification (MNI) [213]. These and similar models assume that expression is approximately described by a system of linear differential equations. That is, each gene X_i is presumed to obey a rate law of the form

$$\frac{dX_i}{dt} = a_{i1}X_1 + a_{i2}X_2 + \cdots + a_{in}X_n$$

for some unknown rate constants a_{ij}. We will further generally assume that there is some perturbing factor specific to each experimental condition, yielding the modified equations

$$\frac{dX_i}{dt} = a_{i1}X_1 + a_{i2}X_2 + \cdots + a_{in}X_n + b_{ik},$$

where b_{ik} is the amount of perturbation presumed on gene i in condition k. For example, if we are interested in comparing wild-type cells to those in which a given gene has been selectively knocked down with RNAi, then b_{ik} will be a negative value for those conditions k in which gene i was suppressed and zero for all other genes and conditions.

We have already seen examples of how one can fit an ODE model to a simple reaction network in chapter 18. Gene network models tend to be much harder to fit, though, because we must generally infer them from sparser data sets relative to the number of parameters. We may have just a few conditions from which to infer connections between many thousands of genes.

Inference in these models is generally accomplished by assuming that the observed expression levels represent steady-state expression values (i.e., stable values over long time periods). By definition, a steady-state expression is one at which the rate of change is zero. Thus,

$$\frac{dX_i}{dt} = 0$$

for all X_i and the ODEs above take the form

$$a_{i1}X_1 + a_{i2}X_2 + \cdots + a_{in}X_n + b_{ik} = 0.$$

Since the X_is are known values and the a_{ij}s and b_{ik}s are unknowns, we can treat this equation as the linear system $A\vec{x} = \vec{b}$, where $\vec{b}$ is the all-zeros vector, $\vec{x}$ is a vector of all of the unknowns,

$$\vec{x}^T = [\, a_{11} \quad a_{12} \quad \cdots \quad a_{1n} \quad b_{11} \quad \cdots \quad b_{1k} \quad a_{21} \quad \cdots \quad a_{2n} \quad b_{21} \quad \cdots \quad b_{2k} \quad \cdots \,],$$

and A is a block matrix in which each row corresponds to one of the above equality constraints. We can therefore in principle learn the model by solving a linear system.

Since this is an even richer model than the general Bayesian network models, one might reasonably infer that it will be impossible to learn for a large number of genes. Indeed, if we assume naïve application of this method to whole-genome expression microarray data, we will likely have millions of variables but only thousands of equations, and thus will have a highly underdetermined linear system to solve. For us to have any hope of solving it, we must reduce the number of variables quite a bit. One will normally apply this technique only to much smaller gene sets, on the order of tens of genes, than one might expect for the preceding methods. One may also improve the tractability a bit by using an approach like the sparse candidate algorithm discussed above [208]. It is worth noting that it is possible to combine Bayesian network inference with fitting kinetic ODE models to network data [214] and get some of the advantages of each approach. We will not attempt to cover such hybrid models here, though.

References and Further Study

Analysis of expression microarrays is an active field right now, and there are many sources of information beyond the primary references cited above. There have been several excellent review articles in the scientific literature on this topic. This chapter was heavily influenced by reviews by D'haeseleer et al. [215], De Jong [216], and Bansal et al. [217]. There are also several textbooks on computational analysis of gene expression microarrays that will provide much greater breadth on the analysis problems we consider and the underlying technologies. Interested readers may consult Causton et al. [218], Berrar et al. [219], or Zhang [220].

The three individual approaches to network inference we considered above draw on much broader literatures than we have considered here. The clustering field was an enormously active area for a number of years, and we have considered just a few of the most popular approaches. Those looking for a concise review of clustering for gene expression in particular can refer to D'haeseleer [221], and those interested in a more in-depth review of clustering methodologies in general can refer to Jain et al. [222].

Bayesian graphical models of the kind we saw above are a widely used tool in statistics and machine learning. While the reviews above will provide a good grounding in their use in expression-modeling specifically, there is much more to know about the field that is likely to be of use to those who plan to work extensively with these models. A general text on statistics, such as Wasserman [112], or on machine learning, such as Mitchell [177], will provide greater depth on Bayesian models and Bayesian inference in general. A number of good texts specifically on Bayesian models, such as Congdon [223], Gelman et al. [224], and Neapolitan [225], are available.

24 Model Validation

In this final chapter, we will examine the problem of how to judge a model we have built. We may imagine that we have taken an imprecise description of some biological system, have made various assumptions and abstractions to end up with a precise mathematical model, and have fitted that model to some data to fill in any unknown parameters. We now want to ask, "Is this model good?" Of course, every model is imperfect in some ways, so we need to be more precise about what we are asking. That is, we need to ask, "Good for what?" and "Compared to what?" The answers to these questions will depend on why we built the model in the first place and what we hope to get out of it. There are some general techniques that are often useful for this task, and we will briefly examine some of them. The goodness of a model is in large measure a philosophical question, however. We will therefore return now to some issues, on which we touched in chapter 1, on reasoning about and evaluating models. We will conclude with a discussion of the broader role of modeling in scientific inquiry.

24.1 Measures of Goodness

One of the first issues we need to consider in more precisely assessing the quality of a model is deciding how we will measure its goodness. This issue is not much different from the issue of defining an objective function for fitting parameters to the model. Almost any function that captures correspondence between a model and some empirical data or our prior expectations may suffice as a measure of goodness. A few kinds of measures show up repeatedly for many different model types, though, and are therefore worth considering here. We have already seen several of these measures in other contexts because a measure of goodness of fit is often useful as an optimization metric and not just an after-the-fact test of fit. Nonetheless, it is worth revisiting them in the present context.

In many cases, we explicitly build a measure of model goodness into the optimization we perform to fit to the model. For example, in a maximum likelihood model, we will often be quite explicit about the universe of possible models from which our model will be drawn and how likely any given model is to be drawn from that universe. We saw examples of how this might work in our coverage of Bayesian models for gene networks, as well as in various applications of expectation maximization. For the gene network inference problem, for example, we can assign a probability $Pr\{G\}$ to any possible network graph G, perhaps favoring simpler graphs. We can also assign a prior probability to possible values of the parameters of the model, λ, perhaps favoring parameter sets close to our previous expectations. Then, when we optimize the other parameters of our model λ, given our input x, we explicitly solve for

$$\max_{\lambda, G} \; Pr\{x \mid \lambda, G\} \; Pr\{\lambda\} \; Pr\{G\}.$$

We can use the same general objective to subsequently validate the model, for example, by comparing it to a data set different from the one used to learn the model. But this process does assume that we have properly constrained the possible universe of models and properly set our prior probabilities. Likelihood can be a good metric, but we need to be careful about its use. In particular, it is generally a mistake to validate a model based on the same criterion used to fit it to the data.

Very often, the success of our modeling problem will be judged in terms of the ability of a set of model predictions to fit a set of data points. We can then judge the goodness of the model at least in part by some metric on the quality of the fit between the observations and the predictions. We have already seen some ways we may measure this fit. We may judge the model by the sum of squares of the differences between the observed data points and the corresponding points predicted by the model. That is, given the observed points $(x_1, y_1), \ldots, (x_n, y_n)$ and the corresponding modeled points $(x_1, \hat{y}_1), \ldots, (x_n, \hat{y}_n)$, we measure the goodness of our model by the metric

$$\sum_{i=1}^{n} (y_i - \hat{y}_i)^2.$$

We can generalize this approach to multiple independent variables by asking for a sum over all observed data points across all independent variables. Similarly, we can generalize to multiple dependent variables by summing least-squares contributions across all variables, perhaps following some normalization.

Another measure of fit we have seen before is the correlation coefficient, which, we may recall, is defined in terms of the variances of two series, Y and $\hat{Y}$, and the covariance of both, by

$$\rho = \frac{Cov(Y, \hat{Y})}{\sqrt{Var(Y) Var(\hat{Y})}}.$$

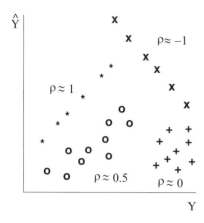

Figure 24.1
Four point sets (*, o, x, and +) showing varying degrees of correlation between their *x* and *y* coordinates.

Figure 24.2
A data set and model that may exhibit no corelation even if the model is a very good representation of the data.

The correlation coefficient, like least-squares, is a generally useful measure of the fit of a model to an observed data set (or in general of any data set to any other). The correlation coefficient essentially measures for the two series Y and $\hat{Y}$ how close to linear the relationship between them is. Figure 24.1 shows examples of point sets that may yield varying amounts of correlation.

When using the correlation coefficient to measure goodness of fit, we have to be careful that we are using it in a way that makes sense for our application. Two data sets can be closely related to one another and yet be completely uncorrelated. For example, suppose we have created a model of diffusion of a particle, as in figure 24.2. If we measure the correlation of the x and y coordinates of the particle over time between the model output and the data, we may conclude that this is a very bad model since the model and the data pursue very different paths. For a model of diffusion, though, we may only care about whether the distribution of particle distances from the origin as a function of time matches between the data and the model. If we measure the correlation between the distances from the starting point of the model and the observation as functions of time, we may conclude that the above is a very good model of particle diffusion. Again, it all depends on what we want the model to capture about our system and how we plan to use the model.

Another alternative that we likewise saw in the context of identifying gene networks is the use of information theoretic measures, such as mutual information. Information theoretic measures provide a more general test of similarity between model predictions and data than does the correlation coefficient, which depends on the relationship being approximately linear. We can use mutual information measures just as we use correlation measures: to test fit between two data streams by fitting each to a common probability model, as we saw with genetic network models. Alternatively, we can use entropy as a direct measure of the goodness of a probability model rather than a particular set of model outputs. If we have a probability model, the entropy of the observed data relative to that model directly provides a measure of the fit of the full model distribution to those data. If the model fits the data well, then the data will have a small amount of information, and thus a low entropy, for that model.

24.2 Accuracy, Sensitivity, and Specificity

When a model is used to predict or classify, several other measures of goodness of fit are frequently used. For instance, suppose we want to predict whether a particular person has a particular disease. We may construct a model for how disease likelihood varies with some symptoms or the outputs of some laboratory assay. We will then want to know how accurate the model is at predicting who has the disease. One simple way to do that is literally to judge our model by its *accuracy*: What fraction of the predictions it makes are correct?

This is likely to be a very poor measure for cases such as disease diagnosis, in which one answer (healthy) is much more common than the other (sick). For example, suppose we have a disease that occurs in 1 percent of people, and we develop two models to predict whether a given person has the disease. Model 1 correctly predicts the disease whenever it is present, but also in an additional 1 percent of the healthy population. Model 2 says that everyone is free of the disease. Which is the better model? Both are equally good by the accuracy measure, since each is right 99 percent of the time and wrong 1 percent of the time. Yet intuitively we are likely to believe that model 1 is the better model. We can formalize this sort of intuition by considering four classes of predictions the model may make:

Table 24.1
Measures of accuracy for two hypothetical disease prediction models

	Model 1	Model 2
True positives	1%	0%
True negatives	98%	99%
False negatives	0%	1%
False positives	1%	0%

1. *True positives*   the model says the person is sick, and the person is sick.
2. *True negatives*   the model says the person is healthy, and the person is healthy.
3. *False negatives* (or *type I errors*)   the model says the person is healthy, but the person is sick.
4. *False positives* (or *type II errors*)   the model says the person is sick, but the person is healthy.

The accuracy of the model is then the fraction of answers that are either true positives or true negatives. For our two hypothetical models above, we will have the rates shown in table 24.1.

Adding true positives and true negatives tells us that each model has 99 percent accuracy. Though that suggests the models are equally good by the accuracy metric, we would generally say the first is much more useful. Of course there may be cases where the second is a better model. If the consequences of being falsely identified as having the disease are worse than actually having the disease, then it may be preferable to use the second, "useless" model. Again, it all depends on what we are trying to do.

Similar issues come up in problems such as motif detection, in which we look for a rare event and cannot consider a single accuracy score a reasonable measure of goodness of a model. If we have a motif that is relatively rare in a genetic sequence, we can create a highly accurate predictor by predicting that there are no motifs, regardless of the data set given. In the case of motif-finding, we may be willing to tolerate some false identifications but really want to get all the true motifs. We will then want a model that has few false negatives, even if that means tolerating many false positives.

In such cases, we can apply two other common measures of goodness. The first is sensitivity:

$$\text{sensitivity} = \frac{|\text{true positives}|}{|\text{true positives}| + |\text{false negatives}|}.$$

Sensitivity measures how often the model identifies what it is supposed to identify. And the second is specificity:

$$\text{specificity} = \frac{|\text{true positives}|}{|\text{true positives}| + |\text{false positives}|}.$$

Specificity asks how often the things the model does identify should have been identified.

For disease diagnosis, we generally want a model with very high specificity. We cannot tolerate many false positives, or else they will overwhelm the true positives. For motif-finding, we may instead want to insist on very high sensitivity; we will tolerate extra motifs but do not want to miss any real ones.

We will often have the option of trading off sensitivity for specificity in a model. For instance, if we construct a hidden Markov model for motif-finding, we can often bias the prior probability of observing a motif to make positive predictions more frequent. We may then increase our rate of true positives at the cost of increasing our rate of false positives. For a Bayesian disease model, we may similarly lower the prior probability of the disease to increase the specificity of the model at the cost of some loss of sensitivity. These trade-offs are captured in a *receiver operating characteristic (ROC) curve*, which shows how sensitivity and specificity of a model vary with some tunable parameter. Figure 24.3(a) shows an example of a hypothetical ROC curve. ROC curves can allow us to distinguish among different models, depending on what model characteristics we need, and to determine which parameter values will give us the best performance for a given application. Figure 24.3(b) shows some hypothetical ROC curves for competing models of a system. The ROC curves

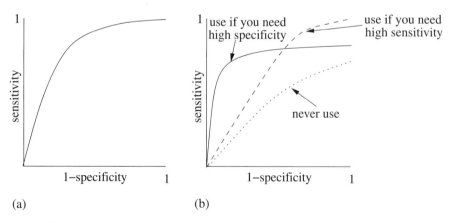

(a)   (b)

Figure 24.3
Receiver operating characteristic (ROC) curves, which plot the trade-off between sensitivity and specificity of a model as a function of some tunable parameter. (a) A hypothetical ROC curve. (b) Three hypothetical curves representing different models for a single application. Our choice of model depends on the needs of a specific application.

provide guidance for selecting which model and which model parameters are most suitable for any given application.

24.3   Cross-Validation

No matter what measure of goodness we use, one problem we have to worry about is *overfitting*. Overfitting means coming up with a model that is too closely tuned to our specific training data, rather than to the full distribution of data the model will encounter in real practice. Overfitting tends to be a problem particularly for overly complex models. For example, if we sample n data points from some distribution and we want to fit a polynomial to them, we can always fit an $n - 1$ degree polynomial and get a perfect fit. But if we then pick one more data point, it is quite likely that our model will fit that extra data point very poorly. If we have chosen instead to use a k-degree polynomial for $k \ll n$ and found a good fit to our n points, then it is likely that same polynomial will also be a good fit to a new point chosen from the same distribution. And if the k-degree polynomial is a poor fit, we will probably realize that from testing on our n points.

There are ways to design protections against overfitting into our modeling procedure. For example, if we are designing a probability model and finding the best fit by a maximum likelihood method, we can build a penalty against complexity into the prior probabilities for the model. One useful version of this strategy is *minimum description length (MDL)* modeling. In an MDL model, we specify a universe of possible models and then seek a model for which we can concisely explain the observed data with the model and the model within its universe. These kinds of methods do not guarantee that we will avoid overfitting, but they do offer some protection against it.

No matter what protections we use, though, it is important to test if our model is overfitting by performing adequate *cross-validation*. Cross-validation means that we train our model using data different from what we use to test the goodness of the model. There are many ways to cross-validate:

• *Twofold*   Split the data into separate testing and training sets, train the model on the training set, then validate on the testing set.
• *k-fold*   Split the data into k sets and perform k tests, each training on $k - 1$ sets and validating on the missing set.
• *Leave-one-out*   a special case of k-fold for $k = n$.
• *Bootstrapping*   Like k-fold, except that for each test we randomly choose n/k elements for testing and train on the remaining $n - n/k$, repeating for multiple random sets.

Which cross-validation method is appropriate mostly depends on the amount of data we have available and how difficult it is to run the tests. If we have very few data points to work with, we want to use as many as possible in each test and may favor leave-one-out validation. If the computations are very costly, we may need to do twofold cross-validation because we cannot afford to train and test our model repeatedly.

Cross-validating adequately is much harder in practice than it may seem. It is not unusual in practice for computational biologists conducting a research project to set aside some training and testing data at the beginning of a study, design and train a model, test it, then go back and tweak the model after the cross-validation, repeating until the method works well. For instance, if we are designing an HMM motif finder, we may get a set of labeled data, design an HMM structure, train the HMM parameters on some of the data, then validate on the rest. If we find the HMM works poorly, we may try adding a new state to our model and repeat the procedure. If this new model does better in fitting the testing data, we may then report the accuracy of this new model on the testing data as a cross-validated assessment of the goodness of the model. Though this practice is common, it is not correct, because we no longer have independence between the model design process and the testing data. As modelers, we must be careful of this danger not only in our own work, but also when reading the work of others. Insufficient cross-validation is one of the most consistent problems one will find when reading computational biology literature critically. This is especially true for problems for which there are widely accepted benchmark data sets available.

24.4   Sensitivity Analysis

When we are dealing with models of many parameters, it is often not enough to say whether a single best-fit parameter set is good or bad. We also need to know whether that parameter set is robust. If our measurements or calculations have a slight bias to them, will the predictions of our model also have a slight bias, or will they be drastically off? We want to know how confident we can be in particular parameter choices and how much the model's behavior depends on the accuracy of those choices. The problem of establishing these properties of a model is called *sensitivity analysis*. Like all of the topics in this final chapter, sensitivity analysis is a broad subject for which many theoretical tools and analysis methods have been developed. Here, we will touch on only the basic principles behind some of the major approaches.

For some model types, sensitivity analysis is something one can perform with empirical experiments or even analytically. For example, suppose we fit a rate constant k to a kinetic rate equation

$$\frac{d[A]}{dt} = -k[A],$$

and ask how well this model predicts the concentration of [A] at some time τ. We can analytically solve for the model as follows:

$$[A](t) = [A](0)e^{-k\tau}.$$

We can then directly determine the effects of a perturbation Δk to the inferred rate constant:

$$\frac{[A](t, k + \Delta k) - [A](t, k)}{[A](t, k)} = \frac{[A](0)(e^{-(k+\Delta k)\tau} - e^{-k\tau})}{[A](0)e^{-k\tau}} = (e^{-\Delta k\tau} - 1).$$

For small τ, the fractional error in the model will be very small. If τ is large, though, the model may be very sensitive to errors. More often, one must do such an analysis empirically, simulating the system with various parameter changes to estimate how much each affects the outputs of the model.

As we saw when looking at gene network inference, we can sometimes use sampling methods to study parameter sensitivity. If we are using a probabilistic class of models for which we can assign some likelihood to each model, then we can create a sampler for the distribution of possible models weighted by their posterior probabilities. The distribution of a given parameter over the full posterior distribution then allows us to establish a best guess and a confidence interval for that parameter. We can similarly use a variant of the bootstrapping method, mentioned above, to establish ranges of parameter values by refitting the model to many random subsets of the data. Parameters that fit in a narrow range under repeated samples are likely to be robust, whereas those that vary widely are likely to be sensitive to input values. Conversely, parameters that fit nonspecifically are likely to be parameters to which the model is not very sensitive, and those that fit to a narrow range are likely to be parameters to which the model is sensitive.

24.5 Modeling and the Scientific Method

We started this text with a discussion about the philosophy of modeling, so it is fitting that we end it by coming back to some more philosophical questions about what a modeler does and how we can think about the problem of modeling. There is an entire field of study devoted to the philosophy of science which is concerned with asking many related questions. There has not traditionally been much practical feedback between philosophers of science and scientists. But a lot of what philosophers of science discuss as an intellectual exercise is actually useful to those of us who practice science, especially when working on problems of model-building. It is therefore worth closing with some considerations of basic issues in the philosophy of science.

Most readers have probably at some point seen the scientific method:

1. Observe
2. Hypothesize
3. Test
4. Theorize.

Philosophers of science have declared the scientific method to be overly simplistic, though. To understand why, let us consider a famous problem of science: understanding the laws of gravity. At one point, various scientists formulated models of different systems we now know to be guided by gravity. For example, Galileo developed models for predicting the motion of falling bodies near the Earth, and Kepler developed rules to describe the motion of objects in space. Later, Newton showed that Galileo's and Kepler's models were special cases of a universal gravitational law. Still later, Einstein showed that Newton's laws were only an approximation to some even more general laws that apply under conditions we cannot easily observe. And most physicists think that there is some more general theory waiting to be discovered that will reconcile theories of gravity with quantum mechanics.

This history can lead us to pose some questions:

• If Newton's unified law of gravity made the same predictions as Galileo's and Kepler's laws collectively, then was Newton's model an improvement over the prior work?
• Did contradictions between Einstein's predictions and Newton's show that Newton's model was wrong?
• Will we ever be able to conclude that Einstein's model, or the quantum gravity model that replaces it, is in any objective sense good?

Thinking of science in terms of the traditional scientific method provides us no way of answering the above questions. For example, the scientific method gives us no way to decide between two theories that make the same predictions (e.g., Galileo+Kepler vs. Newton). Philosophers have proposed a principle called Ockham's Razor for this problem, which essentially says that the simpler model is the better one. This principle provides a rationale for favoring Newton's model over the collective work of Galileo and Kepler. Yet it is not clear why we should accept Ockham's Razor in the first place. The scientific method also does not give us a way to conclude that a model is right. We can say that Einstein's model has led to the rejection of Newton's, but we do not know if Einstein's model will also be rejected when quantum gravity comes along. This notion that we can falsify models but not validate them is associated with the philosopher of science Karl Popper.

And even that is considered simplistic by more modern philosophers of science. We live in an imperfect world, and can never really be completely sure of our data or our analysis of them. According to legend, Galileo showed that all objects fall at the same rate by dropping two different sizes of cannonballs from a tower and observing that they hit the ground at the same time. Historians doubt whether Galileo actually conducted the experiment. If he had, we would actually expect it to have failed, since air resistance will have different effects on two cannonballs unless the experiment is conducted in a vacuum. But suppose we ignore that, and imagine Galileo did conduct this experiment. If he found that the time for the first cannonball to fall was 10 seconds and for the second was 11 seconds, should he have rejected the theory? What if it had been 10.0 seconds versus 10.1? Or 10.0000 vs. 10.0001? And even if he had conducted the experiment and found the same time, could anyone else be sure that his procedure was sound and that he was reporting his results accurately?

Some philosophers of science have concluded from this kind of reasoning that all scientific progress is an illusion, simply an agreement among people who call themselves scientists to support one theory independent of any ground truth. This is known as *social constructivism*. Most adopt a more measured view, saying that science approximates some ground truth, but it is seen through a lens of measurement errors and unconscious biases. In this conception, science generally gets better over time at describing reality. It never exactly gets there, though, and we can never know how close it is. This kind of theory is most closely associated with the philosophers Ludwig Fleck and Thomas Kuhn.

Model-building is all about trying to ask these kinds of philosophical questions in a more rigorous way. We observe some system and want to create a model describing it, just as any scientist observes phenomena and wants to develop a theory describing them. In modeling, though, we explicitly recognize that we are not finding the truth, but rather an imperfect approximation to it. We just want our answer to be "good." We then have to ask:

- In what domains does our answer need to be good?
- By what metrics?
- Compared to what alternatives?

And we need to develop precise formalisms by which we can answer these questions.

We can use these intuitions to try to formulate a "modeler's method," a sort of corrected scientific method in which we explicitly recognize the limits of modeling:

1. Observe process X in domain S.
2. Build a model λ of X in S.

3. Compare λ to reality (or other models) over samples from S, using metric M.
4. Identify those domains $S' \subseteq S$ in which λ is the superior model by metric M.

We can then conclude that λ is the best model available to us when we are working in domains S' and when M is an appropriate metric of goodness. We do not need to have any pretense that λ is the "truth," or even the right model in all circumstances.

All this really is the scientific method, but with more precision about our assumptions. Thus, there is really no distinction between doing modeling and doing science, except that a modeler needs to be more aware of the limits of his or her theories. In short, all of science is model-building. And that, I think, is a good observation on which to end this discussion.

References and Further Study

This chapter, like chapter 1, is more general philosophy than detailed technical methods. References for the subject matter are therefore hard to come by. Though we considered a few common measures of goodness of fit, there is much more one can learn about that topic. The entire field of statistics is essentially the study of how to judge the goodness of fit of models. A general grounding in statistics, particularly with regard to statistics in experimental sciences, is therefore essential for pursuing advanced work in biological modeling. One may look to Wasserman [112] for an introductory treatment of general statistics. Campbell [226] provides a treatment of statistics specifically for biologists.

The concluding discussion on the philosophy of modeling is, to my knowledge, novel to this text. Related issues are, however, broadly discussed in the history and philosophy of science. Interested readers may refer to the classic texts by Fleck [227] and Kuhn [228] for historical treatment of these issues. Those looking to dig deeper may start with Kosso [229] or Rosenberg [230] for a general introduction to the history and philosophy of science.

References

[1] J. Felsenstein. *Inferring Phylogenies*. Sinauer, Sunderland, MA, 2004.

[2] D. Gusfield. *Algorithms on Strings, Trees, and Sequences*. Cambridge University Press, New York, 1997.

[3] C. Semple and M. Steel. *Phylogenetics*. Oxford University Press, New York, 2003.

[4] A. W. F. Edwards and L. L. Cavalli-Sforza. The reconstruction of evolution. *Annals of Human Genetics*, 27 : 105–106 (1963).

[5] H. Taketomi, Y. Ueda, and N. Gō. Studies on protein folding, unfolding and fluctuations by computer simulation. *International Journal of Peptide and Protein Research*, 7 : 445–459 (1975).

[6] H. S. Chan and K. A. Dill. Energy landscape and the collapse dynamics of homopolymers. *Journal of Chemical Physics*, 99 : 2116–2127 (1993).

[7] N. Metropolis, A. W. Rosenbluth, M. N. Rosenbluth, A. H. Teller, and E. Teller. Equation of state calculation by fast computing machines. *Journal of Chemical Physics*, 21 : 1087–1092 (1953).

[8] R. J. Folz and J. I. Gordon. Computer-assisted predictions of signal peptidase processing sites. *Biochemistry and Biophysics Research Communications*, 146 : 870–877 (1987).

[9] K.-C. Chou. Review: Prediction of HIV protease cleavage sites in proteins. *Analytical Biochemistry*, 233 : 1–14 (1996).

[10] P. Saxová, S. Buus, S. Brunak, and C. Keşmir. Predicting proteasomal cleavage sites: A comparison of available methods. *International Immunology*, 15(7) : 781–787 (July 2003).

[11] L. Excoffier and P. E. Smouse. Using allele frequencies and geographic subdivision to reconstruct gene trees within a species: Molecular variance parsiomony. *Genetics*, 136 : 343–359 (1994).

[12] H.-J. Bandelt, P. Forster, B. C. Sykes, and M. B. Richards. Mitochondrial portraits of human populations using median networks. *Genetics*, 141 : 743–753 (1995).

[13] H.-J. Bandelt, P. Forster, and A. Röhl. Median-joining networks for inferring intraspecific phylogenies. *Molecular Biology and Evolution*, 16(1) : 37–48 (1999).

[14] T. H. Cormen, C. E. Leiserson, R. L. Rivest, and C. Stein. *Introduction to Algorithms*, 2nd ed. MIT Press, Cambridge, MA, 2001.

[15] A. V. Aho, J. E. Hopcroft, J. D. Ullman. *The Design and Analysis of Computer Algorithms*. Addison-Wesley, Reading, MA, 1974.

[16] D. Knuth. *The Art of Computer Programming*, vol. 1, *Fundamental Algorithms*, 3rd ed. Addison-Wesley, Reading, MA, 1997.

[17] D. Kozen. *The Design and Analysis of Algorithms*. Springer, New York, 1992.

[18] J. B. Kruskal. On the shortest spanning subtree of a graph and the traveling salesman problem. *Proceedings of the American Mathematical Society*, 7 : 48–50 (1956).

[19] R. C. Prim. Shortest connection networks and some generalizations. *Bell System Technical Journal*, 36 : 1389–1401 (1957).

[20] E. W. Dijkstra. A note on two problems in connexion with graphs. *Numerische Mathematik*, 1:269–271 (1959).

[21] R. Bellman. On a routine problem. *Quarterly Journal of Applied Mathematics*, 16:87–90 (1958).

[22] L. R. Ford and D. R. Fulkerson. *Flows in Networks*. Princeton University Press, Princeton, NJ, 1962.

[23] R. W. Floyd. Algorithm 97 (shortest path). *Communications of the ACM*, 5(6):345 (1962).

[24] D. B. Johnson. Efficient algorithms for shortest paths in sparse networks. *Journal of the ACM*, 24(1):1–13 (1977).

[25] J. Edmonds and R. M. Karp. Theoretical improvements in the algorithmic efficiency for network flow problems. *Journal of the ACM*, 19:248–264 (1972).

[26] H. W. Kuhn. The Hungarian method for the assignment problem. *Naval Research Logistics Quarterly*, 2:83–97 (1955).

[27] H. N. Gabow. *Implementation of Algorithms for Maximum Matching on Non-bipartite Graphs*. Ph.D. thesis, Department of Computer Science, Stanford University, 1974.

[28] S. Needleman and C. Wunsch. A general method applicable to the search for similarities in the amino acid sequences of two proteins. *Journal of Molecular Biology*, 48:443–453 (1970).

[29] T. Smith and M. Waterman. Identification of common molecular subsequences. *Journal of Molecular Biology*, 147(1):195–197 (1981).

[30] P. Weiner. Linear pattern matching algorithms. *Proceedings of the 14th IEEE Symposium on Switching and Automata Theory*, pp. 1–11. IEEE Computer Society Press, Silver Spring, MD, 1973.

[31] E. Ukkonen. On-line construction of suffix trees. *Algorithmica*, 14(3):249–260 (1995).

[32] G. C. L. Johnson, L. Esposito, B. J. Barratt, A. Smith, J. Heward, G. Di Genova, H. Ueda, H. Cordell, I. Eaves, F. Dudbridge, R. C. Twells, F. Payne, W. Hughes, S. Nutland, H. Stevens, P. Carr, E. Tuomilehto-Wolf, J. Tuomilehto, S. C. Gough, D. G. Clayton, and J. A. Todd. Haplotype tagging for the identification of common disease genes. *Nature Genetics*, 29:233–237 (2001).

[33] M. R. Garey and D. S. Johnson. *Computers and Intractability: A Guide to the Theory of NP-Completeness*. W. H. Freeman, San Francisco, 1979.

[34] P. Crescenzi and V. Kann, eds. *A compendium of NP optimization problems*. 1998. http://www.nada.kth.se/~viggo/problemlist/compendium.html.

[35] G. Ausiello, P. Crescenzi, G. Gambosi, V. Kann, A. Marchetti-Spaccamela, and M. Protasi. *Complexity and Approximation*. Springer, New York, 1999.

[36] S. Cook. The complexity of theorem-proving procedures. *Journal of the ACM*, 18:4–18 (1971).

[37] R. M. Karp. Reducibility among combinatorial problems. In *Complexity of Computer Computations: Proceedings*, R. E. Miller anda J. W. Thatcher, eds., pp. 85–103. Plenum Press, New York, 1972.

[38] G. B. Dantzig, W. O. Blattner, and M. R. Rao. All shortest routes from a fixed origin in a graph. In *Theory of Graphs: International Symposium*, pp. 85–90. Gordon and Breach, New York, 1967.

[39] O. Goldschmidt and D. S. Hochbaum. Polynomial algorithm for the k-cut problem. In *Proceedings of the 29th Annual Symposium on the Foundations of Computer Science*, pp. 444–451. Association for Computing Machinery, White Plains, NY, 1988.

[40] M. Yannakakis. *The Node Deletion Problem for Hereditary Properties*. Technical report. Princeton University, Princeton, NJ: 1978.

[41] M. Yannakakis. Node- and edge-deletion NP-complete problems. In *Proceedings of the 10th Annual ACM Symposium on the Theory of Computing*, pp. 253–264. Association for Computing Machinery, San Diego, CA, 1978.

[42] J. M. Lewis. On the complexity of the maximum subgraph problem. In *Proceedings of the 10th Annual ACM Symposium on the Theory of Computing*, pp. 265–274. 1978.

[43] D. Maier. The complexity of some problems on subsequences and supersequences. *Journal of the ACM*, 25:322–336 (1978).

[44] D. Maier and J. A. Storer. *A Note on the Complexity of the Superstring Problem*. Technical report. Princeton University, Princeton, NJ: 1977.

[45] D. J. Rosenkrantz, R. E. Stearns, and P. M. Lewis. An analysis of several heuristics for the traveling salesman problem. *SIAM Journal on Computing*, 6 : 563–581 (1977).

[46] N. Christofides. *Worst-Case Analysis of a New Heuristic for the Traveling Salesman Problem*. Technical report. Carnegie Mellon University, Pittsburgh, PA, 1976.

[47] S. Arora. Polynomial time approximation scheme for Euclidean TSP and other geometric problems. In *Proceedings of the 37th Annual IEEE Symposium on Foundations of Computer Science*, pp. 2–11. IEEE Computer Society, Burlington, VT, 1996.

[48] B. Monien and E. Speckenmeyer. Ramsey numbers and an approximation algorithm for the vertex cover problem. *Acta Informatica*, 22 : 115–123 (1985).

[49] R. Bar-Yehuda and S. Even. A local-ratio theorem for approximating the weighted vertex cover problem. In *Analysis and Design of Algorithms for Combinatorial Problems*, G. Ausiello and M. Lucertiori, eds., pp. 27–46. North-Holland, Amsterdam, 1985.

[50] E. Halperin. Improved approximation algorithms for the vertex cover problem in graphs and hypergraphs. In *Proceedings of the 11th Annual ACM-SIAM Symposium on Discrete Algorithms*. Association for Computing Machinery, San Francisco, CA, 2000.

[51] S. Kirkpatrick, C. D. Gelatt, and M. P. Vecchi. Optimization by simulated annealing. *Science*, 4598 : 671–680 (May 13, 1983).

[52] V. Cerny. A thermodynamical approach to the travelling salesman problem: An efficient simulation algorithm. *Journal of Optimization Theory and Applications*, 45 : 41–51 (1985).

[53] N. A. Barricelli. Esempi numerici di processi di evoluzione. *Methodos*, 6 : 45–68 (1954).

[54] M. Sipser. *Introduction to the Theory of Computation*, 2nd ed. Thompson Course Technology, Boston, 2005.

[55] C. H. Papadimitriou. *Computational Complexity*. Addison-Wesley, Reading, MA, 1994.

[56] D. S. Hochbaum, ed. *Approximation Algorithms for NP-Hard Problems*. PWS, Boston, 1997.

[57] L. Stryer. *Biochemistry*, 4th ed. W. H. Freeman, New York, 1995.

[58] A. M. Maxam and W. Gilbert. A new method for sequencing DNA. *Proceedings of the National Academy of Sciences USA*, 74(2) : 560–564 (1977).

[59] F. Sanger and A. R. Coulson. A rapid method for determining sequences in DNA by primed synthesis with DNA polymerase. *Journal of Molecular Biology*, 94 : 441–448 (1975).

[60] L. M. Smith, J. Z. Sanders, R. J. Kaiser, P. Hughes, C. Dodd, C. R. Connell, C. Heiner, S. B. H. Kent, and L. E. Hood. Fluorescence detection in automated DNA sequence analysis. *Nature*, 321 : 674–679 (June 12, 1986).

[61] H. Drossman, J. A. Luckey, A. J. Kostichka, J. D'Cunha, and L. M. Smith. High-speed separations of DNA sequencing reactions by capillary electrophoresis. *Analytical Chemistry*, 62 : 900–903 (1990).

[62] W. Bender, P. Spierer, and D. S. Hogness. Chromosomal walking and jumping to isolate DNA from the Ace and rosy loci and the bithorax complex in *Drosophila melanogaster*. *Journal of Molecular Biology*, 168(1) : 17–33 (July 1983).

[63] International Human Genome Sequencing Consortium. Initial sequencing and analysis of the human genome. *Nature*, 409 : 860–921 (2001).

[64] P. Pevzner. *Computational Molecular Biology: An Algorithmic Approach*. MIT Press, Cambridge, MA, 2000.

[65] N. C. Jones and P. Pevzner. *An Introduction to Bioinformatics Algorithms*. MIT Press, Cambridge, MA, 2004.

[66] W. Bains and G. C. Smith. A novel method for nucleic acid sequence determination. *Journal of Theoretical Biology*, 135 : 303–307 (1988).

[67] E. Southern. United Kingdom Patent Application GB8810400. 1988.

[68] I. P. Lysov, V. L. Florent'ev, A. A. Khorlin, K. R. Khrapko, and V. V. Shik. Determination of the nucleotide sequence of DNA using hybridization with oligonucleotides. A new method. *Doklady Akademii Nauk SSSR*, Ser. Biol., 303 : 1508–1511 (1988).

[69] R. Drmanac, I. Labat, I. Brukner, and R. Crkvenjakov. Sequencing of megabase plus DNA by hybridization: Theory of the method. *Genomics*, 4:114–128 (1989).

[70] R. Drmanac et al. DNA sequence determination by hybridization: A strategy for efficient large-scale sequencing. *Science*, 260(5114):1649–1652 (1993).

[71] P. A. Pevzner, *l*-tuple sequencing: Computer analysis. *Journal of Biomolecular Structure and Dynamics*, 7(1):63–73 (1989).

[72] R. M. Idury and M. S. Waterman. A new algorithm for DNA sequence assembly. *Journal of Computational Biology*, 2(2):291–306 (1995).

[73] R. D. Fleischmann et al. Whole-genome random sequencing and assembly of *Haemophilus influenzae rd*. *Science*, 269(5223):496–498, 507–512 (1995).

[74] A. Edwards, H. Voss, P. Rice, A. Civitello, J. Stegemann, C. Schwager, J. Zimmerman, H. Erfle, C. T. Caskey, and W. Ansorge. Automated DNA sequencing of the human HPRT locus. *Genomics*, 6:593–608 (1990).

[75] J. L. Weber and E. W. Myers. Whole-genome shotgun sequencing. *Genome Research*, 7(5):401–409 (May 1997).

[76] E. W. Myers et al. A whole-genome assembly of *Drosophila*. *Science*, 287(5461):2196–2204 (March 24, 2000).

[77] D. Huson et al. Design of a compartmentalized shotgun assembler for the human genome. *Bioinformatics*, 17, S1:S132–S139 (2001).

[78] S. Batzoglou et al. ARACHNE: A whole-genome shotgun assembler. *Genome Research*, 12:177–189 (2002).

[79] M. Akeson, D. Branton, J. J. Kasianowicz, E. Brandin, and D. W. Deamer. Microsecond time-scale discrimination among polycytidylic acid, polyadenylic acid, and polyuridylic acid as homopolymers or as segments within single RNA molecules. *Biophysical Journal*, 77(6):3227–3233 (1999).

[80] D. W. Deamer and M. Akeson. Nanopores and nucleic acids: Prospects for ultrarapid sequencing. *Trends in Biotechnology*, 18(4):147–151 (2000).

[81] J. Shendure, R. D. Mitra, C. Varma, and G. M. Church. Advanced sequencing technologies: Methods and goals. *Nature Reviews Genetics*, 5:335–344 (May 2004).

[82] W. H. Press, S. A. Teukolsky, W. T. Vetterling, and B. P. Flannery. *Numerical Recipes in C: The Art of Scientific Computing*. Cambridge University Press, Cambridge, 1992.

[83] J. E. Dennis and R. B. Schnabel. *Numerical Methods for Unconstrained Optimization and Nonlinear Equations*. Society for Industrial and Applied Mathematics, Philadelphia, 2006.

[84] A. P. Ruszczyński. *Nonlinear Optimization*. Princeton University Press, Princeton, NJ, 2006.

[85] R. Fletcher. *Practical Methods of Optimization*. Wiley, New York, 2000.

[86] L. N. Trefethen and D. Bau. *Numerical Linear Algebra*. Society for Industrial and Applied Mathematics, Philadelphia, 1997.

[87] *Wikipedia: The Free Encyclopedia*. http://www.wikipedia.org.

[88] I. Newton. *De Analysi per Aequationes Numero Terminorum Infinitas*. William Jones, London, 1711.

[89] J. Wallis. *A Treatise of Algebra Both Historical and Practical*. J. Playford for R. Davis, London, 1685.

[90] J. Raphson. *Analysis Aequationum Universalis*. A. and I. Churchill, London, 1690.

[91] M. R. Hestenes and E. Stiefel. Method of conjugate gradients for solving linear systems. *Journal of Research of the National Bureau of Standards*, 49:409–436 (1952).

[92] K. Levenberg. A method for the solution of certain non-linear problems in least squares. *Quarterly of Applied Mathematics*, 2:164–168 (1944).

[93] D. Marquardt. An algorithm for least-squares estimation of nonlinear parameters. *SIAM Journal of Applied Mathematics*, 11:431–441 (1963).

[94] S. G. Nash and A. Sofer. *Linear and Nonlinear Programming*. McGraw-Hill, New York, 1996.

[95] G. Dantzig. Maximization of a linear function of variables subject to linear inequalities. In Cowles Commission for Research in Economics, *Activity Analysis of Production and Allocation*. Wiley, New York, 1951.

[96] L. G. Khachiyan. A polynomial algorithm in linear programming. *Doklady Akademiia Nauk SSSR*, 244 : 1093–1096 (1979).

[97] A. S. Nemirovskii and D. B. Iudin. Optimization methods adapting to the "significant" dimension of the problem. *Automatika i Telemekhanika*, 38 : 75–87 (1977).

[98] N. Karmarkar. A new polynomial time algorithm for linear programming. *Combinatorica*, 4 : 373–395 (1984).

[99] *COmputational INfrastructure for Operations Research (COIN-OR)*. http://www.coin-or.org/.

[100] M. Berkelaar, J. Dirks, K. Eikland, and P. Notebaert. *lp_solve*. http://lpsolve.sourceforge.net.

[101] A. Makhorin. *GLPK (GNU Linear Programming Kit)*. http://www.gnu.org/software/glpk.

[102] S. M. Ross. *Introduction to Probability Models*, 8th ed. Academic Press, San Diego, 2003.

[103] S. M. Ross. *Simulation*, 3rd ed. Academic Press, San Diego, 2002.

[104] D. Knuth. *The Art of Computer Programming*. Addison-Wesley, Reading, MA, 1981.

[105] R. Eckhardt. Stan Ulam, John von Neumann, and the Monte Carlo method. *Los Alamos Science*, 15 : 131–137 (1987).

[106] J. von Neumann. Various techniques used in connection with random digits. In *Monte Carlo Methods*, pp. 36–38. Applied Math Series, 12. National Bureau of Standards, Washington, DC, 1951.

[107] G. E. P. Box and M. E. Müller. A note on the generation of random normal deviates. *Annals of Mathematical Statistics*, 29 : 610–611 (1958).

[108] Y. A. Rozanov. *Probability Theory: A Concise Course*. Dover, New York, 1977.

[109] D. T. Gillespie. *Markov Processes*. Academic Press, Boston, 1992.

[110] D. J. Wilkinson. *Stochastic Modelling for Systems Biology*, Chapman & Hall/CRC, Boca Raton, FL, 2006.

[111] A. A. Markov. Extension of the law of large numbers to dependent quantities. *Izvestia Fiziko-Matematicheskikh Obschestva Kazan University*, 15 : 135–136 (1906).

[112] L. Wasserman. *All of Statistics*. Springer, New York, 2004.

[113] S. Geman and D. Geman. Stochastic relaxation, Gibbs distribution, and the Bayesian restoration of images. *IEEE Transactions on Pattern Analysis and Machine Intelligence*, 6 : 721–741 (1984).

[114] G. M. Torrie and J. P. Valleau. Monte-Carlo free energy estimates using non-Boltzmann sampling. *Chemical Physics Letters*, 28 : 578–581 (1974).

[115] A. J. Sinclair and M. R. Jerrum. Approximate counting, uniform generation and rapidly mixing Markov chains. *Information and Computation*, 82(1) : 93–133 (1989).

[116] A. Sinclair. Improved bounds for mixing rates of Markov chains and multicommodity flow. *Combinatorics, Probability and Computing*, 1 : 351–370 (1992).

[117] S. M. Ross. *Stochastic Processes*. Wiley, New York, 1983.

[118] J. L. Doob. *Stochastic Processes*. Wiley, New York, 1953.

[119] D. R. Cox and H. D. Miller. *The Theory of Stochastic Processes*. Wiley, New York, 1965.

[120] S. Karlin. *A First Course in Stochastic Processes*. Academic Press, New York, 1968.

[121] S. I. Resnick. *Adventures in Stochastic Processes*. Birkhäuser, Boston, 1992.

[122] N. G. van Kampen. *Stochastic Processes in Physics and Chemistry*. North-Holland, New York, 1981.

[123] G. B. Benedek and F. M. H. Villars. *Physics with Illustrative Examples from Medicine and Biology: Statistical Physics*, 2nd ed. Springer, New York, 2000.

[124] D. Graur and W.-H. Li. *Fundamentals of Molecular Evolution*, 2nd ed. Sinauer, Sunderland, MA, 2000.

[125] M. Nordborg. Coalescent theory. In *Handbook of Statistical Genetics*, pp. 179–212. Wiley, Chichester, UK, 2001.

[126] D. L. Hartl and A. G. Clark. *Principles of Population Genetics*, 3rd ed. Sinauer, Sunderland, MA, 1997.

[127] T. H. Jukes and C. R. Cantor. Evolution of protein molecules. In *Mammalian Protein Metabolism*, vol. 3, pp. 21–123. Academic Press, New York, 1969.

[128] M. Kimura. A simple method for estimating evolutionary rate of base substitutions through comparative studies of nucleotide sequences. *Journal of Molecular Evolution*, 16:111–120 (1980).

[129] J. F. C. Kingman. The coalescent. *Stochastic Processes and Applications*, 13:235–248 (1982).

[130] D. C. Rapaport. The event scheduling problem in molecular dynamics simulation. *Journal of Computational Physics*, 32:184–201 (1980).

[131] R. Brown. Calendar queues: A fast $O(1)$ priority queue implementation for the simulation event set problem. *Communications of the ACM*, 31(10):1220–1227 (1998).

[132] J. Stoer and R. Bulirsch. *Introduction to Numerical Analysis*, 2nd ed. Springer, New York, 1996.

[133] P. K. Kythe and M. R. Schäferkotter. *Handbook of Computational Methods for Integration*. Chapman & Hall/CRC, Boca Raton, FL, 2004.

[134] R. W. Hamming. *Numerical Methods for Scientists and Engineers*, 2nd ed. Dover, New York, 1986.

[135] R. J. LeVeque. *Numerical Methods for Conservation Laws*, 2nd ed. Birkhäuser Boston, 1992.

[136] D. Gottlieb and S. A. Orszag. *Numerical Analysis of Spectral Methods: Theory and Applications*. Society for Industrial and Applied Mathematics, Philadelphia, 1977.

[137] J. Strickwerda. *Finite Difference Schemes and Partial Differential Equations*. Wordsworth & Brooks, Pacific Grove, CA, 1989.

[138] A. Turing. The chemical basis of morphogenesis. *Philosophical Transactions of the Royal Society of London*, B237:37–72 (1952).

[139] D. J. Higham. An algorithmic introduction to numerical simulation of stochastic differential equations. *SIAM Review*, 43(3):525–546 (2001).

[140] P. E. Kloeden and E. Platen. *Numerical Solution of Stochastic Differential Equations*. Springer, Berlin, 1999.

[141] P. E. Protter. *Stochastic Integration and Differential Equations*. Springer, New York, 2004.

[142] D. Lamberton and B. Lapeyre. *Introduction to Stochastic Calculus Applied to Finance*. CRC Press, Boca Raton, FL, 2000.

[143] B. Øksendal. *Stochastic Differential Equations: An Introduction with Applications*, 6th ed. Springer, New York, 2005.

[144] P. Mendes. GEPASI: A software package for modeling the dynamics, steady states and control of biochemical and other systems. *Computer Applications in the Biosciences [now Bioinformatics]*, 9:563–571 (1993).

[145] M. Tomita, K. Hashimoto, K. Takahashi, T. Simon Shimizu, Y. Matsuzaki, F. Miyoshi, K. Saito, S. Tanida, K. Yugi, J. C. Venter, and C. A. Hutchinson. E-cell: Software environment for whole-cell simulation. *Bioinformatics*, 15:72–84 (1999).

[146] P. Mendes. Biochemistry by numbers: Simulation of biochemical pathways with Gepasi 3. *Trends in Biochemical Sciences*, 22:361–363 (1997).

[147] J. Schaff, C. C. Fink, B. Slepchenko, J. H. Carson, and L. M. Loew. A general computational framework for modeling cellular structure and function. *Biophysical Journal*, 73:1135–1146 (September 1997).

[148] P. Mendes and D. B. Kell. MEG (Model Extender for GEPASI): A program for the modelling of complex, heterogeneous, cellular systems. *Bioinformatics*, 17(3):288–289 (2001).

[149] D. T. Gillespie. A general method for numerically simulating the stochastic time evolution of coupled chemical reactions. *Journal of Computational Physics*, 22(4):403–434 (1976).

[150] N. Le Novère and T. S. Shimizu. StochSim: Modelling of stochastic biomolecular processes. *Bioinformatics*, 17(6): 575–576 (2001).

[151] M. A. Gibson and J. Bruck. Efficient exact stochastic simulation of chemical systems with many species and channels. *Journal of Physical Chemistry*, A104: 1876–1889 (2000).

[152] Y. Cao, H. Li, and L. Petzold. Efficient formulation of the stochastic simulation algorithm for chemically reacting systems. *Journal of Chemical Physics*, 121(9): 4059–4067 (September 2004).

[153] F. Jamalyaria, R. Rohlfs, and R. Schwartz. Queue-based method for efficient simulation of biological self-assembly systems. *Journal of Computational Physics*, 204: 100–120 (2005).

[154] J. R. Stiles, D. Van Helden, T. M. Bartol, Jr., E. E. Salpeter, and M. M. Salpeter. Miniature endplate current rise times < 100 μs from improved dual recordings can be modeled with passive acetylcholine diffusion from a synaptic vesicle. *Proceedings of the National Academy of Sciences USA*, 93: 5747–5752 (June 1996).

[155] J. R. Stiles and T. M. Bartol, Jr. Monte Carlo methods for simulating realistic synaptic microphysiology using MCell. In *Computational Neuroscience: Realistic Modeling for Experimentalists*, pp. 87–127. CRC Press, Boca Raton, FL, 2001.

[156] A. B. Stundzia and C. J. Lumsden. Stochastic simulation of coupled reaction-diffusion processes. *Journal of Computational Physics*, 127: 196–207 (1996).

[157] J. Elf, A. Dončić, M. Ehrenberg. Mesoscopic reaction-diffusion in intracellular signaling. *Proceedings of SPIE*, 5510: 114–124 (2003).

[158] M. Ander, P. Beltrao, B. Di Ventura, J. Ferkinghoff-Borg, M. Foglierini, A. Kaplan, C. Lemerle, I. Tomas-Oliveira, and L. Serrano. SmartCell, a framework to simulate cellular processes that combines stochastic approximation with diffusion and localisation: Analysis of simple networks. *IEEE Systems Biology*, 1: 129–138 (2004).

[159] D. T. Gillespie. Approximate accelerated stochastic simulation of chemically reacting systems. *Journal of Chemical Physics*, 115(4): 1716–1733 (July 22, 2001).

[160] S. Hoops, S. Sahle, R. Gauges, C. Lee, J. Pahle, N. Simus, M. Singhal, L. Xu, P. Mendes, and U. Kummer. COPASI—a COmplex PAthway SImulator. *Bioinformatics*, 22: 3067–3074 (2006).

[161] K. Takahashi, K. Kaizu, B. Hu, M. Tomita. A multi-algorithm, multi-timescale method for cell simulation. *Bioinformatics*, 20(4): 538–546 (2004).

[162] M. L. Blinov, J. R. Faeder, B. Goldstein, and W. S. Hlavacek. BioNetGen: Software for rule-based modeling of signal transduction based on the interaction of molecular domains. *Bioinformatics*, 20: 3289–3291 (June 24, 2004).

[163] L. Lok and R. Brent. Automatic generation of cellular reaction networks with Moleculizer 1.0. *Nature Biotechnology*, 23: 131–136 (2005).

[164] T. Zhang, R. Rohlfs, and R. Schwartz. Implementation of a discrete event simulator for biological self-assembly. In *Proceedings of the Winter Simulation Conference*, pp. 2223–2231. Institute for Operations Research and the Management Sciences, Orlando, FL, 2005.

[165] D. L. Ermak. A computer simulation of charged particles in solution. I. Technique and equilibrium properties. *Journal of Chemical Physics*, 62(10): 4189–4196 (1975).

[166] J. S. van Zon and P. R. ten Wolde. Simulating biochemical networks at the particle level and in time and space: Green's function reaction dynamics. *Physical Review Letters*, 94: 128103 (2005).

[167] M. Hucka, A. Finney, H. M. Sauro, H. Bolouri, J. C. Doyle, H. Kitano, et al. The Systems Biology Markup Language (SBML): A medium for representation and exchange of biochemical network models. *Bioinformatics*, 19(4): 524–531 (2003).

[168] M. Hucka, A. Finney, S. Hoops, S. Keating, and N. Le Novère. *Systems Biology Markup Language (SBML) level 2: Structures and facilities for model definitions.* 2007. http://www.sbml.org.

[169] A. A. Cuellar, P. F. Nielsen, D. P. Bullivant, D. Nickerson, W. Hedley, M. Nelson, and C. M. Lloyd. *CellML specification 1.1. draft.* 2003. http://www.cellml.org.

[170] C. N. Lloyd, M. D. B. Halstead, and P. F. Nielsen. CellML: Its future, present, and past. *Progress in Biophysics and Molecular Biology*, 85(2–3): 433–450 (June–July 2004).

[171] *E-cell.* 2007. http://www.e-cell.org.

[172] *National Resource for Cell Analysis and Modeling.* 2007. http://www.vcell.org.

[173] *COPASI: Complex Pathway Simulator.* 2007. http://www.copasi.org/tiki-index.php.

[174] Center for Quantitative Biological Simulation. *Microphysiology Gateway.* 2007. http://www.mcell.psc.edu.

[175] C. P. Fall, E. S. Marland, J. M. Wagner, and J. T. Tyson, eds. *Computational Cell Biology.* Springer, New York, 2002.

[176] T. Niu, Z. S. Qin, X. Xu, and J. S. Liu. Bayesian haplotype inference for multiple linked single-nucleotide polymorphisms. *American Journal of Human Genetics,* 70 : 157–169 (2002).

[177] T. M. Mitchell. *Machine Learning.* McGraw-Hill, New York, 1997.

[178] L. E. Baum, T. Petries, G. Soules, and N. Weiss. A maximization technique occurring in the statistical analysis of probabilistic functions of Markov chains. *Annals of Mathematical Statistics,* 41 : 164–171 (1970).

[179] A. P. Dempster, N. M. Laird, and D. B. Rubin. Maximum likelihood from incomplete data via the EM algorithm. *Journal of the Royal Statistical Society,* B39 : 1–38 (1977).

[180] E. L. L. Sonnhammer, S. R. Eddy, and R. Durbin. Pfam: A comprehensive database of protein domain families based on seed alignments. *Proteins,* 28 : 405–420 (1997).

[181] L. R. Rabiner. A tutorial on hidden Markov models and selected applications in speech recognition. *Proceedings of the IEEE,* 77 : 257–286 (February 1989).

[182] R. Durbin, S. Eddy, A. Krogh, and G. Mitchison. *Biological Sequence Analysis.* Cambridge University Press, Cambridge, 2003.

[183] D. W. Mount. *Bioinformatics: Sequence and Genome Analysis.* Cold Spring Harbor Laboratory Press, Cold Spring Harbor, NY, 2001.

[184] A. J. Viterbi. Error bounds for convolutional codes and an asymptotically optimum decoding algorithm. *IEEE Transactions on Information Theory,* 13(2) : 260–269 (1967).

[185] L. E. Baum and J. A. Eagon. An inequality with applications to statistical estimation for probabilistic functions of Markov processes and to a model for ecology. *Bulletin of the American Meterological Society,* 73 : 360–363 (1967).

[186] C. Burge and S. Karlin. Prediction of complete gene structures in human genomic DNA. *Journal of Molecular Biology,* 268 : 78–94 (1997).

[187] M. Delorenzi and T. Speed. An HMM model for coiled-coil domains and a comparison with PSSM-based predictions. *Bioinformatics,* 8(4) : 617–625 (2002).

[188] A. Krogh, M. Brown, I. S. Mian, K. Sjölander, and D. Haussler. Hidden Markov models in computational biology: Applications to protein modeling. *Journal of Molecular Biology,* 235 : 1501–1531 (1994).

[189] S. R. Eddy. Hidden Markov models. *Current Opinion in Structural Biology,* 6 : 361–365 (1996).

[190] S. R. Eddy. Profile hidden Markov models. *Bioinformatics,* 14 : 755–763 (1998).

[191] S. R. Eddy. What is a hidden Markov model? *Nature Biotechnology,* 22 : 1315–1316 (2004).

[192] G. Strang. *Linear Algebra and Its Applications,* 3rd ed. Harcourt Brace Jovanovich, San Diego, 1988.

[193] Y. Saad. *Iterative Methods for Sparse Linear Systems.* PWS, Boston, 1996.

[194] P. Kuzmic. Program DYNAFIT for the analysis of enzyme kinetic data: Application to HIV protease. *Analytical Biochemistry,* 237 : 260–273 (1996).

[195] D. F. Walnut. *An Introduction to Wavelet Analysis.* Birkhäuser, Boston, 2004.

[196] J. S. Walker. *A Primer on Wavelets and Their Scientific Applications.* Chapman & Hall/CRC, Boca Raton, FL, 1999.

[197] M. B. Eisen, P. T. Spellman, P. O. Brown, and D. Botstein. Cluster analysis and display of genome-wide expression patterns. *Proceedings of the National Academy of Sciences USA,* 95(25) : 14863–14868 (December 8, 1998).

[198] U. Alon, N. Barkai, D. A. Notterman, K. Gish, S. Ybarra, D. Mack, and A. J. Levine. Broad patterns of gene expression revealed by clustering analysis of tumor and normal colon tissues probed by oligonucleotide arrays. *Proceedings of the National Academy of Sciences USA*, 96 : 6745–6750 (June 1999).

[199] E. Mjolsness, T. Mann, R. Castaño, and B. Wold. *From Coexpression to Coregulation: An Approach to Inferring Transcriptional Regulation Among Gene Classes from Large-Scale Expression Data*. Technical report. Jet Propulsion Laboratory, Pasadena, CA, 1999.

[200] J. MacQueen. Some methods for classification and analysis of multivariate observation. In *Proceedings of the Fifth Berkeley Symposium on Mathematical Statistics and Probability*, vol. 1, pp. 281–297. University of California Press, Berkeley, 1967.

[201] T. Kohonen. The self-organizing map. *Proceedings of the IEEE*, 78 : 1464–1480 (1990).

[202] P. Tamayo, D. Slonim, J. Mesirov, Q. Zhu, S. Kitareewan, E. Dmitrovsky, E. S. Lander, and T. R. Golub. Interpreting patterns of gene expression with self-organizing maps: Methods and application to hematopoietic differentiation. *Proceedings of the National Academy of Sciences USA*, 96 : 2907–2912 (March 16, 1999).

[203] N. Friedman, M. Linial, I. Nachman, and D. Pe'er. Using Bayesian networks to analyze expression data. *Journal of Computational Biology*, 7 : 601–620 (2000).

[204] J. Yu, V. A. Smith, P. P. Wang, A. J. Hartemink, and E. D. Jarvis. Advances to Bayesian network inference for generating causal networks from observational biological data. *Bioinformatics*, 20 : 3594–3603 (2004).

[205] D. Madigan and J. York. Bayesian graphical models for discrete data. *International Statistical Review*, 63 : 215–232 (1995).

[206] B. Efron and R. J. Tibshirani. *An Introduction to the Bootstrap*. Chapman & Hall, New York, 1993.

[207] N. Friedman, M. Goldszmidt, and A. Wyner. Data analysis with Bayesian networks: A bootstrap approach. In *Proceedings of the Fifteenth Conference on Uncertainty in Artificial Intelligence (UAI)*, pp. 206–215. Morgan Kaufmann, San Francisco, 1999.

[208] N. Friedman, I. Nachman, and D. Pe'er. Learning Bayesian network structure from massive datasets: The "sparse candidate" algorithm. In *Proceedings of the Fifteenth Conference on Uncertainty in Artificial Intelligence (UAI)*, pp. 196–205. Morgan Kaufmann, San Francisco, 1999.

[209] S. Tavazoie, J. D. Hughes, M. J. Campbell, R. J. Cho, and G. M. Church. Systematic determination of genetic network architecture. *Nature Genetics*, 22 : 281–285 (1999).

[210] T. Ideker, O. Ozier, B. Schwikowski, and A. F. Siegel. Discovering regulatory and signalling circuits in molecular interaction networks. *Bioinformatics*, 18(supp. 12002) : S233–S240 (2002).

[211] Z. Tu, L. Wang, M. N. Arbeitman, T. Chen, and F. Sun. An integrative approach for causal gene identification and gene regulatory pathway inference. *Bioinformatics*, 22(142006) : e489–e496 (2006).

[212] T. S. Gardner, D. di Bernardo, D. Lorenz, and J. J. Collins. Inferring genetic networks and identifying compound mode of action via expression profiling. *Science*, 301(5629) : 102–105 (July 4, 2003).

[213] D. di Bernardo, M. J. Thompson, T. S. Gardner, S. E. Chobot, E. L. Eastwood, A. P. Wojtovich, S. J. Elliott, S. E. Schaus, and J. J. Collins. Chemogenomic profiling on a genome-wide scale using reverse-engineered gene networks. *Nature Biotechnology*, 23(3) : 377–383 (March 3, 2005).

[214] I. Nachman, A. Regev, and N. Friedman. Inferring quantitative models of regulatory networks from expression data. *Bioinformatics*, 20(supp. 1) : i248–i256 (2004).

[215] P. D'haeseleer, S. Liang, and R. Somogyi. Genetic network inference: From co-expression clustering to reverse engineering. *Bioinformatics*, 16(8) : 707–726 (2000).

[216] H. De Jong. Modeling and simulation of genetic regulatory systems: A review. *Journal of Computational Biology*, 9 : 67–103 (2002).

[217] M. Bansal, V. Belcastro, A. Ambesi-Impiombato, and D. di Bernardo. How to infer gene networks from expression profiles. *Molecular Systems Biology*, 3(78) : 1–10 (2007).

[218] H. Causton, J. Quackenbush, and A. Brazma. *Microarray Gene Expressions Data Analysis: A Beginner's Guide*. Blackwell Science, Malden, MA, 2003.

[219] D. P. Berrar, W. Dubitzky, and M. Granzow, eds. *A Practical Approach to Microarray Data Analysis*. Kluwer Academic, Boston, 2003.

[220] A. Zhang. *Advanced Analysis of Gene Expression Microarray Data*. World Scientific, Singapore, 2006.

[221] P. D'haeseleer. How does gene expression clustering work? *Nature Biotechnology*, 23: 1499–1501 (2005).

[222] A. K. Jain, M. N. Murty, and P. J. Flynn. Data clustering: A review. *ACM Computing Surveys*, 31(3): 264–323 (1999).

[223] P. Congdon. *Applied Bayesian Modelling*. Wiley, Chichester, UK, 2003.

[224] A. Gelman, J. B. Carlin, H. S. Stern, and D. B. Rubin. *Bayesian Data Analysis*. Chapman & Hall/ CRC Press, Boca Raton, FL, 2003.

[225] R. E. Neapolitan. *Learning Bayesian Networks*. Pearson Prentice-Hall, Upper Saddle River, NJ, 2004.

[226] R. C. Campbell. *Statistics for Biologists*, 3rd ed. Cambridge University Press, New York, 1989.

[227] L. Fleck. *Genesis and Development of a Scientific Fact*. University of Chicago Press, Chicago, 1979.

[228] T. Kuhn. *The Structure of Scientific Revolutions*. University of Chicago Press, Chicago, 1962.

[229] P. Kosso. *Reading the Book of Nature: An Introduction to the Philosophy of Science*. Cambridge University Press, New York, 1992.

[230] A. Rosenberg. *The Philosophy of Science: A Contemporary Introduction*. Routledge, New York, 2000.

Index

Acceleration, 211
Accuracy. *See also* Errors
  in Adams-Bashforth methods, 223
  adaptive methods, 224
  centered difference, 84–85, 229–232, 235
  first and second order, 157, 215, 218, 233, 238
  and forward/backward Euler, 218, 222
  in model validation, 358
  of Neville's algorithm, 327
  and Newton-Raphson, 84–85
  and partial differential equations, 233
  and Runge-Kutta methods, 221, 225, 237
  of secant *versus* bisection, 79
  and stability, 221, 223, 251
  and stochastic differential equations, 248–249
  and time, 219, 233
Adams-Bashforth methods, 221–223, 225
Adams-Moulton scheme, 225
Adaptive methods, 224–225
Affine method, 104–107, 110
Aitken's δ^2 process, 338–340
Alleles, 28–33, 280–289
All-pairs shortest path, 21
Amino acids. *See also* Protein folding
  contact energies, 5–7
  and HMMs, 297–299
  and Metropolis method, 145
  proline *cis-trans* isomerization, 180–182
  and proteases, 9
Animation, 333
Annealing. *See* Simulated annealing
Antibiotics, 45
Approximation
  centered difference, 84–85, 89, 229–232, 235
  and extrapolation, 337–340
  forward difference, 84
  and forward Euler, 214, 224
  and interpolation, 327–337
  and reaction-diffusion equations, 237
  for step size, 224
  with Taylor series, 80–81, 85, 232

Approximation algorithms. *See also* Traveling salesman
  and branch-and-bound algorithm, 50
  description, 47–59
  and intractability, 47–49, 50, 55
  reference, 55
  traveling salesman, 36, 48, 54
  and vertex cover, 47, 50–51
Approximation schemes, 48
Automated sequencing, 61–63

Backward algorithm, 298–299
Backward error, 77
Backward Euler, 217–219, 222
Bacteria
  antibiotic sensitivity, 45
  bacterial artificial chromosome (BAC), 64
Barrier methods, 104
Baum-Welch algorithm, 300–307
Bayesian models, 347–350, 353, 356
  additional sources, 353
Bellman-Ford algorithm, 20–21
  background, 33
Best-fit
  in interpolation, 336
  and least-squares, 356
  in parameter-tuning, 275
Bias. *See also* Model validation
  and gene network, 349
  and HMMs, 360
  and importance sampling, 154
  and parameter choices, 362
  unintended, 365
Biconjugate gradient, 319
Bilinear interpolation, 334
Billiard ball model, 206–209
Binary search, 338
Biochemical processes. *See also* Evolution; Reaction networks
  decaying exponentials, 335
  parameters, 267
  whole-cell models, 253–264

BioNetGen, 261
Biophysics, 226
Bipartiteness, 23, 42
Bisection, 76–78, 338
Black-box, 75, 84, 237, 336
Block diagonals, 333
Boltzmann distribution, 7, 142–144, 146
Boltzmann's constant, 7, 142
Bootstrapping, 350, 361, 363
Boundary conditions
  Dirichlet, 230–231
  for multiple dimensions, 234
  Neumann, 231
  and PDEs, 230–233
  and solute diffusion, 230–233
Box-Müller method, 120
  background, 127
Branch-and-bound methods, 49–52
Branching process, 199
Brownian motion, 167, 241–249, 263
Brownian noise, 157
Brute force, 47, 50, 53

Calcium, 260
Calendar queue, 205t, 209
Canonical path, 161–166, 169
  background, 171
Capillary sequencing, 61
Catalysts. See Enzymes
CellML, 264
Cells
  and biochemical networks, 253–264
  cycle synchronization, 323–325
Cell simulation
  and CTMM, 256–259
  electrophysiological components, 264
  hybrid models, 259, 263
  and PDEs, 253–256, 263
  protein expression, 268
  standards and software, 263
  trends, 262
  as very large reaction network, 260–262
Centered difference, 84–85, 229–232, 235
Chain rule, 220
Channel protein, 201–203
Chapman-Kolmogorov equations, 133
Chebyshev polynomials, 340
Chemical reaction. See also Reaction networks
  and interpolation, 336
  and law of mass action, 211
  with noise, 246–248
  and stability, 215–217
Chemical solutions. See Solutions
Chromatic number, 40
Chromosomes
  diploid, 198
  haploid, 192
  haplotypes, 280–286
  tagging SNP selection, 44, 47

Chromosome walking, 63
cis isomer, proline, 180–182
Cliques, 39, 342
  union of, 344–345
Clone-by-clone strategy, 63
Clustering, 342–347, 351
  additional sources, 353
Coalescent
  background, 200
  coalescent simulation, 195
  definition, 193f, 194
  and migration, 198
  and recombinations, 198
  separate populations, 197
  variable population sizes, 196
Coexpression models, 342–347, 351
Collisions, 141, 206–209
Coloring
  in automated sequencing, 61
  in graph problems, 39, 49–50
Compartments, 253–256
Complexity, computational, 55, 260–262, 361. See
  also Intractability; NP-completeness
Computer graphics, 333
Concave functions, 108
Conditional probability, 295
Condition number, of matrix, 319
Conductance method, 166
  background, 171
  bounded random walk, 167–170
Conjugate gradient, 91, 92, 318, 319
Consensus sequence, 32
Constraint satisfaction
  linear program, 96–108
  nonlinear program, 108–110
  parameter-tuning, 269–271
  primal-dual methods, 107
Contact potentials, 5, 267
Continuous distributions
  and importance/umbrella sampling, 154–156
  joint distributions, 119–121, 151–152
  rejection method, 121–124
  transformation method, 116–121, 124f
Continuous optimization. See also Newton-
  Raphson method
  bisection, 76–78, 338
  description, 75
  local versus global optima, 76
  multivariate functions, 85–88
  secant method, 78–80
Continuous systems
  applications, 211–213
  backward Euler, 217–219
  definition, 211
  differential equations, 212
  with discrete event tracking, 206–209, 263
  from discrete points, 323–326 (see also Extrapola-
    tion; Interpolation)
  finite difference, 213, 226

forward Euler (*see* Forward Euler)
leapfrog, 221–223, 225, 236
single-step methods, 219–221, 223–225
Continuous time Markov models (CTMMs)
additional reading, 183
branching process, 199
cell simulation, 256–260
channel protein example, 201–203
and coalescence, 195
description, 173–178
versus discrete event models, 201–204
and DNA base evolution, 187
Kolmogorov equations, 178–182
Moleculizer program, 261
and population dynamics, 212
and protein folding, 180–182
rate inference, 273
and self-transition, 181
waiting time, 173–175
Convection, 237–239
Convection-diffusion, 238
Convergence, 338
order of, 248–249
Convex functions, 108–110
Cooling schedule, 148
COPASI, 260, 264
Correlation coefficients, 343, 356–358
Cross-validation, 361–362
CTMM. *See* Continuous time Markov models
Cubic formula polynomials, 76
Cubic formulas, 76, 329, 333
Curve, receiver operating characteristic (ROC), 360
Curve families, 334–337
Curve generation, 333
Curve linearization, 81, 86, 89, 91
Cut problems
k-cut, 38, 54, 344
maximum cut, 37, 344
minimum cut, 21–23

Data. *See also* Noisy data
ambiguity loss, 30
and Bayesian model, 347
and continuous optimization, 75
fitting, 329–336, 340, 361
gene expression microarray, 341
gene network inference, 352
for HMMs, 299–302
input and output format, 2–3
for intraspecies phylogeny, 29–30
posting time, 205
set relationships, 357
Decision problems, 36
Density
joint, 119
probability, 116–118, 121–122, 154
detailed balance, 143–145, 164, 169
Diagnostics, 358, 360

Differential equations. *See* Finite difference;
Ordinary differential equations; Partial differential equations; Stochastic differential equations
Diffusion
and boundaries, 230–233
and cell simulation, 259
convection-diffusion equation, 238
of particles, in two dimensions, 325
PDE example, 227
reaction-diffusion equations, 234–237, 325
Diffusion term, 234
Dijkstra's algorithm, 20, 21
background, 33
Diploid organisms, 198
Dirichlet boundary, 230–231
Discrete distributions. *See also* Transformation method
and continuous models, 323–326
and Metropolis method, 146
rejection method, 124–126
and transformation method, 124
Discrete event models
artificial event, 208
background, 210
and cell simulation, 260
channel protein case, 201–203
and continuous systems, 206–209, 263, 325
versus CTMMs, 201–204
description, 203
efficiency, 204–206, 208–210
event loop, 204, 207
molecular collisions, 206–209
queuing, 205, 209–210
without queue, 208
Discretization
conversions (multigrid), 325
and gene coexpression, 344
of space, 229, 233, 235, 255, 258
of time, 242
Disease, diagnosis of, 358, 360
Distributions. *See also* Continuous distributions;
Discrete distributions
Boltzmann, 142–144, 146
exponential, 118
gamma, 268
Gaussian, 348
joint, 119–121, 149–152
modified, 156
normal, 120, 123–124
Poisson, 191
prior, 153, 349
probability, 347–349
stationary, 134–138, 149, 153–155, 159, 161
uniform, 115–116
DNA. *See also* String and sequence problems
diploid and haploid, 198
exact set matching, 27
intraspecies phylogeny, 28–33

DNA (cont.)
  motif detection, 152–154, 303–307, 347, 359–360,
    362
  random strings, 129–133
  repetitive, 63
  simulation, 191–195
  tagging SNP selection, 44, 47
DNA bases
  and CTMMs, 187
  evolution, 185–191, 269–271
  frequency analysis, 275–277, 280–286
  and HMMs, 291–293, 303–307
  parameter-tuning, 269–271
DNA microarrays, 64–66, 71, 341
DNA sequencing
  big sequences, 61–63
  computational methods, 64–72
  Eulerian path, 66, 73
  hybridization method, 64–66, 71, 73
  Maxam-Gilbert, 57–59, 61
  nanopore method, 74
  overview, 73–74
  Sanger dideoxy method, 59–61
  shotgun methods, 67–69, 73
  single molecule, 72, 74
Domain recognition, 294, 297
Double-barrel shotgun, 69
  background, 73
Drosophila melanogaster, 74
Duals, 39, 46, 107
Dynafit program, 336

E-Cell system, 260, 264
Edges, graph
  and Bayesian model, 349
  and bipartiteness, 42
  cliques, 39, 342, 344–345
  and CTMMs, 174
  and gene network, 349–351
  in hierarchical clusters, 345
  in intraspecies phylogeny, 29–31, 41
  in Markov model, 143
  and maximum flow, 21
  and mixing, 161–166, 170
  negative weights, 20
  in network structure, 349
  in Steiner trees, 41
  transition probabilities, 160
  in vertex cover, 38, 45, 47, 53
Edit distance, 3–4
Edmonds-Karp algorithm, 22–23, 33
Eigenvalues
  definition, 136
  of Markov models, 136–139, 159, 186
  and matrices, 318, 321, 322
Eigenvectors, 136–139, 186
Einstein, A., 364
Ellipsoid method, 104, 110

Embedded methods, 224
Energy. See also Force field
  and amino acids, 5–7
  and Metropolis method, 143, 147
  potential, 142
  and simulated annealing, 52, 148
  and umbrella sampling, 157
Entropy, 343–344, 358
Enzymatic reactions, 253–256, 324f
Enzymes
  concentration, 325
  and ODEs, 212
  protease, 8–11
Expectation maximization, 345–347
Equilibrium
  and Boltzmann distribution, 142
  in chemical diffusion, 325
  Hardy-Weinberg, 282
Ergodicity
  and canonical path, 164
  definition, 136
  and Markov models, 136, 148, 159, 164, 167, 169
  and Metropolis method, 143
Errors. See also Accuracy
  in differential equation types, 248
  and expectation maximization, 286–287
  and extrapolation, 337–339
  false positives/negatives, 359, 360
  forward and backward, 77, 90
  and intraspecies phylogeny, 30
  in leapfrog method, 222
  Newton-Raphson algorithm, 83–85
  in noisy data, 287
  and physical conservation laws, 226
  and sensitivity analysis, 363
  and steepest descent, 89–90
  and step size, 223
Euclidian distance, 343, 345
Euclidian traveling salesman, 48–49, 54
Eukaryotic genomes
  assembly, 69–70, 73
  DNA sequencing, 63, 67, 73
  gene prediction, 307
  sequence problems, 26
Eulerian path, 66, 73. Euler-Maruyama method,
    246, 249, 250
Event loop, 204, 207
Evolution. See also Continuous time Markov
    models; Molecular evolution
  coalescent model, 193–199
  and data ambiguity, 30
  description, 2–4
  DNA base evolution, 185–191, 269–271
  DNA strand simulation, 191
  genetic algorithms, 52–53
  graph problems, 16–18
  intraspecies phylogeny, 28–33, 41
  Jukes-Cantor model, 185–188

Kimura model, 188–191
and Kolmogorov equations, 187, 190
parameter-tuning, 269–271
tree model, 2–4
Wright-Fisher neutral model, 192
Exact set matching, 27
Exon
and gene structure models, 292–293
length distribution, 129
Expectation maximization
background, 289
and clustering, 345
and goodness of model, 356
haplotype examples, 280–289
and HMMs, 300–307
noisy data, 286–289
reference sources, 289
steps, 277–278, 288–289, 300–302, 305–307
theory, 275, 277–280
weak *versus* strong, 280
Exponential random variables, 118, 175–178, 191
Extrapolation
Aitken's δ^2 process, 337–340
definition, 325
infinite series, 337–340
Richardson method, 225, 337
uses, 323–326, 337

False positives/negatives, 359, 360
Feasible points, 97
Fibonacci heap, 205t, 209
Finite difference iteration, 338
Finite difference methods. *See also* Adams-
Bashforth methods; Runge-Kutta methods
alternatives to, 226
backward Euler, 217–219
definition, 213
forward Euler, 214–217
and independent variables, 239
multistep methods, 221–223
single-step methods, 219–221
stability, 215–217, 218–219, 221
First-order Markov model, 130
First reaction method, 257
Flow problems, 20–22
Floyd-Warshall algorithm, 21
background, 33
Fluorescence, 61–63, 268
Force field, 211
Ford-Fulkerson method, 22–23, 33
Forward algorithm, 298–299
Forward difference, 84
Forward error, 77, 90
Forward Euler. *See also* Euler-Maruyama method
and Brownian motion, 241–246
in convection problem, 238
and coupled differential equations, 229
description, 214–217

and implicitly specified function, 272
with multistep method, 222
reaction-diffusion equations, 235
and step size, 223, 224
Fourier interpolants, 340
Fourier series, 216, 217
Fourier transforms, 226

Galileo, 365
Gamma distribution, 268
Gaussian elimination, 103, 310–316, 318
Gaussian linear model, 348–349
Gauss-Seidel method, 317
Gene expression
additional sources, 353
Bayesian models, 347–350, 353, 356
and cell cycles, 323–325
coexpression models, 342–347, 349, 351
and Gaussian distribution, 348
microarray data, 341
network inference, 341, 347–353, 358
prediction, 309
RNAi, 352
and sampling, 350, 363
General continuous optimization. *See* Continuous
optimization
Generalized minimal residual (GMRES), 319
Gene sequences
Markov models, 129–133
motif detection, 303–307, 347, 359, 362
parameter-tuning, 276
Genetic algorithms, 52
background, 55
Genetic networks. *See* Gene expression
Genetics. *See also* Chromosomes; DNA
gene structure, 276, 292, 299–302
haplotype frequency, 280–286
haplotype inference, 287–289
molecular evolution, 185–192
population genetics, 192–199
tagging SNP selection, 44, 47
Genscan, 307
Geometric series, 337–340
GEPASI program, 253–256, 260
Gibbs sampling, 149–156, 350
background, 158
Gillespie model, 256–260, 263
Global optimum, 52
Gō models, 10
Goodness, measures of, 355–358
Gradient descent, 89
Gradient of objective, 106
Gradient (∇F), 86, 89
Graphing constraints, 96
Graph problems
coloring, 39–40, 49–50
Eulerian path, 66, 73
Hamiltonian path, 37, 65

Graph problems (cont.)
independent set, 38, 42
matching, 23
maximum clique, 39
maximum cut, 37, 344
maximum flow/minimum cut, 21–23
minimum spanning trees, 16–18, 20, 29–31
multigraphs, 16
NP-completeness, 4, 36–42, 47, 344
phylogeny example, 28–33
and set problems, 44
shortest path, 19–21
Steiner trees, 40–41
subgraphs, 42, 54
traveling salesman, 36, 48, 54
and union-of-cliques, 344
vertex cover, 38, 45, 47, 53, 54
Graph properties, 42
Green's function reaction dynamics (GFRD), 263
Grid box, 334
Growth factor, 223
Guilt by association method, 344

Haemophilus influenzae, 73
Hamiltonian path, 36–37, 65
Hamming distance, 41
Haploidy, 198
Haplotypes
frequency estimation, 280–286
inference from noisy data, 286–289
Hard sphere model, 206–209
Hardy-Weinberg equilibrium, 282
Hastings-Metropolis method, 160. *See also*
  Metropolis method
Heat equation, 227
background, 239
Hessian, 86–89, 109
Heuristic methods. *See also* Simulated annealing
background, 53, 158
clustering methods, 344–347
definition, 52
and gene (co)expression, 344–347
genetic algorithms, 52
and Gibbs sampling, 152–154
and intractability, 52
kitchen sink approach, 53
and Metropolis model, 52, 147
and network inference, 349
Hexamers, 260–262
Hidden Markov models (HMMs)
and amino acids, 297–299
background and sources, 289, 307
and DNA bases, 291–293, 303–307
and expectation maximization, 300–307
gene structure, 292, 299–302
motif-finding, 303–307, 359–362
and Newton-Raphson method, 302
and output probability, 297–299

and protein domain, 294
and protein folding, 308
special features, 291
state assignment, 295–297
training, 299–302
transcription factor binding, 293
Hierarchical clustering, 345
HIV, 10
HMM. *See* Hidden Markov models
Huen's method, 224
Hungarian method, 24, 33
Hybridization, sequencing by, 64–66, 71
background, 73
Hydrogen bonds, 158
Hyperplanes, 97

Identity matrix, 310–312, 320
Image analysis, 325, 340
Imino acid, 180
Implicitly specified functions, 271–273
Importance sampling, 154–156, 170
umbrella sampling, 155, 158
Independent set problems, 38–39, 42, 46, 54
Independent variables
and finite difference, 239
multiple, 356
Infeasible points, 97
Infinite series, 337–340
Infinite sites model, 191–192
Information, mutual, 344
Information theory, 343, 358
Inheritable properties, 42
Integer linear programs, 51
Interior point methods, 104–107, 108
Interpolation
best-fit, 336
bilinear, 334
in biochemical reactions, 335
curve families, 334–337
definition, 325
examples, 323–326
Fourier interpolants, 340
Levenberg-Marquardt method, 336
linear, 272
multidimensional, 334
and Newton-Raphson method, 81, 336
and optimization, 335–337
polynomial type, 326–330
rational function, 330
and secant method, 79
splines, 331–334
and steepest descent, 90
Intractability. *See also* NP-completeness
approximation algorithms, 47–49, 50, 55
branch-and-bound methods, 49–52
brute force approach, 47, 53
coping with, 30–32, 35, 46, 49, 53
definition, 24–26, 35

heuristic approaches, 52
trade-offs, 30–32, 46, 49
Isomerization, 180–182
Iterative methods
finite difference, 338
Gauss-Seidel method, 317
Jacobi method, 317
Krylov subspace, 317–320
and Newton-Raphson, 82, 88
Itô integral, 244. *See also* Stochastic integrals;
    Stochastic differential equations
Itô-Taylor series, 249

Jacobian, 86–89, 92
Jacobi method, 317
Johnson's algorithm, 21
background, 33
Joint distributions, 119–121, 149–152
Joint entropy, 344
Jukes-Cantor model, 185–189, 191
background, 200

Karmarkar's method, 104, 108, 110
k-coloring, 40
k-cut problems, 38, 54, 344
k-fold cross validation, 361
Kimura model, 188–191
background, 200
Kinetic models, 351–353
Kolmogorov criterion, 160, 164, 168
Kolmogorov equations
Chapman-Kolmogorov, 133
and CTMMs, 178–182
and discrete event simulation, 201
and evolutionary processes, 187, 190
and implicitly specified functions, 273
Kruskal's algorithm, 17, 31
background, 33
Krylov subspace, 91, 317–320, 333
kth-order Markov model, 130–131

Laplacian, 227
Latent variables, 277, 284, 288–289, 300, 345
Lattice models
background, 10
description, 5–7
and discretized states, 324f, 325
and heuristics, 52
in Markov example, 145
move sets, 10
parameters, 267
and protein folding, 5–6, 145–146
for spatial discretization of PDEs, 258–259
Law of mass action, 211
Lazy queuing, 205
Leapfrog method, 221–223, 225, 236
Least-squares, 320, 336, 349, 356
Leave-one-out cross validation, 361

Levenberg-Marquardt method, 90, 273, 336
background, 93
Likelihood, maximum. *See* Maximum likelihood
Linear congruential generators, 116
Linear interpolation, 272
Linearization, of curve, 81, 86, 89, 91
Linear programming
barrier methods, 104
cost factors, 108
definition, 96
ellipsoid method, 104, 110
primals and duals, 107
relaxation, 51
simplex method, 97–103, 108, 110
software, 107, 111
standard form, 98–99
Linear recurrence, 222
Linear regression, 310
Linear systems
definition, 309
and differential equations, 213
Gaussian elimination, 310–316, 318
and gene networks, 352
and interpolation, 330–334
iterative methods, 316–321
Krylov subspace methods, 317–319
linear regression, 310
and multivariate functions, 87
optimization in, 92
over- and under determined, 320
pivoting, 312–316
preconditioners, 319–320
pseudoinverse, 321
references, 93
and Taylor expansions, 85
Line-by-line method, 256
Local linearizing, 81, 86, 89, 91
Local optimum, 52
LU decomposition, 315

Macromolecular complexes, 260–262, 264
Markov chain Monte Carlo (MCMC), 141–158,
    350
Markov chains
background, 139
definition, 129
and gene network, 350
irreducibility, 136
and mixing times, 163, 166–170
in molecular evolution, 185–188
Markov models
background, 139
branching process, 199
components, 129, 291
conductance, 166–170
continuous time (*see* Continuous time Markov
    models)
and DNA bases, 185–188, 269–271, 291–293

Markov models (cont.)
  and DNA motifs, 153
  eigenvectors, 136–139, 186
  ergodicity, 136, 148, 159, 164, 169
  gene sequence types, 276
  and Gibbs sampling, 149–156
  hidden, 291 (*see also* Hidden Markov models)
  and Metropolis method, 142–148 (*see also* Metropolis method)
  mixing time, 138, 159–160, 166, 170
  and molecular evolution, 185–191
  nonergodic, 137
  order, 130–131
  and prior distribution, 153
  with random walk, 167
  and spatial effects, 258
  stationary distribution, 134–138, 149, 153–155, 159, 161
  and waiting time (*see* Continuous time Markov models)
Mass action, law of, 211
Matching problems
  exact set, 27
  unweighted, 23
  weighted, 24
Mating, 53
Matrices. *See also* Transition matrix
  condition number, 319
  inversion, 87
  over/underdetermined, 310, 320, 330, 333
  permutations, 314
  positive (semi)definite, 92, 318, 319
Maxam-Gilbert method, 57–59, 61
Maximal matching, 47
Maximum a posteriori probability (MAP), 275
Maximum clique problems, 39
Maximum cut problems, 37, 344
Maximum edge loading, 161–166, 170
Maximum flow problems, 21
Maximum likelihood
  background, 289
  and clustering, 345–346
  description, 268
  and expectation maximization, 275, 277–280 (*see also* Expectation maximization)
  in haplotype error correction, 286–287
  in haplotype frequency estimation, 282–283
  and Hardy-Weinberg equilibrium, 282
  and latent variables, 284
  and network inference, 347–351
  and parameter-tuning, 8–10, 268, 275–277, 283
MCell, 258–259, 264
Metropolis criterion, 6–7, 10
Metropolis method
  background, 158
  caveats on use, 146
  efficiency, 154–156
  generalized, 146–147

  and mixing time, 146, 154, 170
  for optimization, 147, 350
  and protein folding, 142, 145, 154
  and simulated annealing, 52, 148
  and thermodynamics, 141–143, 146
  and traveling salesman, 147
Michaelis-Menten reaction, 253–256
Microarrays, 64, 71, 341
Microreversibility, 143–145, 164, 169. *See also* Detailed balance
Midpoint method, 219–222
Migration, 198
Milstein's method, 249, 251
Minimum cut, 21–23
Minimum description length (MDL), 361
Minimum set cover, 45. *See also* Vertex cover
Minimum spanning network, 31
Minimum spanning tree, 16–18, 20, 29–31
Minimum test set, 44
Mixing time
  canonical path method, 161–166, 169, 171
  conductance method, 166–170, 171
  definition, 138, 159–160
  and eigenvalues, 138–139
  and importance sampling, 170
  and Metropolis method, 146, 154
  monomer-dimer systems, 171
Model space, reduction, 351
Model validation
  accuracy, 358 (*see also* Accuracy)
  cross-validation, 362
  goodness measures, 355–358
  overfitting avoidance, 361
  receiver operating characteristic (ROC) curve, 360
  scientific method, 363–366
  sensitivity, 359, 360, 362
  specificity, 359–361
Mode-of-action by network identification (MNI), 351
Modified distribution, 156
Molecular evolution
  coalescent model, 192–198
  DNA strand, 191
  Jukes-Cantor model, 185–188
  Kimura model, 188–191
  and Kolmogorov equations, 187, 190
  one-parameter, 185–188
  and self-transition, 163–164
  two-parameter, 185–188
Molecular modeling
  and continuous optimization, 75
  lattice models, 5–7, 145–146
  macromolecular complexes, 260–262, 264
  and numerical integration, 211
  and stochastic differential equations, 245
  and umbrella sampling, 156–158
Moleculizer program, 261

Monomer-dimer systems, 170–171
Monte Carlo samplers, 350
Motifs
  alignment of, 152
  detection of, 152, 303–307, 347, 359–362
  transcription factor binding, 293
Move sets, for lattice models, 6
Multicommodity flows, 23
Multidimensional curve, 336–337
Multigraphs, 16
Multigrid methods, 324f, 325
Multiple independent variables, 356 (*see also*
  Partial differential equations)
Multiple regression, 351
Multivariate functions, 85–88
Mutations
  in genetic algorithm, 53
  infinite sites model, 191–192
  and Jukes-Cantor model, 187, 191
  and Kimura model, 186f, 188–189
  random, 163–166
  simulation, 4–7, 191
  transitions/transversions, 189
  and Wright-Fisher neutral model, 192
Mutual information, 344

Needleman-Wunsch algorithm, 33
Network identification by multiple regression
  (NIR), 351
Networks
  gene regulatory, 341–353
  inference of, 349–353, 363
  minimum spanning, 31
  reaction networks, 260–264, 323–325, 340
  reduced median, 33
Neumann boundary condition, 231–232
Neville's algorithm, 326–329
Newton-Raphson method
  background, 93
  black-box functions, 84
  and HMMs, 302
  and implicitly specified function, 273
  and interpolation, 336
  and Levenberg-Marquardt method, 90
  multidimentional, 85–88
  and parameter-tuning, 80–84, 269
  and steepest descent, 90
Newton's second law, 211
Next reaction method, 257
Noisy data, 286–289, 323–325, 329, 347
Nonlinear programming, 108–110
Nonlinear systems, 91–92
Nontrivial graph properties, 42
Normal distributions, 120, 123–124
NP-completeness
  background, 53–55
  coping with, 35, 46–53
  and DNA sequencing, 65

linear programming relaxation, 51
  in Steiner tree, 41
  and union-of-cliques graph, 344
NP-hardness. *See* NP-completeness
Numerical integration. *See also* Partial differential
  equations; Stochastic differential equations
  additional readings, 225
  backward Euler, 217–219
  and black box functions, 336
  definition, 213
  and extrapolation, 337
  finite difference method, defined, 213
  forward Euler, 214–217, 222, 223, 224
  implicit, 316
  and interpolation, 336
  and Kolmogorov equations, 273
  leapfrog method, 221–223, 225, 236
  line-by-line method, 256
  midpoint method, 219–221
  multistep methods, 221–223
  and parameter-tuning, 272
  single-step methods, 214–221
  spectral methods, 226
  speed and efficiency, 223–225
  step size selection, 223–225, 233–234
  and transformation method, 118

Objective function, 96, 268–271, 336
ODEs. *See* Ordinary differential equations
Optimization. *See also* Continuous optimization;
  Gibbs sampling; Metropolis method; Parameter-
  tuning
  background, 92
  in bootstrapping, 350, 363
  conjugate gradient, 91, 318
  constrained (*see* Constraint satisfaction)
  and decision problems, 36
  description, 1–4
  discrete, 15
  and gene networks, 349
  and Gibbs sampling, 152–154, 350
  and interpolation, 335–337
  lattice models, 5–7
  Levenberg-Marquardt method, 90, 93
  and Metropolis method, 147–148, 350
  and model goodness, 356
  (non)linear systems, 91–92, 318
  and parameter-tuning, 8, 267–271
  of state assignments, in HMM, 295–297
  steepest descent, 89–90
  without zero-finding, 89–92
Order of convergence, 248
Ordinary differential equations (ODEs)
  backward Euler, 217–219
  and curve fitting, 335
  and errors, 248
  examples, 211–213
  forward Euler, 214–217, 222, 223, 224

Ordinary differential equations (cont.)
and gene networks, 351–353
leapfrog method, 221–223, 225, 236
line-by-line method, 256
living cell simulation, 253–256
midpoint method, 219–221
and reaction network, 323, 352
step size selection, 223–225
Overdetermined systems, 310, 320, 330
Overfitting, 361

Parameter selection, 267, 362
Parameter-tuning. *See also* Expectation
maximization; Hidden Markov models;
Optimization
and biochemical reactions, 267
description, 8–10, 267, 275
DNA base evolution, 269–271
and gene sequences, 276
haplotype frequency, 280–286
haplotype inference, 286–289
implicitly specified functions, 271–273
and linear systems, 309 (*see also* Linear systems)
maximum likelihood, 8–10, 268, 275–277, 283
motif-finding, 303–307
and Newton-Raphson method, 80–84, 269
and noisy data, 286–289
protease example, 8–10
and protein expression, 268
protein folding example, 267
and sensitivity, 363
Parsimony, 3, 29–33
background, 10
Partial differential equations (PDEs). *See also*
Reaction-diffusion equations.
additional information, 239
boundary conditions, 230–233
convection, 237–239
coupled one-dimension, 228–230
diffusion example, 227
initial conditions, 230
line-by-line method, 256
cell simulation, 253–256, 263
multiple spatial dimensions, 233–234
one spatial dimension, 228–230
step size, 233
Particle collisions, 141, 206–209
Particle diffusion, 325
Particle interactions, 177
PDEs. *See* Partial differential equations
Pearson correlation coefficient, 343
Permutation matrix, 314
Pfam protein database, 295
Philosophy of science, 363–366
Phylogeny, intraspecies, 28–33
Pivoting, 312–316
Poisson process, 191
Poisson random variable, 191

Polymerization, 61
Polynomial reduction, 46
Polynomials
Chebyshev, 340
cubic formula, 76, 329, 333
fitting to lower order, 329–331, 340
Neville's algorithm, 326–329
quadratic formula, 76
quartic formula, 76
splines, 331–334
Polytope, 97
Popper, Karl, 364
Population dynamics, 29, 212
Population genetics, 280–286
Posting time, 205
Prediction
cut site, in proteases, 8–11
gene expression, 307, 309
protein expression, 323
Predictor-corrector schemes, 225
Primals and duals, 107
Prim's algorithm, 18, 20
background, 33
Prior distribution. *See* Prior probability
Prior estimate, 303
Priority queue, 18, 205, 209–210
Prior probability, 153, 298, 349
Probability. *See also* Sampling
of best-fit, 275
conditional, and transitioning, 295
distribution, 347–349
fundamental transformation law, 117
maximum a posteriori (MAP), 275
maximum likelihood, 8–10, 268, 275–277, 283,
356
of migration, 198
prior, 298
Proline, 180–182
Proteases
cut site prediction, 8–11
and HIV, 10
and parameter-tuning, 8–10
Proteasomes, 11
Protein expression, 268, 323
Protein folding
and CTMMs, 180–182
and HMMs, 308
importance sampling, 154–156
lattice models, 5–7, 10 (*see also* Lattice models)
Markov model example, 145
Metropolis model, 142, 145, 154
parameters, 267
umbrella sampling, 155–158
Proteins
and Brownian motion, 157
channel protein, 201–203
coiled-coil, 293–295
complexes, 177, 260–262

database, 295
domain recognition, 294, 297
exact set matching, 27
growth rate example, 95
hydrogen bonds, 158
ligand binding, 75
longest common subsequence, 25, 42–43
longest common substring, 26
sampling programs, 261
string and sequence problems, 24–27
structure simulation, 4–7
translation, 268
Pseudoinverse, 321
Pseudorandom numbers, 115
P-value calculators, 343

Quadratic formula, 76
Quadratic programming, 109
Quartic polynomials, 76
Queues, 18, 205, 207–210. *See also* Priority
 queues

Random DNA strings, 129–133
Random mutations, 163–166
Random number generation
 pseudorandom numbers, 115
 rejection method, 121–124
 transformation method, 115–121
Random variables. *See* Distributions
Random walk, 167–170, 324f
Rational function, 330
Rational interpolation, 330
Reaction-diffusion equations, 234–237, 325
 background, 239
Reaction networks, 211, 217, 260–264, 264, 271,
 323–325, 335, 340
 cell simulation, 260–262
 data-fitting, 340
Reaction term, 234
Receiver operating characteristic (ROC) curve,
 360
Recombination, 198
Reduced median network, 33
Rejection method, 121–126
 background, 127
Relaxation, 51
Reversibility, 143–145
Reweighting, 21
Richardson extrapolation, 225, 337
RNAi, 352
Runge-Kutta methods
 and accuracy, 221, 225, 237
 with black box, 237
 and cell simulation, 260
 embedded, 225
 fourth order, 221
 midpoint method, 219–221
 and stability, 221

Run time. *See also* Optimization; Simulation
 and accuracy, 219, 233
 and boundary conditions, 231
 coalescent, 195–197
 and CTMMs, 173–175, 273
 and discrete event models, 204–206, 208–210
 and importance sampling, 155
 and intraspecies phylogeny, 29
 and Krylov subspace methods, 319
 and Metropolis method, 146, 154
 and numerical integration, 225
 and stability, 215
 and step size selection, 217, 233
 and umbrella sampling, 156–158

Sampling. *See also* Gibbs sampling; Importance
 sampling; Markov models; Metropolis method;
 Umbrella sampling
 continuous distributions, 116–124, 156
 discrete distributions, 124–126, 146
 efficiency, 154
 exponential random variable, 118–119
 geometric random variable, 125–126
 joint distributions, 119–121, 149–152
 modified distribution, 156
 and network inference, 350, 363
 normal distributions, 120
 with optimization, 350
 at point in time, 182
 (pseudo)random numbers, 115
 rejection method, 121–124
 and simulation, 7, 115
 transformation method, 116–121
 uniform random variable, 116
Sanger dideoxy method, 59–61
Scaled variables, 105
Science, philosophy of, 363–366
Scientific method, 363–366
Secant method, 78–80
Selfing, 198
Self-transitions
 conversion to, 168
 and CTMMs, 181
 and mixing time bounds, 159
 and molecular evolution, 163–164
Semidefinite programming, 108–110
Sensitivity, 359, 360, 362
Sequences. *See* DNA sequencing; String and
 sequence problems
Set problems
 independent set, 38, 42, 46, 54
 minimum set cover, 45
 minimum test set, 44
Shortest common supersequence, 43
Shortest common superstring, 44
Shortest path, 19–21
Shotgun methods, 67–71
 background, 73

Signal processing, 340
Similarity measures, 342–344
Simplex method, 97–103, 108, 110
Simulated annealing
  background, 54
  and Bayesian models, 349–350
  description, 52
  and Metropolis method, 52, 148
Simulation
  Brownian motion, 241–249
  chemical, in inhomogeneous solution, 234–237
  continuous systems, 211–213 (*see also* Continuous systems)
  of CTMM (pseudocode), 175f
  of discrete events (*see* Discrete event models)
  DNA, haploid, 198
  DNA random string, 129–133
  DNA strand, 191
  DNA whole population, 192–195
  implicit functions, 271–273
  of macromolecular reactions, 260–262
  of mutation, 4–7, 191
  parameter-tuning, 267–271
  of particle collisions, 141, 206–209
  protein structure example, 4–7
  reaction networks, 253–264
  of recombination, 198
  and sampling, 7, 115
Single-molecule sequencing, 72, 74
Single-pair shortest path, 19–21
Single-step methods, 219–221, 223–225
Smith-Waterman algorithm, 33
SNP selection, 44, 47
Social constructivism, 365
Solutions
  convection, 237–239
  diffusion, 227, 230–237, 259, 325
  inhomogeneous, 234
Sparse candidate algorithm, 351, 352
Sparse graphs, 18, 21
Sparse matrices, 315, 316, 322
Spatial models
  discretization, 229, 233, 235, 255, 258
  multidimensional, 85–89, 233, 325
  one dimension, 228–230
  reaction-diffusion equations, 234–236
  three-dimensional, 234
  and time, 233
  two-dimensional, 325
Spearman correlation coefficient, 343
Species tree, 28–33
Specificity, 359–361
Spectral methods. *See also* Eigenvalues; Fourier transforms
  interpolation, 340
  numerical integration, 226, 239
Splines, 331–334

Stability
  and accuracy, 221, 223, 251
  of Adams-Bashforth methods, 223
  additional information, 239
  of backward Euler, 218
  classifications, 215
  disadvantages, 217
  of forward Euler, 215–216
  of leapfrog method, 222
  and mutations, 4–7
  references, 239
  and Runge-Kutta methods, 221
  and step size, 217
  and stochastic differential equations, 249–251
  unconditional, 219
  von Neumann analysis, 215–217
Standards, 264
Standard Weiner process, 241
Stationary distribution, 134–138, 149, 153–155, 159, 161
Steepest descent, 89
Steiner nodes, 32, 41
Steiner trees, 31–32, 40–41
Step sizes, 233, 337
  adaptive methods, 223–225
  predictor-corrector schemes, 225
  and stability, 217
Stochastic differential equations
  accuracy, 248
  additional information, 252
  for Brownian motion, 241–248
  and cell simulation, 256
  Euler-Maruyama method, 246, 249, 250
  and implicit function, 273
  for protein-folding, 157
  stability, 249–251
Stochastic integrals, 244
Stochastic simulation algorithm (SSA), 256–260, 263
StochSim, 256–259
Stratonovich integral, 244
String and sequence problems
  applications, 24
  exact set matching, 27
  haplotype frequency, 280–286
  haplotype inference, 286–289
  HMM, 292
  hybridization, 64–66, 71, 73
  longest common subsequence, 25, 42–43
  longest common substring, 26
  Markov model example, 276
  noisy data, 286–289
  NP completeness, 42–44, 47
  random DNA strings, 129–133
  sequence alignment, 33
  shortest common supersequence, 43
  shortest common superstring, 44
  suffix trees, 26, 27, 33

Subgraphs, 42, 54
Subsequences, 25, 42–43
Subspace. *See* Krylov subspace
Substrings, 26
Successive squaring, 133
Suffix trees, 26, 27, 33
Sum-of-squares. *See* Least-squares
Supersequences, 43
Superstrings, 44
Systems Biology Markup Language (SBML), 264

Tagging SNP selection, 44, 47
Tau leap algorithm, 259
Taylor series
  approximation with, 80–82, 85, 232
  and backward Euler, 218
  and finite difference approximations, 229, 232
  and forward Euler, 215
  and midpoint method, 220
  and multistep methods, 222, 225
  and Newton-Raphson method, 80–82, 84–85
  and Richardson extrapolation, 337
  stochastic. *See* Itô-Taylor series
Temperature. *See* Simulated annealing
Terminal nodes. *See* Steiner trees
Terminator base, 59–61
Thermodynamics
  and CTMMs, 180–182
  and Metropolis method, 141–143, 146
Time. *See* Evolution; Mixing time; Run time
Tractability, 24–26, 35. *See also* Intractability
Transcription factor binding, 293
Transformation method, 116–121, 124
  background, 127
trans isomer, proline, 180–182
Transition, Markov model, 130
Transition matrix
  for CTMMs, 173
  in Jukes-Cantor model, 186
  in Kimura model, 189
  of Markov models, 132, 134–137
Traveling salesman problem (TSP), 36, 48, 54, 147
Trees
  minimum spanning, 16–18, 20, 29–31
  and optimization, 2–4
  Steiner, 31–32, 40–41
  suffix, 26, 27, 33
  and traveling salesman, 48
Triangle traveling salesman, 48–49, 54
True negatives, 359
True positives, 359, 360
Truth, 365
Twofold cross-validation, 361

Umbrella sampling
  background, 158
  and Gibbs sampler, 156–158
  and Metropolis sampler, 155

Unconditional stability, 219
Underdetermined system, 310, 321, 333
Union-of-cliques, 344–347

Variation distance, 160
Vertex cover
  approximation algorithms, 47, 50–51
  description, 38
  and genetic algorithm, 53
  hardness testing, 46
  and independent set, 39
  and minimum set cover, 45
  reference, 54
Virtual Cell, 255, 261, 264
Viterbi algorithm, 296, 299
von Neumann analysis, 215, 219, 220, 250

Waiting time, 173–175
  and coalescence, 198
  and CTMMs, 201–204
  and Poisson process, 191
  and recombination, 199
Wave equation, 238
Wavelets, 223, 226, 340
Weiner process, 241
Whole population sampling, 192–195. *See also* Coalescent
Wikipedia, 93, 110
Wright-Fisher neutral model, 192

Zero, avoiding, 105–107
Zero-finding
  alternative approaches, 89–92
  bisection method, 76–78
  multivariate functions, 85–88
  Newton-Raphson methods, 80–88, 90, 269
  secant method, 78–80
0–1 integer programming, 51